MW01505825

FINITE ELEMENTS:
THEIR DESIGN
AND PERFORMANCE

MECHANICAL ENGINEERING

A Series of Textbooks and Reference Books

Editor

L. L. Faulkner

*Columbus Division, Battelle Memorial Institute
and Department of Mechanical Engineering
The Ohio State University
Columbus, Ohio*

Volume 89. *Finite Elements: Their Design and Performance*
Richard H. MacNeal

*For information about additional volumes in this series,
please contact the publisher:*

Customer Service
Marcel Dekker, Inc.
270 Madison Avenue
New York, NY 10016

800-228-1160

FINITE ELEMENTS: THEIR DESIGN AND PERFORMANCE

RICHARD H. MACNEAL

The MacNeal-Schwendler Corporation
Los Angeles, California

Marcel Dekker, Inc.　　　　New York•Basel•Hong Kong

Library of Congress Cataloging-in-Publication Data

MacNeal, Richard H.
  Finite elements: their design and performance / Richard H. MacNeal.
    p.   cm. -- (Mechanical engineering; 89)
  Includes bibliographical references and index.
  ISBN 0-8247-9162-2
  1. Finite element method. I. Title. II. Series: Mechanical engineering
(Marcel Dekker, Inc.); 89.
  TA347.F5M36   1993
  620'.001'51535--dc20   93-31557
                                                               CIP

The publisher offers discounts on this book when ordered in bulk quantities. For more information, write to Special Sales/Professional Marketing at the address below.

This book is printed on acid-free paper.

Copyright © 1994 by MARCEL DEKKER, INC. All Rights Reserved.

Neither this book nor any part may be reproduced or transmitted in any form or by any means, electronic or mechanical, including photocopying, micro-filming, and recording, or by any information storage and retrieval system, without permission in writing from the publisher.

MARCEL DEKKER, INC.
270 Madison Avenue, New York, New York 10016

Current printing (last digit):
10 9 8 7 6 5 4 3 2 1

PRINTED IN THE UNITED STATES OF AMERICA

To Carolyn

Preface

The preface to a technical book is frequently the only place where an author feels free, given the expectations of scientific etiquette, to express his personal opinions about controversial matters. As the reader will discover, this author labors under no such inhibitions and does not need a preface to escape them. Still, the preface is a good place for the reader to learn a little about the convictions which led an author to write a book.

My experience with finite elements goes back to at least 1965[*] and the beginning of the NASTRAN finite element program as a NASA project. Through the years I have maintained contact with NASTRAN, and later with MSC/NASTRAN, acting more or less in the capacity of resident theoretician. In particular I have authored, or co-authored, the "theory" for nearly all of the

---

[*]To 1947 if you count experience with the precursors of finite elements.

elements, from the primitive ones in early NASTRAN to the p elements currently being coded for MSC/NASTRAN. I have, therefore, abundant experience with the design of commercial finite elements and with users' complaints about them.

Such experience is not, by itself, a sufficient qualification for writing a book about finite elements. I, like most practicing engineers, have long felt inadequacies on the academic side, particularly in mathematical grounding and familiarity with the technical literature. It was not until 1985 and semiretirement that I had the time for study and the inclination to concentrate on such general questions as the causes of element failure and the limits of finite element perfectibility.

The decision to write the book arose, as it so often does, from the preparation of a course of lectures, in my case lectures on the properties of finite elements which I gave to the technical staff at MacNeal-Schwendler in the Spring of 1990. That experience convinced me that my work on the analysis of locking and shape sensitivity could form the core of a book showing how finite element design has been driven, from the earliest days to the present, by the need to suppress these and other flaws.

Whatever else we can say about finite elements, we must admit that the pioneering days are over. In other words, it is high time that we treated the design of finite elements less as an art and more as a science. The pioneers of the 1950s, '60s, and early '70s relied on heuristics, hunches, and experimental data to guide their design choices. The reaction of the academic community has been to supplement the heuristics and hunches with mathematical rigor, based largely on the variational calculus. In my opinion, the addition of mathematical rigor falls well short of the goal of converting finite element design into a science. It does not explain why elements derived rigorously from variational principles sometimes fail miserably in practice. What is needed, and what the book tries to provide, is a rational understanding of the performance of finite elements, including quantitative analysis of their failure modes and of the side effects of proposed remedies. With comprehensive knowledge of this sort, finite element designers can proceed with confidence to design elements which are free from the major flaws – locking, patch test failure, and spurious modes. And finite element analysts can proceed with confidence to use them.

As in any science, the development of understanding begins with the classification of observed effects, in our case the classification of finite element failure modes. To the three which have been mentioned, we will add three more: rigid body failure, induced anisotropy, and shape sensitivity. With a classification of effects in hand, emphasis can shift to a search for underlying causes and then to analyses testing whether the supposed underlying causes explain the observations. Such analyses will also frequently indicate remedies which can be further analyzed.

The treatment of finite element flaws, or disorders as we will call them, will begin side by side with the treatment of fundamentals. For example, in Chapter 3, "Assumed Displacement Fields," we will discuss rigid body failure and induced anisotropy as consequences of inappropriate basis sets and will introduce parametric interpolation as a means to overcome inter-element displacement discontinuity (a frequent cause of patch test failure). In the same chapter we will introduce the practice of inventorying internal degrees of freedom and will discuss the consequences of incompleteness in their polynomial degrees.

The introduction of the patch test in Chapter 5 concentrates on the causes for patch test failure. We will show, in Chapter 6, that the most important of these causes, interpolation failure,[*] is also the underlying cause of locking and shape sensitivity. Chapter 6 is central to the book's purpose. It classifies the types of locking and systematically analyzes locking and shape sensitivity for common types of two- and three-dimensional elements. The introduction of a new tool, *aliasing*, allows us to predict the shapes and magnitudes of locking modes without invoking the cumbersome element numerics developed in Chapter 4.

The analysis of locking repeatedly identifies reduced integration as an obvious remedy. In Chapter 7 we study the benefits and shortcomings of that remedy

---

[*]Failure to interpolate displacement fields correctly from nodal values.

and introduce assumed strain fields and spurious mode stabilization as more sophisticated remedies which avoid some of the undesirable side effects of reduced integration and selective underintegration.

Chapter 8 expands the examination of remedies for locking into the study of bubble functions, drilling freedoms, and direct assumed strain formulations. We then bring many of these tools to the treatment of plate and shell elements in Chapters 9 and 10. That treatment includes thorough analyses of transverse shear locking and membrane locking, and examines various avoidance measures.

The book concludes with a retrospective on the identification, cause, and cure of element disorders and an essay on the limits of finite element perfectibility. It argues not only that the pioneering days are over but that little more is possible concerning the accuracy of lower order elements. If this is true it is time for creative minds to move on, and they largely have. Their product, finite element analysis, remains as a predominant tool in structural mechanics and other fields. It should be studied and understood by its practitioners.

It is traditional for a preface to acknowledge the contributions of others. The book includes short quotations and figures from other publications (by Wiley, Elsevier, McGraw-Hill, Pineridge Press, and Ellis Horwood) which are acknowledged as they occur. I would particularly like to thank John Wiley and Sons for permission to revise and reprint my paper, "On the Limits of Finite Element Perfectibility," as Section 11.4 of the book.

My secretary, Connie Christ, composed the entire book on a Macintosh in finished, camera-ready form. This was not as easy as the ads imply. It required her to master arcane compositional details about fonts and formats and mathematical notation which I hope she can put to good use in the future.

The person to whom I owe the greatest debt of gratitude, although he did not write or revise a single line of the book, is Robert L. Harder. Through the years Bob has supervised the coding of nearly all of the MSC/NASTRAN elements described in the book and has produced all of their test results. The marvelously versatile application of the method of least squares smoothing described in Section 7.7.1 is his work, as are many other details of the MSC/NASTRAN elements.

Finally I must acknowledge the debt owed to the finite element pioneers and to those who followed in their footsteps. To mention some would risk omitting others equally deserving. The development of the finite element method will make a fascinating history when some day it is told.

Richard H. MacNeal

Contents

xi

FINITE ELEMENTS:
THEIR DESIGN
AND PERFORMANCE

1
Introduction

1.1 DOMINANCE OF THE FINITE ELEMENT METHOD

Today the finite element method enjoys a position of predominance among the methods used in structural mechanics and heat transfer. It has occupied that position since at least 1975. The method has also penetrated other fields, but more slowly. In computational fluid dynamics, for example, the method of finite differences still competes with the finite element method, while in electromagnetism, boundary integral methods and equivalent circuit methods remain popular. Worldwide there are, perhaps, 100,000 engineers who use the finite element method. So it is no longer necessary to sell finite elements, at least not to structural engineers and thermodynamicists.

We must go back to 1955 to find real competition among the methods used in the analysis of structures. At the time, new digital computers such as the

IBM 704 were coming on line with all of the features needed for large-scale engineering calculations—storable instructions, higher-order programming languages, substantial high-speed memory (32,000 words), and floating point arithmetic. It was also the heyday of the analog computer, which could solve problems with a few hundred, and ultimately one or two thousand, degrees of freedom very much faster than the digital computer. The era of hand calculation with mechanical desk calculators had reached its zenith and was beginning to fade.

The methods used to analyze structures exhibited similar diversity in 1955. It is useful, in recalling them, to separate methods of formulation from methods of solution. In linear static analysis, for example, the interface between the formulation and solution phases may be taken as the reduction of the problem to a set of linear algebraic equations. From that point, analog computers finessed the solution phase by reducing it to the measurement of response quantities in an analogous physical model (usually an electric circuit). On the digital side, iterative solution methods were preferred, prior to 1955, for any but the smallest sets of equations (ten or less) due to the labor involved in a direct solution.* For example, the number of multiplications required for a direct solution of the linear matrix equation

$$[K]\{u\} = \{P\} \tag{1:1}$$

is, if $[K]$ is a full, square matrix of order N x N,

$$M = \frac{1}{3}N^3 + O(N^2) \tag{1:2}$$

If the matrix $[K]$ is symmetric and banded with semi-bandwidth B, then the required number of multiplications is

$$M = \frac{1}{2}NB^2 + O(NB) \tag{1:3}$$

---

*Today, iterative methods are making a comeback for very much larger problems.[1]

Clearly, mental fatigue sets in at a very small number of equations if a hand calculation of Equation 1:1 is attempted, except in cases where the matrix $[K]$ has a small bandwidth. In similar fashion, considerations of cost and storage capacity limited the direct matrix solution capabilities of 1955-era digital computers to a few hundred equations, or possibly a few thousand for narrowly-banded matrices. With iterative methods of solution, on the other hand, the number of multiplications was reduced to the order of the number of equations, N, times the number of iterations. This seemed particularly advantageous if one was optimistic about the rate of convergence of the iterations.

Engineers also exhibited diversity in their choice of formulation methods. Many problem formulations used finite elements, even though they were not called that since the term "finite element" was not coined until 1960.[2] Truss and framework analyses had employed rod elements since the nineteenth century. Aircraft structural engineers used beam elements to model wings, fuselages, and other airplane parts. Indeed, there was a tendency to force the use of "stick models" into every structural analysis, whether appropriate or not. Even today the MSC/NASTRAN computer program caters to this inclination of aircraft engineers by including a beam element with tapered section properties, offset shear centers, and other refinements.

Crude two-dimensional finite element models also existed for use in cases where stick models were clearly inappropriate. In 1941, Hrennikoff[3] introduced a framework model for plane elasticity. For example, rod elements assembled in a regular lattice of equilateral triangles could simulate plane stress with a Poisson's ratio equal to one-third. By 1955, rod and shear panel models had found widespread use. With the introduction of Garvey's general quadrilateral shear panel[4] in 1951, they had acquired the ability to model arbitrary plane shapes. Even today there are configurations, such as thin sheet metal panels which buckle at low load intensities, where rod and shear panel models excel. MSC/NASTRAN includes a Garvey shear panel to accommodate such cases.

There were other ways of formulating two-dimensional and even three-dimensional problems that commanded support. These included a host of analytical methods based on assumed functions of position which either satisfied the underlying differential equations, as in the case of boundary integral methods,[5] or which did not, as in the case of the method of weighted residuals.[6] Such methods had the advantage that a judicious choice of functions could reduce the number of unknown coefficients needed for accuracy, or could diagonalize their solution matrix. These methods had the disadvantages that they did not cope well with irregular boundaries and that each application required expert mathematical ability.

Many practicing engineers favored finite difference methods.[7] As a simple example, consider Poisson's equation

$$\frac{\partial^2 u}{\partial x^2} + \frac{\partial^2 u}{\partial y^2} = P(x, y) \tag{1:4}$$

In the finite difference method, the partial derivatives are replaced by difference operations, so that Equation 1:4 becomes

$$\frac{u_{m+1}^n - 2u_m^n + u_{m-1}^n}{\Delta x^2} + \frac{u_m^{n+1} - 2u_m^n + u_m^{n-1}}{\Delta y^2} = P_m^n \tag{1:5}$$

The indices (m, n) indicate position in a rectangular grid as shown in Figure 1.1.

An important advantage of Equation 1:5 is that it couples node (m, n) to its nearest neighbors only. Heuristic modifications of Equation 1:5 were devised to accommodate changes in mesh spacing $\left(\Delta x, \Delta y\right)$ and even irregular boundaries. When combined with an iterative solution technique, the finite difference formulation yielded a formidable analysis procedure known as the relaxation method,[8] which could find solutions to two-dimensional problems with a thousand or more node points entirely by hand calculation.

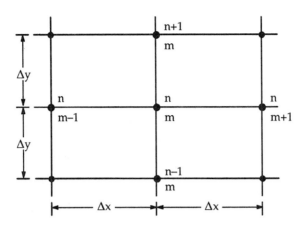

Figure 1.1   Finite Difference Grid.

But the future lay with the digital computer and with the finite element method. In the early 1950s some far-sighted analysts, notably Langefors [9] and Argyris, [10] were laying the groundwork of matrix formulation that would be needed by the coming finite element systems.   In 1956, a landmark paper by Turner, Clough, Martin, and Topp [11] introduced the constant strain triangular element.   Unlike the rod and shear panel model, this element contained all three components of membrane strain in relationships that were correct for any element shape.   The author remembers his reaction to this development—that the mathematical formulation of the new element, which entailed a full 6 x 6 stiffness matrix, was too complex to be incorporated in his analog computer circuits.   But this was precisely the strength of the new IBM 704 and similar digital computers.   Their increased storage capacity allowed a complexity of formulation unmatched by the analog computer and by earlier models of the digital computer.   Added complexity would provide greater accuracy and diminish the user's burden by allowing the computer to take over many of his tasks.   And, as the power of the digital computer increased, larger and larger problems could be solved.

By the early 1960s the elements, the solution methods, and the digital computers had matured to the point where it was possible to contemplate large, general purpose, finite element systems which would be accessible to all engineers. Notable early developments on the element side included the Taig isoparametric membrane quadrilateral[12] (1961), the constant strain tetrahedron [13] (1962), the first paper on plate bending[14] (1961), and the Pian assumed-stress membrane quadrilateral[15] (1964). Element activity attained a peak of intensity at the first Wright Field conference on matrix methods in structural mechanics (1965) which, among other important developments, included presentation of the element library in the new general purpose ASKA system. [16] With Irons' generalization of the isoparametric element concept to higher-order membrane and solid elements [17] (1966), the finite element had reached the stage where it could be applied, with somewhat shaky confidence, to the full range of structural applications.

General purpose systems soon appeared that incorporated the new finite elements. These began in the mid-60s with ASKA in Europe and STARDYNE in the U.S. By 1970 they were joined by ANSYS, MARC, NASTRAN, and SAP in the U.S. and by ASAS, BERSAFE, PAFEC, and SESAM in Europe. These codes were extremely large by then current standards, some with more than 100,000 source statements. It is remarkable that all of them are still in use and that they still constitute, for the most part, the leading commercial finite element systems. With their appearance, the average working structural analyst had access to the best finite element technology. And he still does.

The subsequent history of the finite element method has featured a slow process of incremental improvement, occasionally punctuated by quantum advances in computer architecture or system design. In this book we are primarily concerned with advances in the design of the elements themselves. It has been more than 35 years since the introduction of the constant strain triangle,[11] and to one who has lived through those years progress has indeed been slow. But progress is still needed and is still being made. We now have elements which are much better and much more complicated than the early ones, but few if any are without flaws. We are not satisfied, and perhaps we never will be.

1.2 REASONS FOR THE POPULARITY OF THE FINITE ELEMENT METHOD

Figure 1.2 depicts a few typical finite elements and Figure 1.3 shows their incorporation into the finite element model of a practical structure. To the average structural engineer, Figures 1.2 and 1.3 show about 90% of what he needs to know about finite elements—that they are connected to points in space where displacements and loads are defined, and that, when joined together, they can fill spaces of almost arbitrary complexity. Most of the engineer's time in preparing a finite element analysis is engaged in just this activity—selecting elements and instructing the computer, or perhaps an automatic mesher, to join them together in a meaningful simulation. In so doing, the engineer must decide which structural details he should include in the model and which details he can safely suppress. These may, for example, include such items as holes, fasteners, fillets, and stiffeners. Designing the structural model is typically a big and important task, so that the average engineer may feel little inclination to inquire about the details which are hidden within each element.

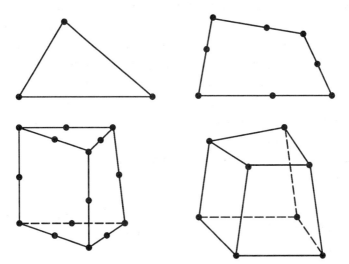

Figure 1.2   Typical Finite Elements.

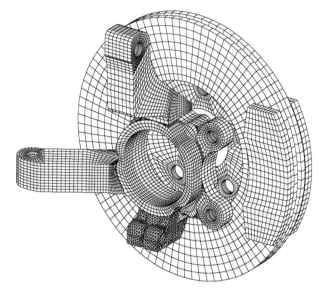

Figure 1.3   Finite Element Model of a Brake Assembly.

Engineers can, in fact, carry out a finite element analysis without knowing much about what goes on within the finite elements. A modern finite element, unlike the earlier rod and shear panel models, has a one-to-one correspondence with a defined region of space and is, therefore, self-contained. Thus, given the locations of connecting node points, the properties of materials, and such other geometric properties as the thicknesses of plate elements, all of the calculations can proceed automatically. This was not generally the case with the methods that the finite element replaced, which often required considerable judgment or mathematical skill. It seems reasonable, therefore, to claim that the popularity of the finite element method rests on the following two propositions:

1. That finite elements can fill spaces of arbitrary complexity. As a result, the method has thorough generality with respect to structural shape.

2. That the use of self-contained element types allows the mathematics of the solution process to be completely hidden from the user.

Taken together, these two propositions constitute a sure-fire recipe for success. They have brought two- and three-dimensional structural analysis, which was formerly the domain of experts, within the reach of the average engineer.

Other methods have cleaned up their act and can now make similar claims, particularly the claim that the solution process is transparent to the user. The boundary integral method, in particular, has been reborn as the boundary element method [18,19] with adherents who claim that its input preparation is easier than that for the finite element method. The boundary element method is not, on the other hand, quite so general with respect to structural shape. No one would, for example, consider using it to analyze a space framework.

1.3 REASONS FOR STUDYING THE DESIGN OF FINITE ELEMENTS

There is one class of engineers for whom finite element design is an obvious course of study, namely those who would like to become finite element designers. For them we have discouraging news. Worldwide, only a few hundred engineers and university professors have designed finite elements which are used by other people, and perhaps only a few dozen have designed commercially successful elements. Further, there is no such thing as a full-time "elementologist." However challenging and important the task may seem, it is at best a part-time job. The reason is not hard to find. Any finite element program needs only a few elements (twenty or less) and the turnover of elements is anything but rapid. For the most part, commercial element design work consists of incremental changes to address customer complaints. A thorough overhaul of the element library occurs perhaps once in ten years.

Element design impacts the general user of finite element systems in oblique ways—through its effect on the cost of analysis and on the accuracy of the results. It may be noted that the finite element is the repository for many of the assumptions made in finite element analysis including, principally, assumptions about spatial discretization and material interactions. Indeed, in linear static analysis the finite element is the repository for all assumptions beyond those which are the user's responsibility (geometry, loading, material constants).

In dynamic analysis or in nonlinear analysis, additional assumptions about the solution procedure (time integration algorithms, convergence criteria, etc.) will affect the cost and accuracy of the results. Even in these cases, the element-embedded assumptions about spatial discretization may well have the most important effect on accuracy.

So it is important that the general user understand something about the effects of elements on accuracy. The first thing he should learn is that all elements are not equal, that some elements are more accurate than others. The next thing he should learn is that a given element does not have equal accuracy in all situations. Both of these points can be demonstrated by the simple beam bending problem illustrated in Figure 1.4.

Note that the QDMEM element in Figure 1.4 gives very poor results in all cases, that the QUAD4 element gives fair results only when the element shape is rectangular, and that QUAD8 and QUADR give good results for both rectangular and nonrectangular shapes. An important qualification is that these results apply only to the case of in-plane bending; nothing can be inferred, for example, about the elements' accuracy for axial extension. We shall assume, for the moment, that it is unimportant to know anything about the design of these elements. We note only that they are all NASTRAN or MSC/NASTRAN elements and that the ones with more recent release dates perform better. Apparently something has been learned with the passage of time.

Consider next a slightly more complicated example. The thin cylindrical shell roof shown in Figure 1.5 is a standard benchmark test problem known as the Scordelis-Lo roof.[20]

The loading is a uniform gravity load, parallel to the z-axis. The output most frequently displayed in benchmark tests is the vertical displacement at the midpoint of the free edge. Symmetry can be used to reduce the analysis to a quarter of the whole roof. Figure 1.6 records results for six elements versus the number of degrees of freedom[*] in the one-quarter model. Degrees of freedom

---

[*]Equal to the sum over nodes of the number of active displacement components per node.

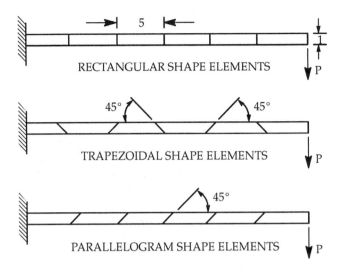

ELEMENT NAME	DATE OF RELEASE	RECTANGULAR ELEMENTS	TRAPEZOIDAL ELEMENTS	PARALLEL-OGRAM ELEMENTS
QDMEM	1970	.032	.016	.014
QUAD4	1974	.904	.071	.080
QUAD8	1980	.987	.946	.995
QUADR	1988	.993	.988	.984
Exact	—	1.000	1.000	1.000

Figure 1.4   **Finite Element Solutions of End-Loaded Cantilever Beam Problems.**

are a better measure of cost than the number of elements because different elements may have different numbers of connected nodes.

It is seen, first of all, that the rates of convergence vary widely and again that the more recent elements perform better. Note also that for the same age class the triangular elements TRIA2 and TRIA3 perform more poorly than the others,

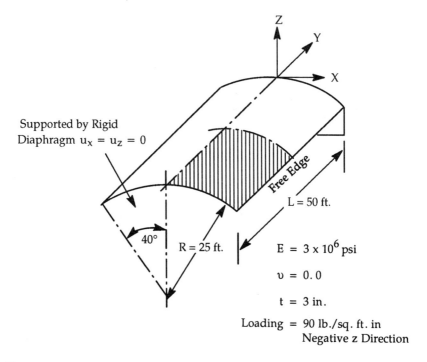

Figure 1.5   Shell Roof Under Gravity Load (the Scordelis-Lo Roof).

which are all rectangular elements. The latter include two versions of the same element, the eight-node Ahmad shell element,[21] which differ only with respect to the order of integration, but which exhibit spectacularly different convergence rates. Evidently some design feature of profound importance is operative here.

Judging from just these two examples, it seems clear that finite element users should be aware of the effects of element selection on accuracy. It is possible to provide an after-the-fact estimate of discretization error[22] so that the user will know the approximate accuracy of his results. This is a good idea but it is not enough. The user also needs to know how to improve the accuracy in an economic way—whether to use more elements, to rearrange them, or to use different types of elements. In practical problems, and particularly in large

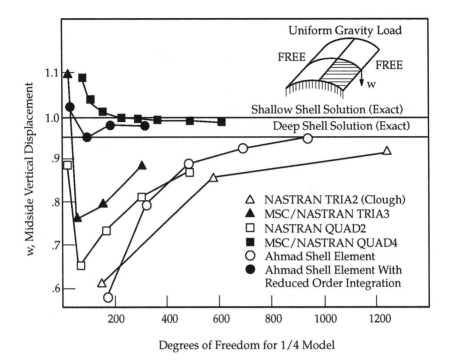

Figure 1.6   **Scordelis-Lo Roof: Performance of Pre-1970 Elements (△, □, ○)**
Versus Post-1970 Elements (▲, ■, ●).

ones, constraints on time and cost frequently limit the number of elements that can be used or the amount of rearranging that can be done.

It is, of course, always better to build a good finite element model the first time. This requires anticipation of expected element behavior. An experienced user will be guided, perhaps intuitively, by guidelines that he has acquired. Examples include the avoidance of constant strain triangles in a bending environment; the avoidance of large aspect ratios, severe tapers, and large skew angles in quadrilateral elements; a preference for eight-node quadrilaterals over four-node quadrilaterals (or vice versa depending on the situation); and avoidance of known defects of particular elements. Some of these guidelines may be wrong or out of date. If learned by rote, there is little chance that the analyst will be able to apply them successfully to new situations or that they will remain current. Someone in every finite element project should have an appreciation for the reasons behind the guidelines. The best way to acquire such an appreciation is to be aware of the principles and limitations of finite element design.

As treated in this book, finite element design means something different than a method of formulation. The word "design" emphasizes choice, while the word "method" restricts it. We will not be looking for a universal method that cranks out finite elements with impeccable academic credentials. Rather we will be studying the reasons why finite elements fail, in order to discover ways to make elements that do not fail or that fail in gentler ways.

It will turn out that finite elements fail in only a small number of different ways. A meaningful analogy can be constructed between the failure modes of finite elements and the diseases of organisms. So we will find it useful to speak of a pathology of finite element diseases (or perhaps *disorders* is a better term) which are complete with symptoms, underlying causes, and indicated therapies. We have already encountered the symptoms of one common finite element disorder in the beam bending problem (Figure 1.4). That disorder is known as *shear locking*. It attacks the bending performance of finite elements and makes them too stiff, sometimes by very large factors, as in the results shown for the QDMEM element.

The cure of a finite element disorder, through design changes, is a matter for specialists, but the general finite element practitioner (i.e., the user) should be able to spot the symptoms, identify the disorder, and take precautions.

Checkups are also useful. These take the form of sets of benchmark test problems. It is helpful to specialists and general practitioners alike if the test problems can identify particular element disorders. Throughout the book we will refer to a set of tests [23] which was selected with this in mind.

To date no two- or three-dimensional finite elements have been found that are disease free, although a few can be claimed to have only mild chronic conditions. We will discover that the prescription of a cure for one disorder very often introduces another. Thus we can speak of side effects.   Finite element design largely consists of the search for cures with minimal side effects.

1.4 CLASSIFICATION OF FINITE ELEMENT CHARACTERISTICS

Finite elements come in many different shapes and have a great variety of other properties, so it is well to say at the outset what will be covered and what will not. Short of creating something like a multi-authored encyclopedia of finite elements, an attempt at comprehensive coverage seems futile. Authors tend to cover best what they know best.

The selection of a set of attributes on which to hang the properties of any complex subject is, to some degree, arbitrary. The classification used here to describe the properties of finite elements includes the following attributes: physical discipline (e.g., elasticity), physical properties (e.g., stiffness), number of dimensions, shape, type and number of degrees of freedom, and method of formulation.

Table 1.1 provides a quick summary of the features that the book will cover within each class of attributes. A more detailed guide can, of course, be found in the Subject Index. In this section we will attempt to explain some of the choices that have been made.

While it is possible to cast finite element theory in a neutral, semantics-free context that is applicable to most of the physical disciplines which employ field theory, the reader's (and also the author's) intuition is lost thereby so that progress is slowed. We have, therefore, decided to develop finite element theory in the language of elasticity, which is the clearest choice based on years and breadth of finite element usage. Heat transfer will also be mentioned occasionally. So will electromagnetism, which is a subject that exhibits, relative to elasticity, some interesting similarities and also some instructive differences.

Most books on finite elements treat linear stiffness to almost the exclusion of other physical properties and this book is no exception. The reason is that most of the troubles, or disorders, of finite elements are connected with the linear stiffness matrix. Thus the design of a new element always begins with the stiffness matrix. The design of the load and mass properties, which is invariably simpler and more trouble free, will be given summary treatment. The simulation of plasticity and other nonlinear properties is a vast but separable subject, best left to experts in that specialty.

Table 1.1

Classification of Finite Element Characteristics

PHYSICAL DISCIPLINES:

ELASTICITY	HEAT TRANSFER	ELECTROMAGNETISM
X	O	O

PHYSICAL PROPERTIES:

STIFFNESS			
LINEAR	NONLINEAR	LOADS	MASS
X	—	O	O

DIMENSION:

ONE DIMENSIONAL	TWO DIMENSIONAL	THREE DIMENSIONAL
—	X	X

SHAPE (2-D):

	TRIANGLE	QUAD
Flat:	X	X
Curved:	X	X
Axisymmetric:	—	—

SHAPE (3-D):

TETRA	PENTA	HEXA
O	O	X

X - emphasized in book
O - covered lightly in book
— - not covered

Table 1.1 (continued)

Classification of Finite Element Characteristics

DEGREES OF FREEDOM (TYPE):	TRANSLATIONS	ROTATIONS	HIGHER-ORDER DERIVATIVES
	X	X	O

DEGREES OF FREEDOM (NUMBER):	MINIMUM	NEXT LOWEST	HIGHER
	X	X	O

METHOD:	ASSUMED DISPLACEMENT	ASSUMED STRESS	ASSUMED STRAIN
	X	O	X

X - emphasized in book
O - covered lightly in book

It will be noted that we do not propose to treat one-dimensional finite elements. The reason is that one-dimensional elements can be designed to arbitrarily high accuracy by any number of methods, so that there is no need to use approximate theories that are better suited to two- and three-dimensional elements. Occasionally we may refer to a one-dimensional example to illustrate a point.

It is interesting that only a few shapes have found favor in finite element design. These include triangles and quadrilaterals and also tetrahedra, pentahedra, and hexahedra. The shape of each may be regular or irregular, with straight or curved edges and flat or curved faces. The quadrilateral and hexahedral shapes will be emphasized because they are used more and have more interesting disorders than the other shapes. We will not treat axisymmetric elements due to the author's lack of recent experience with them.

The use of higher-order derivatives was once common in the design of plate and shell elements but it is no longer. We will touch upon the reasons why higher-order derivatives were considered necessary for a certain class of elements, but otherwise we will stick with translations and rotations as degrees of freedom.

Lowest-order and next-to-lowest-order elements (e.g., four-, eight-, and nine-node quadrilaterals) will be emphasized because they represent the vast majority of elements in use today. Higher-order hierarchical elements[24] will be described because they form the basis of a recent development called the p-method.[25]

Most of the element designs treated in the book will use the assumed displacement method. In some cases, assumed strain fields will be introduced to remedy disorders. The treatment of assumed stress methods, or to use a better term, hybrid stress-displacement methods, will be minimal due to the author's inexperience with this important class of elements.

REFERENCES

1.1   T. J. R. Hughes, I. Levit, and J. Winget, "Element-by-Element Implicit Algorithms for Problems of Structural and Solid Mechanics," *Comput. Methods Appl. Mech. Engrg.*, 36, pp 241-54, 1983.

1.2   R. W. Clough, "The Finite Element Method in Plane Stress Analysis," Proc. 2nd ASCE Conf. on Electronic Computation, Pittsburgh, PA, pp 345-78, 1960.

1.3   A. Hrennikoff, "Solution of Problems in Elasticity by the Framework Method," *J. Appl. Mech.*, 8, pp 169-75, 1941.

1.4   S. J. Garvey, "The Quadrilateral Shear Panel," *Aircraft Engineering*, p. 134, 1951.

1.5   T. von Karman, "Calculation of Pressure Distribution on Airship Hulls," NACA TM 574, 1930.

1.6   B. A. Finlayson and L. E. Scriven, "The Method of Weighted Residuals— A Review," *Appl. Mech. Reviews*, 19, pp 735-8, 1966.

1.7   L. F. Richardson, "The Approximate Arithmetical Solution by Finite Differences of Physical Problems," *Trans. Royal Soc.* (London), A210, pp 307-57, 1910.

1.8   R. V. Southwell, *Relaxation Methods in Theoretical Physics*, The Clarendon Press, Oxford, 1946.

1.9 B. Langefors, "Analysis of Elastic Structures by Matrix Transformation with Special Regard to Semimonocoque Structures," *J. Aeronautical Sci.*, 19, pp 451-8, 1952.

1.10 J. H. Argyris, "Energy Theorems and Structural Analysis," *Aircraft Eng.*, 26 (Oct.-Nov. 1954), 27 (Feb.-May 1955).

1.11 M. J. Turner, R. W. Clough, H. C. Martin, and L. J. Topp, "Stiffness and Deflection Analysis of Complex Structures," *J. Aeronautical Sci.*, 23, pp 803-23, p. 854, 1956.

1.12 I. C. Taig, "Structural Analysis by the Matrix Displacement Method," Engl. Electric Aviation Report No. 5017, 1961.

1.13 R. H. Gallagher, J. Padlog, and P. P. Bijlaard, "Stress Analysis of Heated Complex Shapes," *ARS Journal*, pp 700-7, 1962.

1.14 A. Adini and R. W. Clough, *Analysis of Plate Bending by the Finite Element Method* and Report to Natl. Sci. Foundation/USA, G7337, 1961.

1.15 T. H. H. Pian, "Derivation of Element Stiffness Matrices by Assumed Stress Distribution," *AIAA Intl.*, 2, pp 1332-6, 1964.

1.16 J. H. Argyris, "Continua and Discontinua," Proc. Conf. Matrix Methods in Struct. Mech., Air Force Inst. of Tech., Wright-Patterson AFB, Ohio, 1965.

1.17 B. M. Irons, "Engineering Application of Numerical Integration in Stiffness Methods," *J. AIAA*, 14, pp 2035-7, 1966.

1.18 T. A. Cruse and F. J. Rizzo, "A Direct Formulation and Numerical Solution of the General Transient Elastodynamics Problem," *Intl. J. Math. Anal. Appl.*, 22, pp 244-59, 1968.

1.19 P. K. Banerjee and R. Butterfield, *Boundary Element Methods in Engineering Science*, McGraw-Hill, London and New York, 1981.

1.20 A. C. Scordelis and K. S. Lo, "Computer Analysis of Cylindrical Shells," *J. Amer. Concr. Inst.*, 61, pp 539-61, 1961.

1.21 S. Ahmad, "Pseudo-Isoparametric Finite Elements for Shell and Plate Analysis," Proc. Conf. on Recent Advances in Stress Analysis *J. B.C.&A*, Royal Aero Soc., pp 6-20 to 6-21, 1968.

1.22 O. C. Zienkiewicz and J. Z. Zhu, "A Simple Error Estimator and Adaptive Procedure for Practical Engineering Analysis," *Intl. J. Numer. Methods Eng.*, 24, pp 337-57 1987.

1.23 R. H. MacNeal and R. L. Harder, "A Proposed Standard Set of Problems to Test Finite Element Accuracy," *Finite Elem. Analysis & Design*, 1, pp 3-20, 1985.

1.24   A. G. Peano, "Hierarchies of Conforming Finite Elements for Elasticity and Plate Bending," *Comp. Meth. & Appl.*, pp 211-24, 1976.

1.25   B. A. Szabó and A. K. Mehta, "p-Convergent Finite Element Approximations in Fracture Mechanics," *Intl. J. Numer. Methods Eng.*, 12, pp 551-60, 1978.

2
Formulation Fundamentals

This chapter and the next three chapters outline the basic theory of finite elements based on assumed displacement fields and make brief excursions into the theories of other element types. They constitute, to continue the medical analogy introduced in Chapter 1, the description of element anatomy that is a necessary preliminary to the consideration of element pathology.

2.1 PHYSICS

As indicated in Table 1.1, the finite element coverage in this book emphasizes linear elasticity. In fact, propositions which apply generally will most often be stated in the terminology of engineering mechanics. At the same time, we will also use examples from linear heat transfer and electromagnetism to illustrate points.

An outline of the physics of these three disciplines appears below for ready reference and comparison.

2.1.1 Linear Elasticity

The subject of elasticity deals with the deformations of solid bodies. See References 2.1 and 2.2 for classical engineering treatment of the subject. Let the position of a point in a three-dimensional body be defined by a vector of Cartesian coordinates

$$\{x\} = \begin{Bmatrix} x \\ y \\ z \end{Bmatrix} \qquad (2:1)$$

The *displacement* of the point from its rest position is also represented by a vector

$$\{u\} = \begin{Bmatrix} u \\ v \\ w \end{Bmatrix} \qquad (2:2)$$

where the vector components, u, v, and w are components of displacement in the x, y, and z directions respectively.

Comments about notation are in order at this point. Vectors and matrices will, for the most part, be written with standard matrix notation. Thus curly brackets, $\{\ \}$, will represent a column vector, while $\lfloor\ \rfloor$ or $\{\ \}^T$ will represent a row vector, and $[\ ]$ will represent a matrix. We prefer this notation to the representation of vectors and matrices by special type fonts because it is unambiguous and carries over into handwritten notes. In similar fashion we will employ Σ to designate summation, in preference to a summation convention regarding subscripts. We will, however, borrow the subscripted comma convention from tensor analysis to indicate partial differentiation. Thus, $w_{,xy} = \partial^2 w / \partial x \partial y$. Finally, we will find it convenient, on occasion, to employ the notation of vector analysis for the gradient, ∇, divergence, $\nabla \bullet$, and curl, ∇x, of vectors. In such cases, an arrow or a tilde over a quantity, e.g., $\vec{B}$ or

$\tilde{B}$, will indicate that the quantity is a vector in either two- or three-dimensional space.

To return to the summary of linear elasticity, the *strain* at a point in the body describes the deformations of an infinitesimal cube at that point. It will be represented by the vector

$$\{\varepsilon\} = \begin{Bmatrix} \varepsilon_x \\ \varepsilon_y \\ \varepsilon_z \\ \gamma_{xy} \\ \gamma_{xz} \\ \gamma_{yz} \end{Bmatrix} \qquad (2:3)$$

The strain is, in reality, a second order 3 x 3 tensor, $\left[\varepsilon_{ij}\right]$, but engineering mechanics conventionally treats its six independent components as a vector, or pseudo-vector to be more precise. Note the sequential order of shear strains, $\gamma_{xy}, \gamma_{xz}, \gamma_{yz}$. Different authors prefer different orders. We prefer this order, which is also used by Irons and Ahmad. [3]

In linear elasticity, the strain components are first order derivatives of the displacement components. Thus

$$\{\varepsilon\} = [L]\{u\} \qquad (2:4)$$

where $[L]$ is a matrix of first order derivative operators.

$$[L] = \begin{bmatrix} \dfrac{\partial}{\partial x} & 0 & 0 \\[2ex] 0 & \dfrac{\partial}{\partial y} & 0 \\[2ex] 0 & 0 & \dfrac{\partial}{\partial z} \\[2ex] \dfrac{\partial}{\partial y} & \dfrac{\partial}{\partial x} & 0 \\[2ex] \dfrac{\partial}{\partial z} & 0 & \dfrac{\partial}{\partial x} \\[2ex] 0 & \dfrac{\partial}{\partial z} & \dfrac{\partial}{\partial y} \end{bmatrix} \tag{2:5}$$

Carrying out the indicated operations, we obtain the defining equations for linear Cartesian strain components.

$$\{\varepsilon\} = \begin{Bmatrix} \varepsilon_x \\ \varepsilon_y \\ \varepsilon_z \\ \gamma_{xy} \\ \gamma_{xz} \\ \gamma_{yz} \end{Bmatrix} = \begin{Bmatrix} u_{,x} \\ v_{,y} \\ w_{,z} \\ u_{,y} + v_{,x} \\ u_{,z} + w_{,x} \\ v_{,z} + w_{,y} \end{Bmatrix} \tag{2:6}$$

Stress components can be defined as the self-equilibrating forces acting on the faces of an (infinitesimal) unit cube (see Figure 2.1).

The direct stresses $\left(\sigma_x, \sigma_y, \sigma_z\right)$ are components of traction normal to the faces while the shear stresses $\left(\tau_{xy}, \tau_{xz}, \tau_{yz}\right)$ are components of traction parallel to the faces. The stress components can also be expressed in vector form, i.e.,

$$\{\sigma\} = \begin{Bmatrix} \sigma_x \\ \sigma_y \\ \sigma_z \\ \tau_{xy} \\ \tau_{xz} \\ \tau_{yz} \end{Bmatrix} \tag{2:7}$$

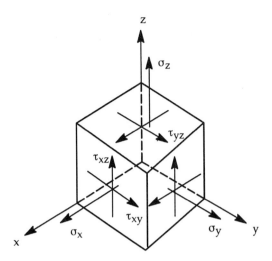

Figure 2.1   Three-Dimensional Stress Components.

The orientation of stress components on the unit cube is such that each component of $\{\sigma\}$ acts directly to increase the corresponding component of the strain vector, $\{\varepsilon\}$. Thus the volumetric density of the incremental work done by $\{\sigma\}$ on increments of strain, $\{\delta\varepsilon\}$, is

$$\delta W = \{\sigma\}^T \{\delta\varepsilon\} \qquad (2:8)$$

The stress vector is also related to the strain vector by a *constitutive* relationship that expresses the elastic properties of the body. The most general form that this relationship can have for linear elastic materials is

$$\{\sigma\} = [D]\{\varepsilon - \varepsilon_0\} + \{\sigma_0\} \qquad (2:9)$$

where $[D]$ is a 6 x 6 matrix of elastic moduli, $\{\varepsilon_0\}$ is a vector of initial strains, most often due to thermal expansion, and $\{\sigma_0\}$ is a vector of initial stresses. The elasticity matrix $[D]$ is symmetric so that, in the most general case of anisotropy, it may have twenty-one independent terms. For isotropic materials, all of the terms of $[D]$ depend on just two elastic constants. The elastic constants most frequently employed in engineering are E, the elastic modulus, and υ, Poisson's ratio. For the isotropic case

$$[D] = \frac{E}{2(1 + \upsilon)} \begin{bmatrix} \dfrac{2(1 - \upsilon)}{1 - 2\upsilon} & \dfrac{2\upsilon}{1 - 2\upsilon} & \dfrac{2\upsilon}{1 - 2\upsilon} & 0 & 0 & 0 \\ & \dfrac{2(1 - \upsilon)}{1 - 2\upsilon} & \dfrac{2\upsilon}{1 - 2\upsilon} & 0 & 0 & 0 \\ & & \dfrac{2(1 - \upsilon)}{1 - 2\upsilon} & 0 & 0 & 0 \\ & & & 1 & 0 & 0 \\ & \text{SYMMETRIC} & & & 1 & 0 \\ & & & & & 1 \end{bmatrix}$$

(2:10)

alternative pairs of isotropic elastic constants are sometimes used; for example, the shear modulus,

$$G = \frac{E}{2(1 + \upsilon)} \qquad (2:11)$$

and the bulk modulus,

$$K = \frac{E}{3(1 - 2\upsilon)} \qquad (2:12)$$

The shear modulus and the bulk modulus are more fundamental constants than the elastic modulus and Poisson's ratio because they relate, respectively, to the two fundamental types of deformation: shear (deviatoric strain) and volumetric expansion (dilatation). For nearly incompressible materials, K is very much larger than G, so that υ approaches one-half.

The stresses $\{\sigma\}$ are in equilibrium with the vector of body force density, $\{p\}$, applied to the unit cube. The equations of equilibrium are

$$\{p\} = \begin{Bmatrix} p_x \\ p_y \\ p_z \end{Bmatrix} = - \begin{Bmatrix} \sigma_{x,x} + \tau_{xy,y} + \tau_{xz,z} \\ \sigma_{y,y} + \tau_{xy,x} + \tau_{yz,z} \\ \sigma_{z,z} + \tau_{xz,x} + \tau_{yz,y} \end{Bmatrix} \qquad (2:13)$$

Body forces can arise from electromagnetic effects, i.e., the interaction of magnetic and electric fields with currents and charges in the body, or from inertia effects. In the latter case

$$\{p^i\} = \begin{Bmatrix} p_x^i \\ p_y^i \\ p_z^i \end{Bmatrix} = -\rho \begin{Bmatrix} u_{,tt} \\ v_{,tt} \\ w_{,tt} \end{Bmatrix} \qquad (2:14)$$

where ρ is the mass density and t is time.

On the surface of the body, the stresses are in equilibrium with the vector of tractions, $\{t\}$, acting on the surface. The equations of equilibrium (boundary conditions) are

$$\{t\} = \begin{Bmatrix} t_x \\ t_y \\ t_z \end{Bmatrix} = \begin{Bmatrix} \sigma_x n_x + \tau_{xy} n_y + \tau_{xz} n_z \\ \sigma_y n_y + \tau_{xy} n_x + \tau_{yz} n_z \\ \sigma_z n_z + \tau_{xz} n_x + \tau_{yz} n_y \end{Bmatrix} \qquad (2:15)$$

where $\left(n_x, n_y, n_z \right)$ are the direction cosines of the outward normal to the surface.

The foregoing summarizes the physics of classical linear three-dimensional elasticity. Important two-dimensional cases include plane stress, plane strain, and axisymmetry. For plane stress, the stress components σ_z, τ_{xz}, and τ_{yz} are assumed to be zero while for plane strain, the strain components ε_z, γ_{xz}, and γ_{yz} are assumed to be zero. If isotropy is assumed, the homogeneous constitutive equation for plane stress $\left(\sigma_z = 0 \right)$ becomes

$$\left\{ \begin{array}{c} \sigma_x \\ \\ \sigma_y \\ \\ \tau_{xy} \end{array} \right\} = \frac{E}{1-\upsilon^2} \left[\begin{array}{ccc} 1 & \upsilon & 0 \\ \\ \upsilon & 1 & 0 \\ \\ 0 & 0 & \dfrac{1-\upsilon}{2} \end{array} \right] \left\{ \begin{array}{c} \varepsilon_x \\ \\ \varepsilon_y \\ \\ \gamma_{xy} \end{array} \right\} \qquad (2{:}16)$$

while for plane strain $\left(\varepsilon_z = 0 \right)$

$$\left\{ \begin{array}{c} \sigma_x \\ \\ \sigma_y \\ \\ \tau_{xy} \end{array} \right\} = \frac{E}{1+\upsilon} \left[\begin{array}{ccc} \dfrac{1-\upsilon}{1-2\upsilon} & \dfrac{\upsilon}{1-2\upsilon} & 0 \\ \\ \dfrac{\upsilon}{1-2\upsilon} & \dfrac{1-\upsilon}{1-2\upsilon} & 0 \\ \\ 0 & 0 & \dfrac{1}{2} \end{array} \right] \left\{ \begin{array}{c} \varepsilon_x \\ \\ \varepsilon_y \\ \\ \gamma_{xy} \end{array} \right\} \qquad (2{:}17)$$

The two-dimensional (plane) form of the other important relationships, e.g., Equations 2:6, 2:13, 2:14, and 2:15, is obtained by removing all quantities with z subscripts.

2.1.2   Linear Heat Conduction (and Storage)

As defined in physics, heat describes the average kinetic energy of particles in random motion.   Heat can be transferred from one place to another by conduction (transfer between adjacent particles with fixed mean positions), by convection (transfer by the directed motion of particles), or by electromagnetic radiation. Convection and radiation tend to be highly nonlinear and will not be considered here. Heat can also be stored by raising the random kinetic energy of particles. The parameter which measures the random kinetic energy is the temperature, u. The rate at which heat is transferred within a conducting medium is proportional to the (negative) gradient of the temperature. Thus if we define $\{q\}$ to be the vector of *heat flow* (energy transferred across a unit area per unit time),

$$\{q\} = \begin{Bmatrix} q_x \\ q_y \\ q_z \end{Bmatrix} = -[K] \begin{Bmatrix} u_{,x} \\ u_{,y} \\ u_{,z} \end{Bmatrix} = -[K]\{\nabla u\} \qquad (2{:}18)$$

where $[K]$ is the *conductivity* matrix.

At every point, a balance exists between the rate at which heat is generated, Q (energy input per unit volume per unit time), and heat outflow. Thus

$$\nabla \bullet q = q_{x,x} + q_{y,y} + q_{z,z} = Q \qquad (2{:}19)$$

Heat can be generated in a solid medium by electrical effects (ohmic heating) and by chemical or nuclear effects. It can also be stored as random thermal energy and released. The value of Q due to the release of stored heat is proportional to the (negative) rate of change of temperature. Thus

$$Q^s = -\rho c u_{,t} \qquad (2{:}20)$$

where ρ is the mass density and c is a constant.

Heat may also be input at the boundary of a conducting body. If q^b is the rate of heat input per unit area at a point on the boundary, then the components of heat flow at the same point are

$$\{q\} = -\{n\}q^b \qquad (2{:}21)$$

where the components of $\{n\}$ are the direction cosines of the outward normal.

2.1.3 Magnetostatics

Engineers have used finite elements for quite a long time to analyze problems in electromagnetism,[4] but with nowhere near the same intensity as in structural mechanics. Recent interest centers in the development of techniques that are suitable for three-dimensional field problems. We consider here the application of a construct known as the *magnetic vector potential*[5] because of its power in three-dimensional applications and its strong analogy with displacement in structural mechanics. It should also be said that there is, at present, no general consensus among electrical engineers on the best general approach to field problems. The situation is not unlike that which existed when the force and displacement methods of structural mechanics were contending in the 1950s and 1960s.

We will restrict our attention here to magnetostatics, which considers the interaction between magnetic fields and electrical currents. It is a subject of much importance to the designers of electrical machinery. The quantity of greatest interest is the vector of *magnetic induction*, $\vec{B}$. It is the quantity which induces mechanical forces on currents in conductors according to the relationship

$$\vec{F} = \vec{J} \times \vec{B} \qquad (2{:}22)$$

where $\vec{J}$ is the *current density*. It also induces voltages in conducting loops according to

$$\nabla \times \vec{E} = -\dot{\vec{B}} \qquad (2{:}23)$$

where $\vec{E}$ is the *electric field strength* (volts/meter).

The magnetic induction is itself generated by current, but indirectly through another vector quantity, $\vec{H}$, called the *magnetic field strength*. $\vec{B}$ is related to $\vec{H}$ through a matrix of magnetic material properties $[\mu]$ known as the *permeability* matrix. Thus

$$\vec{B} = [\mu]\,\vec{H} \qquad (2{:}24)$$

Permeability is very much larger in iron, nickel, and other ferromagnetic materials than it is in air. It tends to be strongly nonlinear and anisotropic in ferromagnetic materials.

The magnetic field strength, $\vec{H}$, is related to current density by

$$\nabla \times \vec{H} = \vec{J} \qquad (2{:}25)$$

An important property of the magnetic induction is that it has no sources or sinks. Thus

$$\nabla \cdot \vec{B} = 0 \qquad (2:26)$$

We note that, if the current density is given, $\vec{B}$ is obtained as the solution of a *first* order differential equation

$$\nabla \times [\mu]^{-1} \vec{B} = \vec{J} \qquad (2:27)$$

subject to a differential constraint (Equation 2:26).

This is a rather odd situation with which to confront a numerical analyst, who would much prefer a second order differential equation and no constraint. That happier condition can be realized by expressing $\vec{B}$ as the curl of another vector, called the *magnetic vector potential*, $\vec{A}$. Thus, let

$$\vec{B} = \nabla \times \vec{A} \qquad (2:28)$$

which automatically satisfies the divergence condition (Equation 2:26), since the divergence of the curl of a vector is identically zero. In addition, Equation 2:27 is replaced by

$$\nabla \times [\mu]^{-1} \nabla \times \vec{A} = \vec{J} \qquad (2:29)$$

We note that $\vec{A}$ has a physical interpretation because, by comparing Equations 2:23 and 2:28,

$$\vec{E} = -\dot{\vec{A}} + \nabla \psi \qquad (2:30)$$

where ψ is some scalar quantity.

Thus $\vec{A}$ is the (negative) time integral of the *induced* electric field strength. ($\nabla \psi$ is the electric field strength due to electrostatic charge.)

Another important quantity is the density of energy in the magnetic field

$$V = \frac{1}{2} \vec{B} \cdot \vec{H} \qquad (2:31)$$

From the point of view of finite element design, the important results are the following.

$$\vec{B} = \nabla \times \vec{A} \qquad (2{:}32a)$$

$$\vec{H} = [\mu]^{-1}\vec{B} \qquad (2{:}32b)$$

$$\nabla \times \vec{H} = \vec{J} \qquad (2{:}32c)$$

$$V = \tfrac{1}{2}\,\vec{B} \bullet \vec{H} \qquad (2{:}32d)$$

These equations are equivalent, respectively, to the following relationships in elasticity: the definition of strain, a constitutive relationship, an equilibrium equation, and the definition of strain energy.

2.1.4   Similarities and Differences

The variables and relationships of classical linear elasticity, heat transfer, and magnetostatics have analogies which allow these disciplines to be analyzed with the same procedures. Table 2.1 indicates the analogies and provides a generic term to describe each type of quantity or relationship. Thus the "potential functions" for elasticity, heat conduction, and magnetostatics are, respectively, the displacement vector $\{u\}$, the temperature u, and the magnetic vector potential $\{A\}$, while the "stresses" are, respectively, the stress vector $\{\sigma\}$, the heat flow $\{q\}$, and the magnetic field strength $\{H\}$. The analogy extends to the relationships which are most important for finite element design, namely the definition of strain, the constitutive relationship, and the equation of equilibrium.

Different notation conventions exist side by side in Table 2.1. This is done out of respect for the conventions most commonly used in each discipline but also for conciseness. Thus the equilibrium equation listed for elasticity in Table 2.1 treats stress as a second order tensor (see Equation 2:15 for the expanded vector equivalent).

Table 2.1

Comparison of Analogous Properties in Three Disciplines

GENERIC TERM	DISCIPLINE		
	ELASTICITY	HEAT CONDUCTION	MAGNETO-STATICS
Potential Function	$\{u\}$	u	$\{A\}$
"Strain"	$\{\varepsilon\}$	$-\{\nabla u\}$	$\{B\}$
"Stress"	$\{\sigma\}$	$\{q\}$	$\{H\}$
Load Density	$\{p\}$	$\{Q\}$	$\{J\}$
Boundary Load	$\{t\}$	$\{q^b\}$	$\bar{H} \times \bar{n}$
Definition of Strain	$\{\varepsilon\} = [L]\{u\}$	$-\{\nabla u\}$	$\{B\} = \nabla \times \{A\}$
Constitutive Relationship	$\{\sigma\} = [D]\{\varepsilon\}$	$\{q\} = -[K]\{\nabla u\}$	$\{H\} = [\mu]^{-1}[B]$
Potential Energy	$\frac{1}{2}\{\sigma\}^T\{\varepsilon\}$	—	$\frac{1}{2}\{H\}^T\{B\}$
Equilibrium	$\sum_j \sigma_{ij,j} = -P_i$	$\nabla \bullet \{q\} = Q$	$\nabla \times \{H\} = \{J\}$
Inertia Loading	$\{p^i\} = -\rho\{u_{,tt}\}$	$Q^s = -\rho c u_{,t}$	—

While strong similarities exist between the mathematical formulations of the three disciplines, there are also important differences. Let the student beware who believes that, by learning an analogy, he has mastered a new discipline. Note, for example, that the quantities of greatest practical importance in analogous disciplines need not belong to the same generic type. Thus the quantity of greatest practical importance in elasticity is the stress $\{\sigma\}$, while in heat conduction it is the temperature u, and in magnetostatics it is the magnetic induction $\{B\}$. In heat conduction the quantity corresponding to "strain" does not even have a special name; it is simply the (negative) temperature gradient. Further, there is no name in heat transfer for the quantity corresponding to "potential energy" in mechanics. That is because heat transfer is a higher order abstraction than mechanics or electromagnetism which deals directly with energy density (temperature) and energy transfer (heat flow) as first order response quantities. We can, of course, create a functional corresponding to potential energy for heat transfer, but that is entirely a matter of mathematical convenience without physical standing.

In electromagnetism, the magnetic vector potential $\{A\}$ has so little standing that most of those electrical engineers who are even familiar with it consider it to be a mathematical construct without physical significance. They use it reluctantly and only where necessary, as in certain types of three-dimensional analysis. Perhaps that situation will change with the increased availability of finite element codes based on vector potential formulations. [6]

The definitions of "strain" for elasticity and magnetostatics exhibit interesting similarities and differences. Each is composed of first order spatial derivatives of a vector potential with three components. There are, in all, nine such derivatives, but observe below how each discipline selects different components.

ELASTICITY   MAGNETOSTATICS

$$\{\varepsilon\} = \begin{Bmatrix} u_{,x} \\ v_{,y} \\ w_{,z} \\ u_{,y} + v_{,x} \\ u_{,z} + w_{,x} \\ v_{,z} + w_{,y} \end{Bmatrix} \qquad \{B\} = \begin{Bmatrix} A_{z,y} - A_{y,z} \\ A_{x,z} - A_{z,x} \\ A_{y,x} - A_{x,y} \end{Bmatrix}$$

Since $\{A\}$ is the analog of $\{u\}$, we see that the components of $\{B\} = \nabla x\{A\}$ are the analogs of components of rotation (with only a factor of 1/2 missing). Thus, in an analog sense, the three components of $\{B\}$ are orthogonal to the six components of $\{\varepsilon\}$; or in other words, magnetostatics keeps the derivatives that elasticity throws away, and vice versa! This fact makes for some interesting differences between finite element analysis in the two disciplines. For example, while elastic bodies have six rigid body modes (three translations and three rotations), magnetic fields have nine (three sets of constant values for $\{A\}$ and six constant derivatives of $\{A\}$ in the form of the elastic strains).

2.2 BASIC ASSUMPTIONS

An instructive way to define the term "finite element" is to state a set of rules that finite elements should satisfy. This approach generates discussion about exceptions, thereby further refining the finite element concept. The following is a set of such rules, or assumptions, about finite element formulation. These rules are observed by nearly all of the elements treated in this book.

Rule 1:   *Each finite element fills a well-defined region of space and represents all of the relevant physics within the space. A finite element which satisfies this rule is self contained.*

A trivial example which does not satisfy Rule 1 is a scalar spring. Since a scalar spring has no stated geometric properties, its stiffness must be supplied by the user. It is not self contained.

Another example which does not satisfy the rule is the representation of a membrane plate by a triangular lattice of rod elements (in the manner of Hrennikoff[7]). While the properties of each rod may be derived from the geometry of the array and the material properties of the plate, each rod represents strain in one direction only and is not, therefore, self contained. This becomes quite apparent near the edges of the plate.

Rule 2:   *Two finite elements interact with each other only through the common values*
             of a finite set of variables located in their mutual boundary.

In elastic finite elements, the boundary variables are most frequently
the components of displacement at specific locations. More generally,
they may simply be parameters which fix the magnitudes of
distributed displacement states, as in the case of hierarchical
elements. [8] They may also be tractions but such elements are not,
generally speaking, well behaved. [9] Note that this rule establishes
the "finiteness" of finite elements. Note further that the finite elements
are isolated from each other by the boundary variables. This makes it
possible to reduce all element properties to a set of equations written
in terms of the boundary variables only. Each such boundary variable
is then directly coupled only to the variables in the boundaries of the
elements it touches (see Figure 2.2).

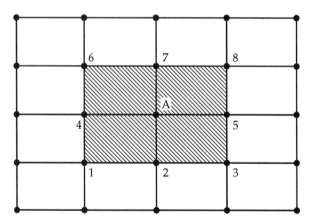

**Figure 2.2   Illustration of Node Coupling.   Node A is Coupled
Only to Nodes 1 Through 8.**

Finite element analysis based on the force method [10] does not make
the same use of this rule, with the result that the *flexibility* matrix that
couples redundant forces tends to be much fuller than the typical
stiffness matrix of the displacement matrix. This is the principal
reason why the force method has fallen into disfavor.

An important exception to Rule 2 occurs when radiation is present in heat transfer analysis. Radiated heat is transferred between noncontiguous heat conduction elements according to a set of *view factors.* [11] The rule can be finessed by considering the entire space bounding the heat conduction elements to be a special "radiation" element.

Rule 3:   *In the interior of finite elements, response variables (e.g., displacements, stresses, and strains) vary according to functions of position selected by the element designer. The degree of smoothness of these functions must be sufficient to ensure integrability of the (strain) energy.*

Other conditions are frequently placed on the assumed fields, such as the requirement that displacements be continuous at inter-element boundaries, that stresses be in equilibrium, or even that all of the differential equations be satisfied. Such conditions give rise to distinct classes of elements.

An example which violates even the relaxed requirements of Rule 3 is an element whose stiffness matrix is specified by the user, such as the NASTRAN general element, GENEL. Violations of Rules 1 and 3 tend to go together.

These three basic assumptions seem to be enough. We could state more assumptions but they would tend to exclude important types of elements. For example, we could insist that all element properties be derived according to strict variational principles, but that would eliminate selective underintegration and other successful "variational crimes."

2.3   THE BASIC STIFFNESS FORMULATION

Nearly all finite element programs in use today express the system equations in terms of a vector of displacement variables, $\left\{ u_i \right\}$, whose components fix the magnitudes of distributed displacement states within the elements. Usually, but not necessarily, the components of $\left\{ u_i \right\}$, correspond to the components of displacement at a discrete set of points. The general form of the matrix system equation is

$$\left[K_{ii} \right] \left\{ u_i \right\} = \left\{ P_i \right\} \tag{2:33}$$

where $\left[K_{ii} \right]$, the stiffness matrix, has constant elements and $\left\{ P_i \right\}$, the load vector, includes all forces except those due to linear elasticity. Thus $\left\{ P_i \right\}$ may, in general, include nonlinear and time-dependent terms. For linear static analysis, $\left\{ P_i \right\}$ is a vector of constants.

The system equation is obtained by summing the contributions of individual elements. Thus,

$$\left[K_{ii} \right] = \sum_e \left[K_{ii}^e \right] \tag{2:34}$$

$$\left\{ P_i \right\} = \sum_e \left\{ P_i^e \right\} + \left\{ P_i^d \right\} \tag{2:35}$$

where the summations extend over all elements. The load vector contains a part associated with the elements $\sum_e \left\{ P_i^e \right\}$ and a part associated with the displacement variables, $\left\{ P_i^d \right\}$.

The basic assumptions outlined in Section 2.2 form the basis of a general procedure for computing $\left[K_{ii}^e \right]$ and $\left\{ P_i^e \right\}$. We recall that the assumptions assert, in brief, that a finite element occupies a well-defined region of space, that it interacts with a neighboring element through common values of variables located in their mutual boundary, and that interior response quantities vary according to designer-supplied functions of position.

Figure 2.3 shows such an element. We see that it is not a practical element because it has five sides. Being impractical, it can act as a neutral representative of all two- and three-dimensional elements. Note that the edges are straight, but they could as well be curved. The forces acting on the element include a body force density, $\left\{ p \right\}$, and surface (or edge) tractions, $\left\{ t \right\}$. The vector $\left\{ u_i \right\}$ contains the boundary variables, which are displacement-like. As represented in Figure 2.3, they are associated with discrete nodal points at the corners of the element, but that is not required.

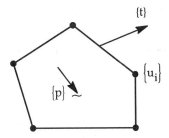

Figure 2.3 A Representative Finite Element.

It is useful, in the present context, to replace the surface tractions $\{t\}$, which represent forces exerted by adjacent elements, by generalized forces $\left\{F_i^e\right\}$, which are forces exerted on the element by the boundary displacements $\left\{u_i\right\}$. (The forces exerted by the element on $\left\{u_i\right\}$ are elements of the vector $-\left\{F_i^e\right\}$.) This substitution is permitted by the assumption that adjacent elements interact only through common values of the boundary displacements. The consequences of this substitution will be explored later.

We also need a relationship between the strains $\{\varepsilon\}$ at interior points and the boundary displacement set $\left\{u_i\right\}$. Let this relationship be expressed as

$$\{\varepsilon\} = [B]\left\{u_i\right\} \tag{2:36}$$

The matrix $[B]$, called the *strain-displacement* matrix, is derived from the designer-supplied functions of position. It can, for example, be evaluated from an assumed displacement distribution

$$\{u\} = [N]\left\{u_i\right\} \tag{2:37}$$

by using the definition of strain (see Section 2.1.1)

$$\{\varepsilon\} = [L]\{u\} = [L][N]\left\{u_i\right\} \tag{2:38}$$

If we restrict our attention to three-dimensional elasticity, then $\{u\}$ in Equation 2:37 has three components. $\{u_i\}$, on the other hand, has a much larger number of components. It is immaterial whether we consider $\{u_i\}$ to include all the boundary variables or only those in the boundaries of a particular element.

We will show that the matrix $[B]$ can also be evaluated from an assumed strain distribution, from an assumed stress distribution, or even from a combination of two assumed distributions. For the present, we do not concern ourselves with how $[B]$ is evaluated. We only assume that it exists.

Our goal is to compute the element stiffness matrix $\left[K_{ii}^e\right]$ and the associated applied load vector $\left\{P_i^e\right\}$. The easiest routes to this goal employ energy principles. We can, for example, use the principle of virtual work. This principle assumes the existence of virtual (infinitesimal) displacements $\{\delta u\}$ and virtual (infinitesimal) strains $\{\delta\varepsilon\}$. It then equates the virtual (incremental) work done by applied forces, δW_a, to the increment in stored potential energy, δW_s. Thus

$$\delta W_s = \delta W_a \tag{2:39}$$

The virtual work done by the applied forces, including the generalized forces exerted by the boundary displacements, is

$$\delta W_a = \int_{V_e} \{p\}^T \{\delta u\} dV + \left\{F_i\right\}^T \left\{\delta u_i\right\} \tag{2:40}$$

where the integral extends over the element's volume, V_e.

The increment in stored energy is, using Equation 2:8

$$\delta W_s = \int_{V_e} \{\sigma\}^T \{\delta\varepsilon\} dV \tag{2:41}$$

so that, from Equation 2:39,

$$\int_{V_e} \{\sigma\}^T \{\delta\varepsilon\} dV = \int_{V_e} \{p\}^T \{\delta u\} dV + \{F_i\}^T \{\delta u_i\} \qquad (2:42)$$

Using Equations 2:36 and 2:37, we can make the substitutions

$$\{\delta\varepsilon\} = [B]\{\delta u_i\} \qquad (2:43)$$

$$\{\delta u\} = [N]\{\delta u_i\} \qquad (2:44)$$

which gives

$$\left(\int_{V_e} \{\sigma\}^T [B] dV\right) \{\delta u_i\} = \left(\int_{V_e} \{p\}^T [N] dV + \{F_i\}^T\right) \{\delta u_i\} \qquad (2:45)$$

Since the elements of $\{\delta u_i\}$ are independent, we can remove $\{\delta u_i\}$ from both sides of Equation 2:45. Thus we obtain, taking the transpose of both sides,

$$\int_{V_e} [B]^T \{\sigma\} dV = \int_{V_e} [N]^T \{p\} dV + \{F_i\} \qquad (2:46)$$

This result has fundamental importance. It shows how to compute the generalized nodal forces, $\{F_i\}$, from the stresses and the applied load density.

Equation 2:46 applies whether the strain-displacement matrix is linear or nonlinear. In the latter case $[B] = \left[\varepsilon_{/u_i}\right]$, the matrix of partial derivatives of strain with respect to nodal displacements. Note that the matrix which carries the assumed displacement distribution, $[N]$, need not be consistent with the strain-displacement matrix, $[B]$. As a crude example, the designer could simply transfer the body forces to node points in an intuitive manner. Equation 2:46 also shows that, given an assumed relationship between strains and nodal displacements, a separate assumed relationship between stresses and nodal forces is not necessary. Indeed, a separate relationship would not be wise because it might destroy the symmetry of the element's stiffness matrix.

To complete the computation of the stiffness matrix, we use the homogeneous part of the constitutive relationship, Equation 2:9, and write

$$\{\sigma\} = [D]\{\varepsilon\} \tag{2:47}$$

Later we will see how to use the inhomogeneous part of $\{\sigma\}$ to include the effects of thermal expansion. Substitution into the left-hand side of Equation 2:46 gives

$$\int_{V_e} [B]^T\{\sigma\}dV = \int_{V_e} [B]^T[D]\{\varepsilon\}dV = \left[K_{ii}^e\right]\left\{u_i\right\} \tag{2:48}$$

where

$$\left[K_{ii}^e\right] = \int_{V_e} [B]^T[D][B]dV \tag{2:49}$$

We note also that the element-associated vector of nodal loads is

$$\left\{P_i^e\right\} = \int_{V_e} [N]^T\{p\}dV \tag{2:50}$$

so that Equation 2:46 can be written as

$$\left[K_{ii}^e\right]\left\{u_i\right\} = \left\{P_i^e\right\} + \left\{F_i^e\right\} \tag{2:51}$$

The final step is to sum Equation 2:51 over all elements connected to $\left\{u_i\right\}$. We note that equilibrium requires

$$\sum_e \left\{F_i^e\right\} = \left\{P_i^d\right\} \tag{2:52}$$

where $\left\{P_i^d\right\}$ is the vector of loads applied directly to $\left\{u_i\right\}$. Thus the summation yields the system equation (Equation 2:33) with the definitions of $\left[K_{ii}\right]$ and $\left\{P_i\right\}$ indicated by Equation 2:34 and 2:35. It is immaterial whether we keep the part of $\left\{P_i\right\}$ which is applied to the elements separate from the part which is applied directly to $\left\{u_i\right\}$. If we consider that both parts are applied directly to $\left\{u_i\right\}$, we can then write Equation 2:46 as

$$\left\{F_i\right\} = \int_{V_e} [B]^T \{\sigma\} dV \qquad (2:53)$$

which is homogeneous. In general, we will find it convenient to separate homogeneous and inhomogeneous relationships, or in other words, to separate the stiffness and load calculations.

As an alternative to the principle of virtual work, we could have used the principle of minimum potential energy. The potential energy of the element is usually defined as

$$\Pi_p = \int_{V_e} \left(\left(\tfrac{1}{2}\right)\{\sigma\}^T\{\varepsilon\} - \{p\}^T\{u\}\right)dV - \int_{S_e} \{t\}^T\{u\}dS \qquad (2:54)$$

where $\{t\}$ is the vector of the boundary tractions and the second integral extends over the element's external surface. The assumption that elements interact only through common boundary displacement sets allows the substitution

$$\int_{S_e} \{t\}^T\{u\}dS \rightarrow \left\{F_i^e\right\}^T\left\{u_i\right\} \qquad (2:55)$$

If we use, in addition, the constitutive relationship and the assumed relationships of strains and interior displacements to boundary displacements, we obtain

$$\Pi_p = \tfrac{1}{2}\left\{u_i\right\}^T\left[K_{ii}^e\right]\left\{u_i\right\} - \left\{P_i^e\right\}^T\left\{u_i\right\} - \left\{F_i^e\right\}^T\left\{u_i\right\} \qquad (2:56)$$

where $\left[K_{ii}^e\right]$ and $\left\{P_i^e\right\}$ are defined by Equations 2:49 and 2:50.

Setting the partial derivatives with respect to $\{u_i\}$ to zero, in order to minimize Π_p, gives Equation 2:51, which is the result obtained by application of the principle of virtual work. The only significant difference between the two procedures is that the principle of minimum potential energy requires the constitutive relationship to be linear and homogeneous while, in general, the principle of virtual work does not. (The assumption of homogeneity was made in the latter case to obtain a stiffness matrix after an expression for $\{F_i\}$ had been derived.)

To summarize, the computation of the element stiffness matrix includes the following three steps:

1. Write a strain-displacement matrix, $[B]$ (Equation 2:36).

2. Form the triple matrix product $[B]^T[D][B]$.

3. Integrate over the element's volume (Equation 2:49).

The second and third steps appear to be routine with little, if any, potential for a display of ingenuity by the element designer. Later we will see that the third step is not quite so routine as it appears. Still, the first step is by far the most important. It is not an exaggeration to call it the central problem of finite element design. We will consider many ways to evaluate $[B]$.

It is interesting to note that the only physical entity which is explicitly present in the computation of the stiffness matrix is the modulus matrix $[D]$. The dual physical principles which express the definition of strain (Equation 2:6) and the equilibrium of stresses (Equation 2:13) are not explicitly present. One or the other will, however, enter into the calculation of $[B]$. For example, $[B]$ can be calculated from an assumed displacement distribution using the definition of strain (Equation 2:38) or $[B]$ can be inferred from an assumed relationship between $\{\sigma\}$ and $\{F_i\}$ (Equation 2:53).

There is no requirement that an element must satisfy the equilibrium of stresses at all points within its interior. Indeed most finite elements derived from an assumed displacement distribution do not satisfy internal equilibrium. External, nodal equilibrium (Equation 2:33) is substituted for internal equilibrium.

Note also that the boundary tractions $\{t\}$ do not explicitly enter the calculation of stiffness. This is made possible by a principle of energy conservation which permits generalized forces on boundary displacements to be related directly to stresses within the element. Methods do exist, however, which employ boundary tractions in the computation of stiffness (see Section 2.5).

2.4 THE STRAIN-DISPLACEMENT MATRIX FOR THE CONSTANT STRAIN TRIANGLE EVALUATED BY THREE METHODS

As emphasized in the preceding section, evaluation of the strain-displacement matrix $[B]$ is the central problem of finite element design. $[B]$ depends directly on the assumed form of the internal field(s) supplied by the designer. We consider here three distinct types of assumed field—an assumed displacement field, an assumed strain field, and an assumed stress field—and examine how each may be used to compute the $[B]$ matrix.

The constant strain membrane triangle makes an excellent example to display these techniques because of its simplicity, its practical importance, and its place in the history of the finite element method.[12] Figure 2.4 illustrates the element. Its edges are straight and it has three connected nodes at each of which two components of in-plane motion (u_i, v_i) are defined. Thus the element has six external degrees of freedom.

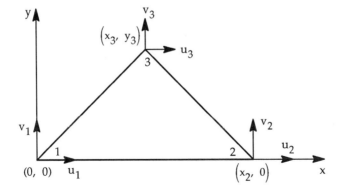

Figure 2.4   The Constant Strain Membrane Triangle.

The external degrees of freedom are balanced by six internal degrees of freedom which, in principle at least, may be selected at the discretion of the designer. We have, however, already given the designer's choices away by calling the element a *constant strain* triangle. Thus there are six candidates for selection as internal degrees of freedom which have overwhelming credentials. They are the three rigid body motions (two components of constant translation plus in-plane rotation) and three independent constant values for the components of in-plane strain $\left(\varepsilon_x , \varepsilon_y , \text{and } \gamma_{xy} \right)$. The choice of any other internal degree of freedom, say $\varepsilon_x = ax$, would require deletion of one of the rigid body motions or one of the constant strain states, which are clearly more important.

The balance between the external and the (preferred) internal degrees of freedom may be presented as a little table or balance sheet.

External Degrees of Freedom	
3 nodes x 2 displacement components:	6

Internal Degrees of Freedom	
Rigid body modes:	3
Constant strain states:	3
Total:	6

While this table seems trivial for the constant strain triangle, similar tables for more complex elements can provide valuable insights.

2.4.1   The Assumed Displacement Method

With the assumed displacement method, the vector of internal displacements, $\{u\}$, relates to the nodal displacement vector $\left\{ u_i \right\}$ by

$$\{u\} = [N]\{u_i\} \qquad (2:57)$$

where $[N]$ is a matrix of designer-supplied *shape functions* of position. The vector of in-plane strains is

$$\left\{\begin{array}{c} \varepsilon_x \\ \varepsilon_y \\ \gamma_{xy} \end{array}\right\} = \{\varepsilon\} = [L]\{u_i\} = [L][N]\{u_i\} \qquad (2:58)$$

where $[L]$ is a linear derivative operator. Thus, since $[B]$ is defined by $\{\varepsilon\} = [B]\{u_i\}$, we have

$$[B] = [L][N] \qquad (2:59)$$

or, in other words, $[B]$ is obtained by applying a linear derivative operator to the matrix of shape functions.

In order to obtain constant strain states, it is evident that the displacement states need to be linear in x and y. Thus, let the design choice be

$$u = a_1 + a_2 x + a_3 y$$
$$\qquad (2:60)$$
$$v = b_1 + b_2 x + b_3 y$$

Evaluation of u at each node gives, if we employ the nodal coordinate data shown in Figure 2:4,

$$\left\{\begin{array}{c} u_1 \\ u_2 \\ u_3 \end{array}\right\} = \left[\begin{array}{ccc} 1 & 0 & 0 \\ 1 & x_2 & 0 \\ 1 & x_3 & y_3 \end{array}\right] \left\{\begin{array}{c} a_1 \\ a_2 \\ a_3 \end{array}\right\} \qquad (2:61)$$

The solution of Equation 2:61 is easily obtained because the matrix is triangular. The solution is

$$
\left\{ \begin{array}{c} a_1 \\ a_2 \\ a_3 \end{array} \right\} =
\begin{bmatrix}
1 & 0 & 0 \\
\dfrac{-1}{x_2} & \dfrac{1}{x_2} & 0 \\
\dfrac{(x_3 - x_2)}{x_2 y_3} & \dfrac{-x_3}{x_2 y_3} & \dfrac{1}{y_3}
\end{bmatrix}
\left\{ \begin{array}{c} u_1 \\ u_2 \\ u_3 \end{array} \right\}
\qquad (2\text{:}62)
$$

The solution for the b's in terms of the v_i's has the same 3×3 matrix of coefficients. The in-plane strains are defined by

$$
\left\{ \begin{array}{c} \varepsilon_x \\ \varepsilon_y \\ \gamma_{xy} \end{array} \right\} =
\left\{ \begin{array}{c} u_{,x} \\ v_{,y} \\ u_{,y} + v_{,x} \end{array} \right\}
\qquad (2\text{:}63)
$$

or, using the assumed displacement state, Equation 2:60,

$$
\left\{ \begin{array}{c} \varepsilon_x \\ \varepsilon_y \\ \gamma_{xy} \end{array} \right\} =
\left\{ \begin{array}{c} a_2 \\ b_3 \\ a_3 + b_2 \end{array} \right\}
\qquad (2\text{:}64)
$$

Substitution of the values of a_2 and a_3 from Equation 2:62 and the values of b_2 and b_3 from the corresponding equation for the b's then gives

$$
\left\{ \begin{array}{c} \varepsilon_x \\ \varepsilon_y \\ \gamma_{xy} \end{array} \right\} =
\left[\begin{array}{ccc|ccc}
-\dfrac{1}{x_2} & \dfrac{1}{x_2} & 0 & 0 & 0 & 0 \\
0 & 0 & 0 & \dfrac{x_3 - x_2}{x_2 y_3} & -\dfrac{x_3}{x_2 y_3} & \dfrac{1}{y_3} \\
\dfrac{x_3 - x_2}{x_2 y_3} & -\dfrac{x_3}{x_2 y_3} & \dfrac{1}{y_3} & -\dfrac{1}{x_2} & \dfrac{1}{x_2} & 0
\end{array} \right]
\left\{ \begin{array}{c} u_1 \\ u_2 \\ u_3 \\ \hline v_1 \\ v_2 \\ v_3 \end{array} \right\}
$$

$$(2\text{:}65)$$

The 3×6 matrix in this result is the desired strain-displacement matrix, $[B]$.

Notice that we did not need to form explicit values for the elements of $\lfloor N \rfloor$ in order to evaluate $[B]$. We can, however, easily obtain these shape functions by substituting Equation 2:62 and the corresponding equation for the b's into Equation 2:60. If this is done, it will be seen that

$$u = \lfloor N \rfloor \{ u_k \}$$

$$v = \lfloor N \rfloor \{ v_k \} \qquad k = 1, 2, 3 \tag{2:66}$$

where

$$\lfloor N \rfloor = \lfloor 1, \ x, \ y \rfloor [A] \tag{2:67}$$

and $[A]$ is the 3×3 matrix appearing in Equation 2:62. Thus, if we write $\lfloor N \rfloor = \lfloor N_1, N_2, N_3 \rfloor$, the specific values for the shape functions are

$$N_1 = 1 - \frac{x}{x_2} + \frac{(x_3 - x_2)y}{x_2 y_3}$$

$$N_2 = \frac{x}{x_2} - \frac{x_3 y}{x_2 y_3} \tag{2:68}$$

$$N_3 = \frac{y}{y_3}$$

Note that, in this example, u and v use the same shape functions. Other important properties of shape functions are described in Chapter 3.

2.4.2 The Assumed Strain Method

In a pure form of the assumed strain method, the designer equates the strains to assumed functions of position. He then integrates the strains to obtain relationships with the displacements at nodal points. Line integrals can be employed for this purpose. [13] Thus, if ε is the strain directed along a straight line, ab, and $\bar{u}_a$, $\bar{u}_b$ are the displacements in the direction of the line at its end points (see Figure 2.5), we can write

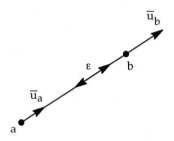

Figure 2.5   Integration of Strain Along a Line Member.

$$\bar{u}_b - \bar{u}_a = \int\limits_a^b \varepsilon \, d\ell \qquad\qquad (2{:}69)$$

Equation 2:69 gives a relationship between $\bar{u}_a$, $\bar{u}_b$ and the coefficients of the assumed strain field. Integration along enough line segments will provide enough relationships to evaluate all of the coefficients in terms of nodal point displacements; for example, in the constant strain triangle, line integrals along each of the three edges suffice.

In some cases it may be preferable to form indefinite integrals of the given strain fields in order to obtain displacement fields that can then be related to nodal point displacements. In the case of the constant strain triangle, the assumed strain field may be taken to be

$$\left\{ \begin{array}{c} \varepsilon_x \\ \varepsilon_y \\ \gamma_{xy} \end{array} \right\} = \left\{ \begin{array}{c} u_{,x} \\ v_{,y} \\ u_{,y} + v_{,x} \end{array} \right\} = \left\{ \begin{array}{c} a \\ b \\ c \end{array} \right\} \qquad\qquad (2{:}70)$$

where a, b, and c are constants. Thus

$$u = ax + F(y)$$
$$v = by + G(x) \qquad\qquad (2{:}71)$$

and, from the definition of γ_{xy},

$$F'(y) + G'(x) = c \qquad (2:72)$$

The only forms for $F(y)$ and $G(x)$ that satisfy Equation 2:72 are

$$F(y) = d + fy, \quad G(x) = e + (c - f)x \qquad (2:73)$$

Substitution into Equation 2.71 gives the form of the displacement field

$$\begin{aligned} u &= d + ax + fy \\ v &= e + (c - f)x + by \end{aligned} \qquad (2:74)$$

where (a, b, c) are the coefficients of the assumed strain field and (d, e, f) are integration constants. There are just enough coefficients (six) to evaluate (u, v) in terms of nodal point displacements.

In this example, the assumed strain method is seen to be trivially related to the assumed displacement method. That is not the case for more complicated elements.

In practice, assumed strain fields often appear in conjunction with assumed displacement fields. This topic is treated briefly in Section 2.5.2. It will be brought up again in later chapters.

2.4.3 The Assumed Stress Method

With the assumed stress method, the designer first selects the form of the spatial distribution of the stress field. He then finds the tractions on the boundary of the element (see Equation 2:15) and relates them to the generalized nodal forces. In the case of the constant strain triangle, the relationship between stresses and nodal forces yields the elements of $[B]^T$ (see Equation 2:53). More complex elements require additional steps (see Section 2.5.1).

For our constant strain triangle, we must, quite clearly, assume that the in-plane stresses are constant. Thus

$$\begin{Bmatrix} \sigma_x \\ \sigma_y \\ \tau \end{Bmatrix} = \begin{Bmatrix} \sigma_{xo} \\ \sigma_{yo} \\ \tau_o \end{Bmatrix} = \{\sigma_o\} \qquad (2:75)$$

Equation 2:53 gives, since both $[B]$ and $\{\sigma_o\}$ are constant,

$$\{F_i\} = \int\limits_{V_e} [B]^T \{\sigma\} dV = \tfrac{1}{2} x_2 y_3 [B]^T \{\sigma_o\} \qquad (2:76)$$

where $\tfrac{1}{2} x_2 y_3$ is the area of the triangle in Figure 2.4 and it is assumed that the element has unit thickness.

If we have an independent way to evaluate $\{F_i\}$ in terms of $\{\sigma_o\}$, we have, by the same token, a way to evaluate the elements of $[B]^T$. We can attempt to find such a relationship through the mechanism of static equilibrium.

Consider the force distributions shown in Figure 2.6. The triangular element is enclosed in a rectangle to facilitate visualization of the forces acting on its edges. The edge forces are in equilibrium and so must be the nodal forces that replace them. But there still seems to be some latitude in the way the edge forces may be concentrated at the nodes. For example, in the case of the vertical component of stress, σ_y, we could elect to concentrate the entire force, $x_2\sigma_y$, at node 3, in which case we would have to put $-(x_2 - x_3)\sigma_y$ at node 1 and $-x_3\sigma_y$ at node 2 to satisfy equilibrium. Or we could elect to put no part of the force at node 3, in which case there would be no net forces at nodes 1 and 2. The point is that we need to make a decision and that this requires an additional assumption or an appeal to an additional principle, which is much the same thing.

We could assume, for example, that the tractions on each edge are distributed equally to the adjacent corners. Thus we could extend the principle of equilibrium for the element as a whole to the requirement that each edge be in equilibrium. The resulting relationship between nodal forces and stresses is, by inspection of Figure 2.6,

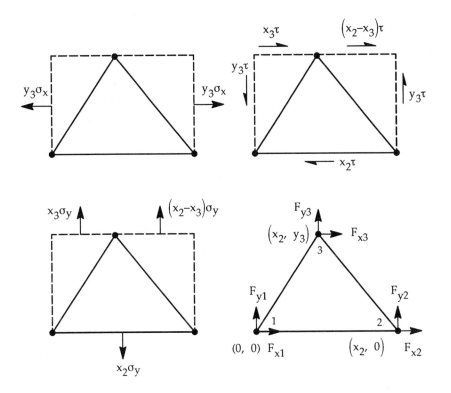

Figure 2.6   Boundary Forces and Nodal Forces for the Constant Strain
Triangle.

$$
\begin{Bmatrix} F_{x1} \\ F_{x2} \\ F_{x3} \\ \text{---} \\ F_{y1} \\ F_{y2} \\ F_{y3} \end{Bmatrix} = \frac{1}{2} \left[\begin{array}{ccc} -y_3 & 0 & x_3 - x_2 \\ y_3 & 0 & -x_3 \\ 0 & 0 & x_2 \\ \text{---} & \text{---} & \text{---} \\ 0 & x_3 - x_2 & -y_3 \\ 0 & -x_3 & y_3 \\ 0 & x_2 & 0 \end{array} \right] \begin{Bmatrix} \sigma_{xo} \\ \sigma_{yo} \\ \tau_o \end{Bmatrix} \qquad (2:77)
$$

Comparing this result with Equation 2:76, we see that the 6×3 matrix in Equation 2:77 is equal to $x_2 x_3 [B]^T$. This produces the same $[B]$ matrix as that given by the assumed displacement method (Equation 2:65).

Instead of just assuming that the forces on each edge are distributed equally to adjacent corners, we could appeal to the principle that the work done by the generalized forces $\{F_i\}$ be equal to the work done by edge tractions. Thus we might require that, from Equation 2:55,

$$[F_i]^T \{u_i\} = \int_{S_e} \{t\}^T \{u\} dS \qquad (2:78)$$

We see that we have not avoided making an assumption because now we must assume a relationship between $\{u\}$ and $\{u_i\}$. In other words, if we let

$$\{u\} = [N]\{u_i\} \qquad (2:79)$$

so that

$$\{F_i\} = \int_{S_e} [N]^T \{t\} dS \qquad (2:80)$$

then we still must decide how the shape functions in $[N]$ vary along the edges of the element. In the case of the constant strain triangle, the choice is an easy one because the only displacement field that is consistent with constant stresses is one that varies linearly with position. Thus we should use the shape functions already derived by the assumed displacement method (Equation 2:68). It turns out that the resulting distribution of nodal forces is exactly the same as that obtained by distributing the edge tractions equally to the adjacent corners.

In summary, each of the three assumed field methods, including two variations of the assumed stress method, give exactly the same strain-displacement matrices. This is no accident because the three assumed fields—linear displacement, constant strain, and constant stress—are fully compatible with each other. In the case of the constant strain triangle, no other plausible choices are available to the designer. The number of external degrees of freedom (six) allows a complete linear displacement field and a complete constant strain field. Raising the number of connected nodes from three to four, i.e., replacing the triangle by a rectangle, allows a complete linear displacement field plus two extra terms. Finding the best form for the two extra terms has occupied the attention of element designers for more than a generation. [14]

2.5 HYBRID METHODS

We have seen in the preceding section that the assumed stress method requires an additional assumption, and that one way to frame the additional assumption is as an assumed displacement field. Thus the assumed stress method is, in reality, a method that requires *two* assumed fields. We will also encounter examples where independent displacement and strain fields are assumed. Such methods of finite element formulation are called *hybrid* methods, *mixed* methods, or *two-field* methods.

Variational principles have been devised which include two or even three [15] independent fields. Such principles may be used to formulate finite elements or to show that elements formulated by other means satisfy a variational principle. We have shown, for example, that the principle of minimum potential energy (Equation 2:54) can be used as a basis for the basic, single-field stiffness formulation.

For the record, the two-field variational principle, which is commonly referred to as the Hellinger-Reissner principle [16,17] can be stated as follows. Let superscripts u, ε, and σ, refer to fields derived, respectively, from an assumed displacement field, an assumed strain field, and an assumed stress field. The Hellinger-Reissner principle comes in two forms—one for the case of assumed displacement and strain fields, and the other for the case of assumed displacement and stress fields.*

---

*This form is also called the modified Hu-Washizu principle. [15]

The functionals to be minimized are

$$\Pi_R(u, \varepsilon) = \int\limits_{V_e} \left[-\frac{1}{2}\{\sigma^\varepsilon\}^T\{\varepsilon^\varepsilon\} + \{\sigma^\varepsilon\}^T\{\varepsilon^u\} - \{p\}^T\{u^u\} \right] dV$$

$$\hspace{4cm} (2{:}81)$$

$$- \int\limits_{S_e} \{t\}^T\{u^u\}\, dS$$

$$\Pi_R(u, \sigma) = \int\limits_{V_e} \left[-\frac{1}{2}\{\sigma^\sigma\}^T\{\varepsilon^\sigma\} + \{\sigma^\sigma\}^T\{\varepsilon^u\} - \{p\}^T\{u^u\} \right] dV$$

$$\hspace{4cm} (2{:}82)$$

$$- \int\limits_{S_e} \{t\}^T\{u^u\}\, dS$$

It is seen that the two forms differ only in the interchange of σ and ε in the superscripts.* For comparison, the functional for the single field principle of minimum potential energy is

$$\Pi_p = \int\limits_{V_e} \left[\frac{1}{2}\{\sigma\}^T\{\varepsilon\} - \{p\}^T\{u\} \right] dV - \int\limits_{S_e}\{t\}^T\{u\} dS \hspace{1cm} (2{:}83)$$

The functionals are minimized by setting to zero the partial derivatives with respect to the free parameters that define the fields. In the case of the principle of minimum potential energy, the parameters are the nodal displacements $\{u_i\}$. With two assumed fields, there will be additional equations for the partial derivatives with respect to the parameters of the assumed strain or assumed stress field.

2.5.1   The Pian Assumed Stress Hybrid Element

As has been noted, the form of the assumed stress method described in Section 2.4.3 requires additional work to be applicable to elements that are more complex than the constant strain triangle. T. H. H. Pian was the first to

---

*The constitutive matrix $[D]$ must be invertible to allow the functionals to be stated as shown.

describe an assumed stress hybrid element of general applicability.[18] His
starting point is to state an assumed stress distribution in the form

$$\{\sigma\} = \left[C_\sigma\right]\{\beta\} \tag{2:84}$$

where the elements of $\left[C_\sigma\right]$ are functions of position and $\{\beta\}$ is a vector of
undetermined stress coefficients. For the case of the constant strain triangle,
$\left[C_\sigma\right]$ is an identity matrix.

The next step is to relate $\{\beta\}$ to the generalized nodal forces through the
agency of boundary tractions. In two dimensions, for example, the tractions are
related to stresses by

$$\{t\} = \begin{Bmatrix} t_x \\ t_y \end{Bmatrix} = \begin{Bmatrix} \sigma_x n_x + \tau_{xy} n_y \\ \tau_{xy} n_x + \sigma_y n_y \end{Bmatrix} \tag{2:85}$$

where n_x and n_y are direction cosines of the normal to the boundary.

In general, we can compute

$$\{t\} = \left[C_t\right]\{\beta\} \tag{2:86}$$

without making additional assumptions.

Then, using Equation 2:80 to relate boundary tractions to generalized nodal
forces, we obtain

$$\{F_i\} = \int_{S_e} [N]^T \{t\} dS = [T]\{\beta\} \tag{2:87}$$

where

$$[T] = \int_{S_e} [N]^T \left[C_t\right] dS \tag{2:88}$$

Thus the calculation of $[T]$ involves the displacement shape functions $[N]$. A
separate assumed displacement field is required and the term "hybrid" is
properly applied.

The matrix $[T]$, which has been called the *leverage* matrix,[19] occupies a position of central importance for the assumed stress formulation, similar to that of the $[B]$ matrix for the basic stiffness formulation.

The $[T]$ matrix can be expressed in a different way that directly involves $[B]$. Using Equation 2:53, we can write

$$\{F_i\} = \int_{V_e} [B]^T\{\sigma\}dV = [T]\{\beta\} \qquad (2:89)$$

where

$$[T] = \int_{V_e} [B]^T\left[C_\sigma\right]dV \qquad (2:90)$$

Comparing Equations 2:87 and 2:89, we observe that

$$\int_{V_e} [B]^T\{\sigma\}dV = \int_{S_e} [N]^T\{t\}dS \qquad (2:91)$$

Equation 2:91 is an application of the divergence theorem which can be derived using integration by parts.[20] Its validity requires that $[B]$ and $[N]$ be consistent, i.e., that $[B] = [L][N]$.

In Section 2.4.3, we simply used Equation 2:89 to evaluate $[B]^T$ for the constant strain triangle and called the analysis complete. To complete the derivation of Pian hybrid elements, on the other hand, we must continue on and derive an expression for the stiffness matrix $\left[K_{ii}\right]$ in terms of $\left[C_\sigma\right]$ and $[T]$. We begin by defining a generalized "displacement" vector $\{\alpha\}$ that is associated with the vector of undetermined force coefficients $\{\beta\}$. The strain energy can then be expressed as

$$W = \tfrac{1}{2}\{\beta\}^T\{\alpha\} = \tfrac{1}{2}\{F_i\}^T\{u_i\} = \tfrac{1}{2}\{\beta\}^T[T]^T\{u_i\} \qquad (2:92)$$

where Equation 2:87 has been used in the last equality. Since the elements of $\{\beta\}$ are independent, we have

$$\{\alpha\} = [T]^T\{u_i\} \tag{2:93}$$

Taken together, Equations 2:87 and 2:93 express the reciprocal static and kinematic properties of a rigid transformation.

The strain energy can also be expressed as a volume integral

$$W = \tfrac{1}{2}\int_{V_e}\{\sigma\}^T\{\varepsilon\}dV = \tfrac{1}{2}\int_{V_e}\{\sigma\}^T[D]^{-1}\{\sigma\}dV \tag{2:94}$$

or, substituting $\{\sigma\} = [C_\sigma]\{\beta\}$,

$$W = \tfrac{1}{2}\{\beta\}^T[f]\{\beta\} \tag{2:95}$$

where the *flexibility matrix*

$$[f] = \int_{V_e}[C_\sigma]^T[D]^{-1}[C_\sigma]dV \tag{2:96}$$

Comparing Equation 2:95 with the first form in Equation 2:92, we observe that

$$\{\alpha\} = [f]\{\beta\} \tag{2:97}$$

We can then replace Equation 2:92 with

$$W = \tfrac{1}{2}\{\alpha\}^T[f]^{-1}\{\alpha\} = \tfrac{1}{2}\{u_i\}^T[K_{ii}^e]\{u_i\} \tag{2:98}$$

so that, using Equation 2:93 for $\{\alpha\}$, we finally obtain the element stiffness matrix

$$[K_{ii}^e] = [T][f]^{-1}[T]^T \tag{2:99}$$

In practical applications, the designer must select specific forms for $[C_\sigma]$ and $[N]$. In his original paper, Pian designed a four-node rectangular element. He assumed that the shape functions in $[N]$ varied linearly along the edges and that $\{C_\sigma\}$ included up to ten independent polynomial functions of x and y. The minimum number of functions is five, corresponding to the element's five independent strain states. The form that Pian chose for the minimum set of stress functions is

$$
\begin{Bmatrix} \sigma_x \\ \sigma_y \\ \tau \end{Bmatrix} = \begin{bmatrix} 1 & y & 0 & 0 & 0 \\ 0 & 0 & 1 & x & 0 \\ 0 & 0 & 0 & 0 & 1 \end{bmatrix} \begin{Bmatrix} \beta_1 \\ \beta_2 \\ \beta_3 \\ \beta_4 \\ \beta_5 \end{Bmatrix}
\tag{2:100}
$$

This is an excellent choice. The linear variations of σ_x in the y direction and σ_y in the x direction allow in-plane bending along the x- and y-axes. We shall see later that the achievement of similar stress states with the assumed displacement method requires special procedures.

2.5.2 Assumed-Strain Hybrid Elements

It will be demonstrated that strains based on assumed displacement fields frequently have accurate values only at particular points. In such situations the strains can be interpolated to other points, such as Gauss integration points, by means of assumed interpolation functions. Elements designed in this way have an assumed displacement field and an assumed strain field which are collocated at particular points.

In other situations, strains derived from an assumed displacement field may have undesirable higher-order terms or even discontinuities. Such effects can be eliminated with a least-squares fit between the displacement-based strain field and an assumed lower-order strain field.

In both of these cases, the resulting elements may be classified as assumed-strain hybrids. For the moment, we will not comment further on them. They will receive full treatment as situations which warrant them arise.

REFERENCES

2.1 I. S. Sokolnikoff, *Mathematical Theory of Elasticity* (2nd ed.), McGraw-Hill, New York, 1956.

2.2 S. Timoshenko and J. N. Goodier, *Theory of Elasticity* (3rd ed.), McGraw-Hill, New York, 1969.

2.3 B. M. Irons and S. Ahmad, *Techniques of Finite Elements*, Ellis Horwood, Chichester, p. 462, 1980.

2.4 P. P. Silvester and R. L. Ferrari, *Finite Elements for Electrical Engineers*, Cambridge University Press, London, 1983.

2.5 J. D. Jackson, *Classical Electrodynamics*, John Wiley & Sons, New York, 1962.

2.6 B. E. MacNeal, J. R. Brauer, and R. N. Coppolino, "A General Finite Element Vector Potential Formulation of Electromagnetics Using a Time-Integrated Electric Scalar Potential," IEEE Trans. MAG-26, p. 1768, 1990.

2.7 A. Hrennikoff, "Solution of Problems in Elasticity by the Framework Method," *J. Appl. Mech.*, 8, pp 169-75, 1941.

2.8 O. C. Zienkiewicz, J. P. de S.R. Gago, and D. W. Kelly, "The Hierarchical Concept in Finite Element Analysis," *Comput. Struct.*, 16, pp 53-65, 1983.

2.9 B. Fraeijs de Veubeke, "Displacement and Equilibrium Models in the Finite Element Method," Chap. 9, *Stress Analysis*, O. C. Zienkiewicz and G. Holister (eds.), John Wiley & Sons, 1965.

2.10 P. H. Denke, "A Matrix Method of Structural Analysis," Proc. 2nd U.S. Natl. Congress of Applied Mech., p. 445, 1954.

2.11 M. Jacob, *Heat Transfer*, Vol. II, John Wiley & Sons, pp 1-24, 1957.

2.12 M. J. Turner, R. W. Clough, H. C. Martin, and L. J. Topp, "Stiffness and Deflection Analysis of Complex Structures," *J. Aeronautical Sci.*, 23, pp 803-23, p. 854, 1956.

2.13 R. H. MacNeal, "Derivation of Element Stiffness Matrices by Assumed Strain Distributions," *Nucl. Eng. Design*, 70, pp 3-12, 1982.

2.14 R. H. MacNeal, "Toward a Defect-Free Four-Node Membrane Element," *Finite Elem. Analysis & Design*, 5, pp 31-7, 1989.

2.15 K. Washizu, *Variational Methods in Elasticity and Plasticity*, 2nd ed., Pergamon Press, Oxford, 1975.

2.16 E. Hellinger, "Der Allgemeine Ansatz der Mechanik der Kontinua," *Encycl. Math. Wissensch.*, 4(4), p. 602, 1914.

2.17   E. Reissner, "On a Variational Theorem in Elasticity," *J. Math. Phys.*, 29, pp 90-5, 1950.

2.18   T. H. H. Pian, "Derivation of Element Stiffness Matrices by Assumed Stress Distribution," *AIAA Intl.*, 2, pp 1332-6, 1964.

2.19   B. M. Irons and S. Ahmad, *Techniques of Finite Elements*, Ellis Horwood, Chichester, p. 256, 1980.

2.20   J. Barlow, "A Different View of the Assumed Stress Hybrid Method," *Intl. J. Numer. Methods Eng.*, 22, pp 11-6, 1986.

3

Assumed Displacement Fields

The starting point for most finite element designs is an assumed displacement field. The design of assumed stress and assumed strain hybrid elements also requires an assumed displacement field (see Section 2.5). The only exception we have encountered is the assumed constant strain element described in Section 2.4.2.

In this chapter we will examine the choices for displacement fields that are available to the designer. We will see that those choices are restricted by the number and arrangement of the boundary variables and by the desire to achieve or to avoid particular effects.

3.1 SHAPE FUNCTIONS AND BASIS FUNCTIONS

The terms *shape function* and *basis function* are used interchangeably in the finite element literature. We shall make a distinction between these terms as part of a particular way of describing displacement fields. We have already encountered *shape functions* in Chapter 2 (see Equation 2:37) to describe the displacement field $\{u\}$ in terms of the set of boundary displacement variables $\{u_i\}$. Thus

$$\{u\} = [N]\{u_i\} \tag{3:1}$$

where $[N]$ is a matrix whose elements are functions of position that we will call shape functions (not basis functions). Considering a particular component of $\{u\}$, we can write

$$u = \lfloor N \rfloor \{u_i\} = \sum_{i=1}^{n} N_i u_i \tag{3:2}$$

where n is the number of boundary variables.

In most finite element designs, the same shape functions are used to relate each of the components of $\{u\}$ to its boundary values. Later we will encounter exceptions in connection with the design of plate and shell elements. If $\{u_i\}$ is partitioned into subsets corresponding to the coordinate directions, then for three-dimensional elements that use the same shape functions for all three components,

$$u = \sum_{i=1}^{n} N_i u_i \quad ; \quad v = \sum_{i=1}^{n} N_i v_i \quad ; \quad w = \sum_{i=1}^{n} N_i w_i \tag{3:3}$$

Note that n now refers to the number of degrees of freedom in each subset rather than to the total number of boundary variables. Note also that the relationships between field components (u, v, w) and nodal components are invariant to rotation of the coordinate axes.

In developing the constant strain triangle, we started by assuming the form of the displacement field without reference to the boundary variables (see Equation 2:60). Thus, instead of Equation 3:2 or Equation 3:3, we had

$$u = \lfloor X \rfloor \left\{ a_j \right\} = \sum_{j=1}^{n} X_j a_j \qquad (3:4)$$

where, for the constant strain triangle,

$$\lfloor X \rfloor = \lfloor \ 1, \ x, \ y \ \rfloor \qquad (3:5)$$

The elements of $\lfloor X \rfloor$ will be called *basis functions* and the elements of $\left\{ a_j \right\}$ will be called *basis coefficients*. If we assume that the number of basis functions equals the number of boundary variables in the subset associated with u, then the relationship between the basis coefficients $\left\{ a_j \right\}$ and the boundary variable subset $\left\{ u_i \right\}$ can be expressed as

$$\left\{ a_j \right\} = \left[A_{ji} \right] \left\{ u_i \right\} \qquad (3:6)$$

where $\left[A_{ji} \right]$ is a square matrix of constants.[*]

Substitution of Equation 3:6 into Equation 3:4 shows that

$$\lfloor N \rfloor = \lfloor X \rfloor \left[A_{ji} \right] \qquad (3:7)$$

Thus the row vector of shape functions, $\lfloor N \rfloor$, can be factored into a row vector of basis functions, $\lfloor X \rfloor$, and a square matrix of constant coefficients. In the case of the constant strain triangle (see Equation 2:68),

---

[*]If the size of $\left\{ a_j \right\}$ is chosen to be greater than the size of $\left\{ u_i \right\}$, we can simply add some of the members of $\left\{ a_j \right\}$ to $\left\{ u_i \right\}$ in order to make the sizes equal. The extra members of $\left\{ u_i \right\}$ then become *internal* degrees of freedom, not shared with neighboring elements. If the size of $\left\{ a_j \right\}$ were chosen to be less than the size of $\left\{ u_i \right\}$, the matrix $\left[A_{ji} \right]$ could be evaluated by some kind of least squares fit. This choice appears to offer no practical advantage.

$$[A_{ji}] = \begin{bmatrix} 1 & \vdots & 0 & \vdots & 0 \\ -\dfrac{1}{x_2} & \vdots & \dfrac{1}{x_2} & \vdots & 0 \\ \dfrac{x_3 - x_2}{x_2 y_3} & \vdots & \dfrac{-x_3}{x_2 y_3} & \vdots & \dfrac{1}{y_3} \end{bmatrix} \tag{3:8}$$

We can characterize a set of shape functions as a set of basis functions with particular values for the coefficients $\{a_j\}$. Our study of the general properties of assumed displacement fields will not often require knowledge of these coefficients. Accordingly, we will frequently find it more convenient to use elementary basis functions such as those shown in Equation 3:5.

The calculation of an element's stiffness matrix requires, of course, a way to evaluate the matrix $[A_{ji}]$. Two cases are distinguished which depend on the way $\{u_i\}$ is formed. In the first case the components of $\{u_i\}$ are assumed to be the values of u at nodal points. In this case it is clear, from the form of Equation 3:3, that the shape function N_i must have unit value at node (i) and zero value at all other nodes. Evaluation of Equation 3:4 at nodal points then gives, since u_i is the value of u at node (i),

$$\{u_i\} = [X_{ij}]\{a_j\} \tag{3:9}$$

where elements in the ith row of $[X_{ij}]$ equal the values of the basis functions at node i. Assuming $[X_{ij}]$ to be nonsingular, we obtain

$$\{a_j\} = [X_{ij}]^{-1}\{u_i\} \tag{3:10}$$

so that

$$[A_{ji}] = [X_{ij}]^{-1} \tag{3:11}$$

and

$$\lfloor N \rfloor = \lfloor X \rfloor [X_{ij}]^{-1} \tag{3:12}$$

In order for $\left\lfloor X_{ij} \right\rfloor$ to be non-singular, the basis functions must be linearly independent. This means that, if X_1 and X_2 are independent basis functions, the only coefficients $\left(C_1, C_2 \right)$ which satisfy

$$C_1 X_1 + C_2 X_2 = 0 \qquad (3{:}13)$$

at all nodes are the null set $C_1 = C_2 = 0$.

Equation 3:12 shows that the shape functions are entirely determined by the functional forms of the basis functions and by their values at nodes. Furthermore, the forms of the basis functions can be changed, within limits, without changing the shape functions. To see this, let

$$\left\lfloor \overline{X} \right\rfloor = \left\lfloor X \right\rfloor [C] \qquad (3{:}14)$$

where $[C]$ is a square matrix of constants. Equation 3:14 expresses each element of $\left\lfloor \overline{X} \right\rfloor$ as a linear combination of the basis functions in $\left\lfloor X \right\rfloor$. Equation 3:4 gives, if $[C]$ is nonsingular,

$$u = \left\lfloor X \right\rfloor \left\{ a_j \right\} = \left\lfloor \overline{X} \right\rfloor [C]^{-1} \left\{ a_j \right\} = \left\lfloor \overline{X} \right\rfloor \left\{ \overline{a}_j \right\} \qquad (3{:}15)$$

where $\left\{ \overline{a}_j \right\} = [C]^{-1} \left\{ a_j \right\}$ is the vector of basis coefficients associated with $\left\lfloor \overline{X} \right\rfloor$. Substitution for $\left\lfloor X \right\rfloor$ and $\left\lfloor X_{ij} \right\rfloor$ from Equation 3:14 into Equation 3:12 gives

$$\left\lfloor N \right\rfloor = \left\lfloor X \right\rfloor \left\lfloor X_{ij} \right\rfloor^{-1} = \left\lfloor \overline{X} \right\rfloor \left\lfloor \overline{X}_{ij} \right\rfloor^{-1} \qquad (3{:}16)$$

Equation 3:16 shows that shape functions are invariant to a non-singular recombination of basis functions. For example, the sets of basis functions, $\left\lfloor 1, x, x^2 \right\rfloor$ and $\left\lfloor 1, x, (1-x)^2 \right\rfloor$, are equivalent in that they produce the same shape functions. Basis functions are, as we see, marked by a certain indefiniteness in their representation of the displacement field. Shape functions, on the other hand, have unique and explicit forms which suit them better for computation than for speculation about the properties of assumed displacement fields.

The second way of forming $\{u_i\}$ is a slight generalization of the first. It allows $\{u_i\}$ to consist partly of the values of u at nodes and partly of coefficients which are not related to nodes. Thus

$$\{u_i\} = \left\{ \begin{array}{c} u_j \\ \hline a_k \end{array} \right\} \tag{3:17}$$

where the elements of $\{u_i\}$ are the values of u at nodes, and the elements of $\{a_k\}$ are other "generalized" displacement variables. The vectors of shape functions and basis functions are similarly partitioned, so that

$$u = \left\lfloor N_j \vdots N_k \right\rfloor \left\{ \begin{array}{c} u_j \\ \hline a_k \end{array} \right\} = \left\lfloor X_j \vdots X_k \right\rfloor \left\{ \begin{array}{c} a_j \\ \hline a_k \end{array} \right\} \tag{3:18}$$

Note that we have conveniently chosen the generalized displacement variables, $\{a_k\}$, to be a subset of the vector of basis coefficients. Evaluation of u at nodes gives

$$\{u_j\} = \left[X_{jj} \vdots X_{jk} \right] \left\{ \begin{array}{c} a_j \\ \hline a_k \end{array} \right\} \tag{3:19}$$

so that, assuming $\left[X_{jj} \right]$ to be nonsingular,

$$\{a_j\} = \left[X_{jj} \right]^{-1} \{u_j\} - \left[X_{jj} \right]^{-1} \left[X_{jk} \right] \{a_k\} \tag{3:20}$$

Substitution into Equation 3:18 yields

$$u = \left\lfloor X_j \right\rfloor \left[X_{jj} \right]^{-1} \{u_j\} + \left\lfloor X_k - X_j\, X_{jj}^{-1}\, X_{jk} \right\rfloor \{a_k\} \tag{3:21}$$

Equation 3:21 is simplified if $\left[X_{jk} \right] = 0$ or, in other words, if the generalized basis functions $\left\lfloor X_k \right\rfloor$ are null at the node points. If this is required, then

$$u = \lfloor X_j \rfloor \lfloor X_{jj} \rfloor^{-1} \{u_j\} + \lfloor X_k \rfloor \{a_k\}$$

$$= \lfloor N_j \rfloor \{u_j\} + \lfloor N_k \rfloor \{a_k\} \qquad (3:22)$$

Note that $\lfloor N_k \rfloor = \lfloor X_k \rfloor$ only if $\lfloor X_{jk} \rfloor = 0$. Note, in addition, that the physical interpretation of u_j as the value of u at node (j) is preserved by requiring $\lfloor X_{jk} \rfloor = 0$.

Equation 3:22 underlies the design of *hierarchical* elements.[1] Such elements have a few nodes, usually located at corners, and a variable number of generalized displacements, $\{a_k\}$. The addition of one nodal displacement variable changes *all* the nodal shape functions, $\lfloor N_j \rfloor = \lfloor X_j \rfloor \lfloor X_{jj} \rfloor^{-1}$, but the addition of one generalized displacement variable, a_{k+1}, does not change any of the other shape functions. This fact can be exploited in the design of a sequence of hierarchical elements with increasing numbers of generalized displacements (see Section 4.5).

We have not yet addressed the question of what kinds of functions make suitable basis functions for finite elements. The terms of a simple power series are by far the most common choice. The one-dimensional basis functions in this class are $\lfloor 1, x, x^2, x^3, \cdots \rfloor$, and the two-dimensional basis functions are $\lfloor 1, x, y, x^2, xy, y^2, x^3, \cdots \rfloor$. It is quite natural to allow x and y to be Cartesian position coordinates. They could also be polar coordinates or coordinates in some other system or even, as we shall see, parameters that are functions of position.

The terms in a trigonometric series are sometimes used as basis functions for finite elements.[2,3] The motivation is to use functions that give a good approximation to the displacement solution, even if only a few terms are used. Thus trigonometric basis functions are appropriate for wave propagation because the solutions tend to be (slowly) decaying sine waves. In static stress analysis, on the other hand, the most important solutions are constant and linearly varying strain fields. In this case an appropriate set of displacement basis functions would be a finite power series with terms that are complete through second degree. A power series in three dimensions is said to be complete to second degree if it includes all terms of the form $x^a y^b z^c$, where a, b, and c are nonnegative integers such that $a + b + c \leq 2$. If x, y, and z are Cartesian position coordinates, the element might then be able to display exact solutions for any constant or linearly varying strain field.

The notion of completeness is an important one in finite element design. We have just used it to describe a power series that is complete through the terms of second degree. We will encounter many other examples of this kind but we should stop, for a moment, to explore some general implications of the concept. Assume, for example, that we wish to approximate a function u by a series of terms $u' = u_1 + u_2 + u_3 \cdots u_n$. The mean squared error in the approximation over the volume of the element is

$$E = \frac{1}{V} \int_V (u - u')^2 dV \qquad (3{:}23)$$

We hope and expect that the error, E, should tend to zero as n increases. This will be possible if the magnitudes of the added terms tend to zero as n increases. Consider, however, what happens if the terms $u_i, i = 1, 2, \cdots n$, are orthogonal and if one of the terms, u_m, is left out. Then

$$\int_V (u_m u') dV = 0 \qquad (3{:}24)$$

so that if u consists entirely of u_m,

$$\begin{aligned} E &= \frac{1}{V} \int \left(u_m^2 - 2u_m u' + (u')^2 \right) dV \\ &= \frac{1}{V} \int \left(u_m^2 + (u')^2 \right) dV \end{aligned} \qquad (3{:}25)$$

This expression, being positive definite, does not tend to zero as n increases. In fact, $u' = 0$ gives the best approximation. The implication is that a series of orthogonal functions cannot converge to an arbitrary function if the series is incomplete, i.e., if any of the terms in the series are left out. Of course, in practical finite element design we can use only a finite number of terms so that errors due to the neglected higher degree terms will always be present.

It is important to ask whether the completeness requirement for convergence applies to basis functions which are the terms in a power series $\left(1, x, x^2, \cdots\right)$. The answer is that it does apply because we can construct an orthogonal set of functions, such as Legendre polynomials, by linear combination of the terms in a power series. We know further from Equation 3:16 that such a recombination of basis functions will not alter the element's shape functions.

3.2 NODE LOCATIONS AND THE SELECTION OF BASIS FUNCTIONS

If the components of $\left\{u_i\right\}$ are assumed to be the values of u at nodes, then by prior agreement (see Equation 3:6) we take the *number* of basis functions to be equal to the number of nodes. We will assume, in this section, that the basis functions are terms from a power series of Cartesian position coordinates. Usually, but not always, the functions will be chosen to be complete to the highest degree possible with a few scattered higher degree terms as needed to make the total number of basis functions equal the number of nodes.

3.2.1 Triangular Elements

Consider the three triangular elements shown in Figure 3.1. They differ only in the number and location of nodes. The triangle in Figure 3.1(a) has three nodes located at the corners; the element in Figure 3.1(b) has six nodes located at the corners and on the edges; and the element in Figure 3.1(c) has ten nodes: one at each corner, two along each edge, and one inside the element. We note in passing that the edges need not be straight and the edge nodes need not be centered. The triangular elements in Figure 3.1 are seen to form a sequence with an increasing number of edge nodes. The significance of the center node in the third element is that it allows the element's basis functions to be a complete set.

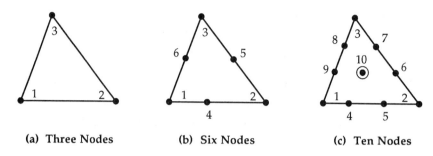

(a) **Three Nodes**   (b) **Six Nodes**   (c) **Ten Nodes**

Figure 3.1   Three Triangular Elements.

Figure 3.2 shows the terms in a two-dimensional power series arranged in a Pascal triangle. The complete (cumulative) number of terms through the first, second, and third degrees is, respectively, three, six, and ten. Therefore, to have complete sets of basis functions the three elements must have, respectively, three, six, and ten nodes. The boundary nodes of the first two elements conveniently satisfy this requirement for completeness, but the third element needs an extra interior node. It causes no harm. It can, indeed, be removed during problem execution by static condensation of the element's stiffness matrix.[4]

								Cumulative No. of Terms
1								1
x	y							3
x^2	xy	y^2						6
x^3	x^2y	xy^2	y^3					10
x^4	x^3y	x^2y^2	xy^3	y^4				15
--	--	--	--	--	--	--	--	--
x^n	$x^{n-1}y$	$x^{n-2}y^2$	--	x^2y^{n-2}	xy^{n-1}	y^n		$\dfrac{(n+1)(n+2)}{2}$

Figure 3.2   Pascal's Triangle.

It might well be asked why we are so keen to preserve completeness. Convergence is not a decisive issue because we intend to keep only a few terms. In the case of triangles, an argument for completeness can be advanced in terms of isotropy. We would much prefer that the element's stiffness matrix not depend on the orientation of the element's internal coordinate system. For example, the stiffness matrix should not change when the direction of the x-axis is changed from side 1-2 to side 2-3 of the triangle in Figure 3.1(c).[*] This requirement will be met if we get the same set of basis functions, except for linear recombination, when the coordinate system is rotated through an arbitrary angle. Complete polynomials clearly satisfy the requirement, but observe what happens when we leave one term out of the cubic polynomial so as to achieve an element with nine nodes. The only combinations of cubic or lower degree terms that are invariant to coordinate rotation are the constant term (unity) and $x^2 + y^2$. The elimination of the first of these terms would be unacceptable in any element (it would destroy the rigid body property). The elimination of the second of these terms would be unacceptable in a cubic element (the lower order quadratic element would be better). Later we will see, in connection with triangular plate elements, the lengths to which element designers are willing to go to preserve isotropy or, to put it another way, to avoid *induced anisotropy*.

Another important question is whether any difficulty will arise in forming the shape functions for the three triangular elements shown in Figure 3.1. The answer is that difficulty does occur in special circumstances. Consider the six-node "triangle" shown in Figure 3.3. All of its nodes lie on a circle. To form the shape functions, we need to invert $\left[X_{ij} \right]$, the matrix of basis functions

---

[*]This allows the coordinate axes for triangles to be chosen, without penalty, by an arbitrary rule. A common rule is to direct the x-axis along the side joining the two corner nodes mentioned first in the element connection data. Non-arbitrary rules can also be devised to accommodate anisotropy, such as a rule directing x along the longest side. This and all similar rules become ambiguous for particular cases.

evaluated at nodes (see Equation 3:12). Inversion is not possible if a linear combination of the columns exists which is null. The vector of basis functions is $\lfloor X \rfloor = \lfloor 1, x, y, x^2, xy, y^2 \rfloor$. Since $x^2 + y^2 = R^2$, the radius of the circle, the sum of columns 4 and 6 of $\left[X_{ij} \right]$ is equal to R^2 times column 1. Thus $\left[X_{ij} \right]$ is singular and the shape functions cannot be formed. Irons and Ahmad[5] apply the term *neutral function* to a combination of basis functions which vanishes at all nodes. They give several examples. In the present example, $x^2 + y^2 - R^2$ is a neutral function.

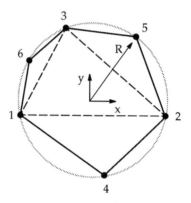

Figure 3.3   Six-Node "Triangle" Whose Nodes Lie on a Circle.

3.2.2 Tetrahedral Elements

Tetrahedral elements can be regarded as the extension to three dimensions of triangular elements. Figure 3.4 shows the first three members of a sequence of tetrahedral elements with zero, one, and two nodes per edge respectively. The basis functions are conveniently taken to be terms in a three-dimensional power series. The complete terms through cubic degree are shown in Table 3.1.

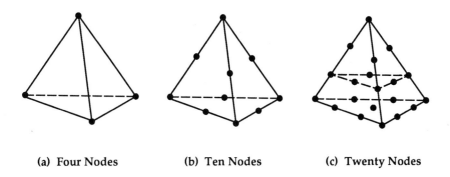

(a) Four Nodes   (b) Ten Nodes   (c) Twenty Nodes

Figure 3.4   Three Tetrahedral Elements.

Table 3.1

Complete Three-Dimensional Polynomial Sets Through Third Degree

DEGREE	TERMS	CUMULATIVE NO. OF TERMS
0	1	1
1	x, y, z	4
2	$x^2, y^2, z^2, xy, xz, yz$	10
3	$x^3, y^3, z^3, x^2y, x^2z, xy^2, xz^2, yz^2, z^2y, xyz$	20

Since a tetrahedron has four corners and six edges, we see that the element with corner nodes only can have basis functions that are complete through the linear terms and that the element with corner nodes and one node per edge can have basis functions that are complete through the quadratic terms. The element with two nodes per edge, however, falls short of completeness by four terms. These can be made up by nodes at the centers of the triangular faces, as shown in Figure 3.4(c).

3.2.3 Rectangular Elements

Although rectangular elements were formerly important,[6,7] they are encountered nowadays only as special cases of general quadrilateral elements. They have, however, certain symmetry properties which provide useful insights. Later we will see how to extend those properties to general quadrilateral elements.

Figure 3.5 shows five rectangular elements with 4, 8, 9, 12, and 16 nodes respectively. We see right away that the number of nodes does not agree, for any of the five elements, with the number of terms in a complete polynomial. (The sequence of the number of such terms is, from Figure 3.2, 1, 3, 6, 10, 15, 21, etc.) Nor does there appear to be any symmetrical way of adding nodes that will allow the basis functions to form complete polynomials. The four-node rectangle in Figure 3.5(a) would, for example, require two additional nodes which cannot be added without destroying the double symmetry of the node pattern with respect to the x- and y-axes. Why is such symmetry important? The reason is to preserve the isotropy of the element with respect to its axes. If, for example, two extra nodes were placed on the x-axis, the stiffness matrix would depend on which of the element's two principal axes was selected as the x-axis.

So we are stuck with incomplete polynomials as basis functions for quadrilateral elements. That is not necessarily a bad thing but it does raise the issue of how to select terms from an incomplete polynomial degree. In the case of the four-node element, we could pick either x^2, xy, or y^2 as the fourth basis function to go along with 1, x, and y. The correct choice is xy. This can be shown by either of two separate arguments. Consider that x^2 is the choice. Then, in a rotation of coordinate axes $\left(\overline{x} = y, \overline{y} = -x\right)$, it is seen that the new set of basis functions $\lfloor 1, \overline{x}, \overline{y}, \overline{x}^2 \rfloor = \lfloor 1, y, -x, y^2 \rfloor$ is not a linear recombination of the original set $\lfloor 1, x, y, x^2 \rfloor$. Thus the choice of x^2 as a basis function will give a different set of shape functions and a different stiffness matrix depending on which principal axis is selected as the x-axis. The choice of x^2 as a basis function also causes $\left[X_{ij} \right]$ to be singular. This occurs because the value of x^2 at each of the four nodes is a^2. Thus the fourth column of $\left[X_{ij} \right]$ is equal to a^2

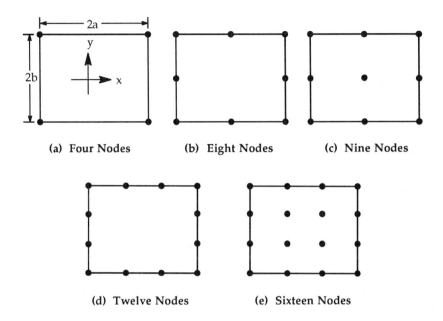

Figure 3.5   Five Rectangular Elements.

times the first column. In contrary distinction, the choice of xy as the fourth
basis function passes both the isotropy test and the singularity test.

We do not often need to employ such lengthy arguments to arrive at a suitable
choice of basis functions. In fact, the job can usually be done by inspection.
Consider, again, the four-node element in Figure 3.5(a). There are two nodes on
the top edge of the element which happens to be parallel to the x-axis. The
function of minimum complexity which can match the values of displacement
at the two nodes is $u = a_1 + a_2 x$. The bottom edge is also parallel to the x-axis
but it has a different value of y. To get correct values of u at all four nodes we
must, therefore, allow the expression for u to vary in the y direction.
Consequently, $u = \left(a_1 + a_2 x\right)\left(b_1 + b_2 y\right)$ is the function of lowest degree in x
and y that can match the values of displacement at all four nodes.

Turning to the eight-node element, Figure 3.5(b), we observe that $\left(1, x, x^2\right)$ is
the basis function set of minimum complexity that can match the values of u at

nodes along the top edge. To match values of u along the bottom edge as well, we need to add the set $\left(1, x, x^2\right)y$. Finally, to match values of u at the midpoints of the left and right edges, we need to add the set $(1, x)y^2$. Combining all terms we obtain the complete vector of basis functions for an eight-node rectangle $\lfloor X^8 \rfloor = \lfloor 1, x, y, x^2, xy, y^2, x^2y, xy^2 \rfloor$.

Note that $\lfloor X^8 \rfloor$ is complete through the quadratic terms and has xy symmetry (the same set of terms is obtained if x and y are interchanged). These properties would be retained if x^2y and xy^2 were replaced by x^3 and y^3, but the resulting $\lfloor X_{ij} \rfloor$ would be singular. Thus the set selected for $\lfloor X^8 \rfloor$ is the set of lowest degree which is both isotropic and non-singular for the eight-node rectangle.

The selection of basis functions for the nine-node rectangle is even easier. Here again we note that the set $\left(1, x, x^2\right)$ is needed to match u at nodes on the top edge. The three nodes on the top edge are repeated at two other values of y so that the complete required set can be expressed as $\left(1, x, x^2\right)\left(1, y, y^2\right)$. The vector of basis functions, $\lfloor X^9 \rfloor = \lfloor 1, x, y, x^2, xy, y^2, x^2y, xy^2, x^2y^2 \rfloor$, contains all the terms in $\lfloor X^8 \rfloor$ plus the quartic term, x^2y^2. Again, this is the basis set of lowest degree for the nine-node rectangle which is both isotropic and non-singular.

We can proceed in like manner with the twelve-node and sixteen-node rectangles. The basis functions for all five rectangles are superimposed on Pascal triangles in Figure 3.6. The five rectangular elements separate into two series. In one series called *serendipity*[*] elements, all of the nodes lie on the boundary of the element. In the other series, called *Lagrange*[**] elements, the

---

[*]The word serendipity means "an aptitude for making fortunate discoveries accidentally." The term was coined by Horace Walpole (c. 1754) after his tale *The Three Princes of Serendip*.

[**]Because their shape functions can be described by Lagrange interpolation polynomials (see Equation 4:2).

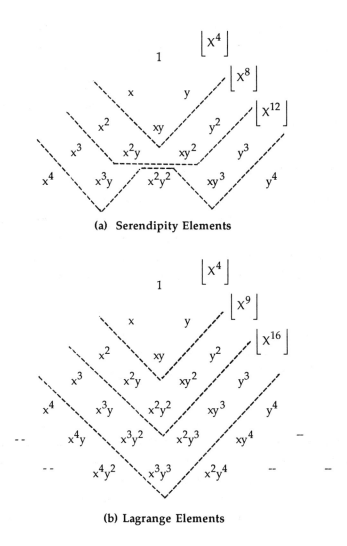

(a) Serendipity Elements

(b) Lagrange Elements

Figure 3.6   Basis Functions for Rectangular Elements.

node locations make a rectangular grid on the surface of the element. The lowest order members of both series are the same element, the four-node rectangle. The eight-node and nine-node elements are similar enough that they compete heavily with each other. We will, in later chapters, have occasion to discuss their advantages and disadvantages. It is seen, from Figure 3.6, that the basis functions for serendipity elements are more compact in the sense that there are fewer incomplete higher degree terms. Serendipity elements should, therefore, be more efficient than Lagrange elements. It will turn out, however, that the extra higher degree terms of Lagrange elements have some surprising benefits for accuracy.

3.2.4 Rectangular Brick Elements

The basis functions for rectangular brick elements follow quite easily from the inspection procedure devised for plane rectangular elements. Figure 3.7 illustrates the three lowest order elements with isotropic node patterns. They have, respectively, eight, twenty, and 27 nodes. Comparing them with the complete three-dimensional polynomial sets shown in Table 3.1, we observe that only the twenty-node brick has the correct number of nodes to produce basis functions that form a complete polynomial. Alas, even this is an illusion, as a careful application of the inspection method of deriving basis functions will show. Basis functions for each of the three elements are recorded in Table 3.2. These are, again, the sets of lowest degree which are nonsingular and which satisfy isotropy.

The twenty-node brick has nodes only at the eight corners and at the midpoints of the twelve edges. It may, therefore, be classed as a serendipity element. The nodes of a 27-node brick form a regular lattice, which makes it a Lagrange element. Eight- and twenty-node bricks dominate practical applications. 32-node serendipity elements (i.e., elements with two nodes per edge) are occasionally seen, but there appears to be little practical use of Lagrange brick elements.

3.2.5 Pentahedral Elements

Figure 3.8 illustrates two elements whose shapes are right triangular prisms. Such elements sometimes appear in relatively thick shell analyses as substitutes for triangular shell elements. We will call them pentahedral (five-sided) elements to include their extension to non-prismatic shapes.

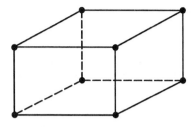

(a) Eight-Node Brick

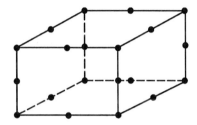

(b) Twenty-Node Brick

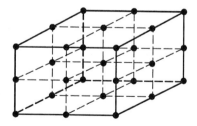

(c) 27-Node Brick

Figure 3.7   Rectangular Brick Elements.

Table 3.2

Basis Functions for Rectangular Brick Elements

(a) Eight-Node Brick

DEGREE	TERMS
0	1
1	x, y, z
2	xy, xz, yz
3	xyz

(b) Twenty-Node Brick

DEGREE	TERMS
0	1
1	x, y, z
2	x^2, y^2, z^2, xy, xz, yz
3	x^2y, x^2z, xy^2, xz^2, y^2z, yz^2, xyz
4	x^2yz, xy^2z, xyz^2

(c) 27-Node Brick

DEGREE	TERMS
0	1
1	x, y, z
2	x^2, y^2, z^2, xy, xz, yz
3	x^2y, x^2z, xy^2, xz^2, y^2z, yz^2, xyz
4	x^2yz, xy^2z, xyz^2, x^2y^2, x^2z^2, y^2z^2
5	x^2y^2z, x^2yz^2, xy^2z^2
6	$x^2y^2z^2$

(a) Six Nodes

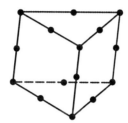

(b) Fifteen Nodes

Figure 3.8   Pentahedral Elements.

Table 3.3 records the basis functions for the six-node and fifteen-node pentahedra. They were obtained by multiplying the basis functions for a three-node or six-node triangle by appropriate functions of z.

Other node patterns suggest themselves. We could, for example, omit the three nodes on the vertical legs of fifteen-node elements or the nodes at the midpoints of the top and bottom edges. Neither of these variations alter the symmetry of the basis functions which are symmetrical with respect to x, y, but not with respect to x, z or y, z.

Table 3.3

Basis Functions for Pentahedral Elements

(a) Six-Node Pentahedron

DEGREE	TERMS
0	1
1	x, y, z
2	xz, yz

(b) Fifteen-Node Pentahedron

DEGREE	TERMS
0	1
1	x, y, z
2	x^2, y^2, z^2, xy, xz, yz
3	x^2z, xyz, xz^2, y^2z, yz^2

3.2.6 Node Deletion

The node patterns just described as variations of pentahedral elements employed the technique of node deletion to modify a "basic" element. Most finite element systems allow the user to delete edge nodes at program execution time. This practice can accommodate a variety of purposes, such as the juxtaposition of higher and lower order elements or simply the achievement of economy by removing higher order terms from a particular direction where they are not needed.

A corresponding number of basis functions must also be removed. Consider, for example, the eight-node rectangle with top and bottom edge nodes deleted, as shown in Figure 3.9(a).

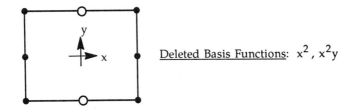

Deleted Basis Functions: x^2, x^2y

(a) Top and Bottom Edge Nodes Deleted.

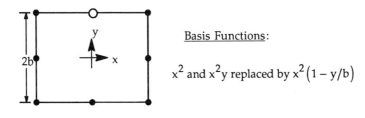

Basis Functions:

x^2 and x^2y replaced by $x^2\left(1 - y/b\right)$

(b) Top Edge Node Deleted.

Figure 3.9 **Eight-Node Rectangle with Deleted Nodes.**

The basis functions that must be deleted in this case are, from symmetry, x^2 and x^2y. The situation becomes somewhat more complicated when only one edge node is deleted, as shown in Figure 3.9(b). Here we elect to replace two basis functions, x^2 and x^2y, by the single function $x^2\left(1 - y/b\right)$. Since the replacement function vanishes at $y = b$, the variation of u along the top edge is linear. The justification for this choice does not come from a desire to preserve symmetry or non-singularity of the basis function matrix. It comes, rather, from a desire to maintain inter-element displacement continuity, which is the subject of Section 3.4.

Edge node deletion can have significant effects on the states of strain within elements. In the examples just given, the quadratic displacement states and, correspondingly, the linear strain states were complete and independent prior to the removal of edge nodes, but not afterward. The absence of completeness in linear strains states can cause locking, as will be shown.

One final remark about the selection of basis functions seems appropriate before passing on to other topics. In all of the examples given, except the last one, there was always one obvious choice for the best set of basis functions. All other choices either induced anisotropy, gave singular matrices, or substituted higher order terms for lower order ones. The designer's real options were, as a rule, confined to the selection of node locations. Once the node locations were selected, the basis functions logically followed. This condition probably applies generally.

3.3 INVENTORIES OF DEGREES OF FREEDOM

The degrees of freedom of a finite element have two aspects: external and internal. The external degrees of freedom are defined as the boundary displacement variables whether they are actually located on the boundary or in the interior. If these variables are confined to be components of displacement at nodes, then the number of external degrees of freedom equals the sum over the nodes of the number of displacements components at each node. The internal degrees of freedom include rigid body modes and strain states. Later we will encounter a third class of internal degrees of freedom—spurious modes. The number of internal and external degrees of freedom must be equal: they are just two different aspects of the same reality.

The internal degrees of freedom of an element are, so to speak, the designer's raw material. He must know how many there are and what they are before he can decide which ones to leave alone, to modify, or to delete. That is, in a nutshell, the principal subject matter of modern finite element design.

If the assumed displacement method is used, the spatial dependence of the internal degrees of freedom follows immediately from the form of the assumed basis functions. All we need do to obtain strains and rotations is to take spatial derivatives of the basis functions. In this way we can readily compile an inventory of the internal degrees of freedom for each of the elements described in Section 3.2. In order to reduce burdensome detail we will, for the present, list only the *number* of strain states in each polynomial degree (constant strains, linear strains, etc.). Later we will be concerned with the functional form of individual strain states.

Tables 3.4 to 3.8 list degree-of-freedom inventories for all of the elements described in Section 3.2. Classical elasticity is assumed for the physics of the elements. The numbers, but not the concepts, change for other disciplines.

Each element is identified by a name that describes the shape of the element and the number of nodes. Thus TRIA3 identifies the three-node triangle while HEXA27 identifies the 27-node hexahedral (brick) element. The number of strain states in a particular polynomial order is simply computed as the number of displacement basis functions in the next higher order times the number of displacement components. Constant strain states are an exception because linear displacements also give rise to rigid rotations.

The number of strain states needed to achieve completeness in a particular polynomial order is also listed in the tables. An element which has a complete set of strains states of a given order can, at least in principle, give correct results for any imposed stress or strain field of that order. If the strain states of a given order are incomplete, the element can give correct results for only certain types or certain orientations of imposed fields of the same order. Consider, for example, the four-node quadrilateral, QUAD4, in Table 3.6. It has only two linear strain states rather than the six needed for completeness. That has not stopped a generation of users from employing the four-node quadrilateral for in-plane bending applications. The user should, of course, be aware of the element's limitations and carefully select its orientation relative to the strain field. (How many users actually do this!)

The number of quadratic and higher order strain states needed for completeness is actually less than might be expected due to the existence of certain identities. In two dimensions, for example, we might expect the number of quadratic strain states to be (3 components of strain) x (3 variations $\left(x^2, xy, y^2\right)$) = 9 states. The actual number is eight, due to the identity

$$\varepsilon_{x,y^2} + \varepsilon_{y,x^2} = \gamma_{xy,xy} \tag{3:26}$$

which follows from the definition of strains in terms of displacement components.

Table 3.4

Degree-of-Freedom Inventory for Triangular Elements

ELEMENT TYPE	TRIA3	TRIA6	TRIA10	
DOF at Boundary Nodes	6	12	18	NEEDED FOR COMPLETE-NESS
DOF at Interior Nodes	--	--	2	
External DOF	6	12	20	
Rigid Body Modes	3	3	3	3
Constant Strain States	3	3	3	3
Linear Strain States	–	6	6	6
Quadratic Strain States	--	--	8	8
Internal DOF	6	12	20	

Table 3.5

Degree-of-Freedom Inventory for Tetrahedral Elements

ELEMENT TYPE	TETRA4	TETRA10	TETRA20	
DOF at Boundary Nodes	12	30	60	NEEDED FOR COMPLETE-NESS
DOF at Interior Nodes	--	--	--	
External DOF	12	30	60	
Rigid Body Modes	6	6	6	6
Constant Strain States	6	6	6	6
Linear Strain States	–	18	18	18
Quadratic Strain States	--	--	30	30
Internal DOF	12	30	60	

Table 3.6

Degree-of-Freedom Inventory for Rectangular Elements

ELEMENT TYPE	QUAD4	QUAD8	QUAD9	QUAD12	QUAD16	
DOF at Boundary Nodes	8	16	16	24	24	NEEDED FOR COMPLETE-NESS
DOF at Interior Nodes	--	--	2	--	8	
External DOF	8	16	18	24	32	
Rigid Body Modes	3	3	3	3	3	3
Constant Strain States	3	3	3	3	3	3
Linear Strain States	2	6	6	6	6	6
Quadratic Strain States	–	4	4	8	8	8
Cubic Strain States	–	--	2	4	6	10
Quartic Strain States	–	–	–	–	4	12
Quintic Strain States	--	--	--	--	2	14
Internal DOF	8	16	18	24	32	

Table 3.7

Degree-of-Freedom Inventory for Brick Elements

ELEMENT TYPE	HEXA8	HEXA20	HEXA27	
DOF at Boundary Nodes	24	60	78	NEEDED FOR COMPLETE-NESS
DOF at Interior Nodes	--	--	3	
External DOF	24	60	81	
Rigid Body Modes	6	6	6	6
Constant Strain States	6	6	6	6
Linear Strain States	9	18	18	18
Quadratic Strain States	3	21	21	30
Cubic Strain States	–	9	18	42
Quartic Strain States	–	–	9	54
Quintic Strain States	--	--	3	66
Internal DOF	24	60	81	

Table 3.8

Degree-of-Freedom Inventory for Pentahedral Elements

ELEMENT TYPE	PENTA6	PENTA15	
DOF at Boundary Nodes	18	45	NEEDED FOR COMPLETE-NESS
DOF at Interior Nodes	--	--	
External DOF	18	45	
Rigid Body Modes	6	6	6
Constant Strain States	6	6	6
Linear Strain States	6	18	18
Quadratic Strain States	--	<u>15</u>	30
Internal DOF	18	45	

Another way to look at the completeness issue is to recall that, in the assumed displacement method, quadratic strain states are obtained by differentiating cubic displacement states. Since, in two dimensions, there are four cubic basis functions $\left(x^3, x^2y, xy^2, y^3 \right)$ and two displacement components, there can be only eight independent quadratic strain states. We may also note, in passing, that the strain identity, or strain compatibility condition (Equation 3:26), is not automatically satisfied when the element design employs an assumed strain or an assumed stress field.

In three dimensions, there are six strain identities similar to Equation 3:26 so that the expected number of quadratic strain states (36) is reduced by six. At cubic or higher orders, the expected numbers of strain states are reduced by the numbers of possible derivatives of the strain identities.

If completeness is our objective, we can boil Tables 3.4 to 3.8 down to a list which gives the order of strain to which each element is complete. That list, shown in Table 3.9, indicates that the lowest order element of each type is complete through constant strains only. The next higher order elements (those with one node per edge regardless of the presence or absence of interior nodes) are complete through linear strains, and the elements with two nodes per edge are complete through the quadratic strain states.

Table 3.9
Orders of Completeness of Various Elements

Constant Strain:	TRIA3, TETRA4, QUAD4, HEXA8, PENTA6
Linear Strain:	TRIA6, TETRA10, QUAD8, QUAD9, HEXA20, HEXA27, PENTA15
Quadratic Strain:	TRIA10, TETRA20, QUAD12, QUAD16

We have seen that strain compatibility reduces the required number of quadratic and higher order strain states needed for completeness. We might also ask whether the stress equilibrium equations (see Equation 2:13) have a similar effect. The answer is that they most certainly do if a stress field is assumed and that they very well might if a displacement field is assumed.

In two dimensions the stress equilibrium equations are

$$\sigma_{x,x} + \tau_{xy,y} + p_x = 0$$
$$\sigma_{y,y} + \tau_{xy,x} + p_y = 0$$

$$(3:27)$$

Since a complete set of linear strain states has six terms, the equilibrium equations can potentially reduce the required number of terms to four. We shall use the term *quasi-complete* to refer to a set of strain states that meets this reduced requirement. Computation of the reduction in the number of terms needed for quasi-completeness follows the procedure used to compute the effect of strain compatibility. At quadratic and higher orders, the derivatives of Equation 3:27 produce additional conditions on the stress states. Table 3.10 indicates the resulting number of strain states needed for quasi-completeness in both two and three dimensions. It is interesting to note that, in two dimensions, only four terms are needed in each order above the constant terms.[*]

Table 3.10

Number of Terms Needed for

Quasi-Completeness of Sets of Strain Components

With Ascending Polynomial Degrees

	NUMBER OF TERMS NEEDED FOR QUASI-COMPLETENESS	
ORDER OF STRAINS	IN 2-D	IN 3-D
Constant	3	6
Linear	4	15
Quadratic	4	21
Cubic	4	24
Quartic	4	24
Quintic	4	21

Recognition of quasi-completeness raises the order of "competence" of some elements but not all. Of the elements listed in Tables 3.4 to 3.8, the following have their order of competence raised when quasi-completeness is substituted for completeness.

---

[*]It is possible, in the higher orders, that some of the constraints imposed by strain compatibility and stress equilibrium may not be independent. This has not been factored into Table 3.10.

QUAD8:   Linear to Quadratic
QUAD9:   Linear to Quadratic
QUAD12:   Quadratic to Cubic
QUAD16:   Quadratic to Quartic
HEXA20:   Linear to Quadratic
HEXA27:   Linear to Quadratic

These higher levels represent ideals which can only rarely be achieved. As we shall see, many factors enter which degrade the performance of finite elements.

It is also worth noting that, while the four-node membrane element, QUAD4, does not have linear quasi-completeness, the reduction of the required number of linear terms from six to four can be exploited if, somehow, the number of external degrees of freedom is raised. The so-called *drilling freedoms*[8] provide this possibility (see Section 8.2).

3.4   THE ISSUE OF INTER-ELEMENT CONTINUITY

Finite elements interact with each other through the common values of discrete displacement variables in their mutual boundaries. This does not imply that the displacements of two adjacent elements are necessarily equal along their entire mutual boundary. Indeed, we have deliberately avoided making the continuity of boundary displacements a requirement because many successful elements lack such continuity. On the other hand, inter-element displacement continuity brings important benefits which must be considered and understood. Later we can decide, in individual cases, whether the abandonment of inter-element continuity is worth the loss of these benefits.

The chief benefit of inter-element displacement continuity is that it allows convergence proofs to be constructed. Later, in Chapter 5, we will develop one such proof in detail. For now we will only attempt to show that inter-element displacement continuity makes convergence plausible.

Consider an element that is subjected to known tractions along its edges. Virtual work may be used to transfer the tractions to nodes (see Equation 2:80).

The element responds by assigning values to the coefficients of its displacement basis functions. As the number of basis functions is increased, the element's strain energy increases and, if the basis functions satisfy a completeness criterion, the strain energy of the element will approach the strain energy of an elastic continuum occupying the same space. Looked at a little differently, the removal of any basis function imposes a constraint on the element's displacement field by denying it the basis function's form. The imposition of each such constraint stiffens the element. The process of adding basis functions is, therefore, a Ritz process[9] which should converge to the exact solution for the continuum if the basis functions are complete.

Consider next that an elastic domain is subdivided into elements and subjected to known tractions along its outer boundary. Each element will respond in the manner described above as the number of basis functions is increased within each element. Let us examine, in addition, what happens at the interior boundaries between elements. If the displacements of adjacent elements are equal at their boundaries, nothing remarkable happens as tractions are transferred smoothly across the boundaries between elements. The process of adding basis functions to each element can then be regarded as just part of a process of removing constraints from the entire domain. The strain energy increases with the addition of each basis function and approaches the correct value for the elastic domain.

If, on the other hand, the displacements of adjacent elements are not equal, gaps open up along the boundaries between elements as loads are applied and the tractions on the boundaries do work against the relative displacements.[*] Because the gaps soften the structure, we no longer know for certain that the solution for a finite number of basis functions is too stiff. Neither can we be sure of convergence. For example, do the gaps, or slits, between elements tend to widen or contract as the number of basis functions is increased? Each case must be examined on its own merits.

---

[*]We can think of the tractions on the edges of one element as being transferred first to its boundary nodes and then to the edges of adjacent elements.

In summary, we can be certain, if the displacements of adjacent elements are equal at their common boundary, but not otherwise, that the finite element model is stiffer than the actual structure and we can expect that the solution will converge to the true solution as basis functions are added to the elements. These conclusions also apply when each element is subdivided into similar elements because that also adds to the total number of internal degrees of freedom.

Inter-element displacement continuity is frequently described by other terms. Elements which have this property are said to *conform*, to be *conforming*, or to have *conformability*. They are also said to have C^0 continuity, or possibly C^1 continuity. Here we should define terms. An element has C^0 continuity if its displacement field is continuous with the fields of adjacent elements and has continuous *first* derivatives at interior points. The continuity level is raised to C^1 if the *slope* of the displacement field is continuous along element boundaries and if the field has continuous *second* derivatives at interior points.

Displacement fields with C^1 continuity have application in plate elements. If transverse shear strain is neglected, the slopes of the normal displacement, $w_{,x}$ and $w_{,y}$, become degrees of freedom at node points. Inter-element continuity of these degrees of freedom then requires C^1 continuity of w.

First, second, and higher order derivatives have, in the past, been used as degrees of freedom for membrane and solid elements. In contrast with the higher-order elements described in Section 3.2, which all have edge nodes, such elements frequently have all of their degrees of freedom, such as $\left(u, u_{,x}, u_{,y}, u_{,xy}, \text{etc.} \right)$, concentrated at corner nodes. This is a more efficient formulation in terms of computer time (matrix bandwidth is reduced). Unfortunately, it may also impose an unwanted degree of continuity on the displacement field. Real engineering structures have plenty of discontinuities, such as free edges, changes in thickness, and changes in material properties. The first derivatives of displacement fields, i.e., strains, are likely to be discontinuous at such places, so that the provision of continuous first derivatives may be an unwelcome gift. In recognition of this fact, modern element designs have avoided the use of displacement derivatives as degrees of freedom. Plate and shell elements are an exception but even there the trend is to design elements with C^0 continuity only.

Given that inter-element displacement continuity is desirable, at least to order C^0, the next logical question is how to determine whether it is present. To answer this we refer to the expansion of the displacement field in terms of shape functions

$$u = \lfloor N \rfloor \{u_i\} = \sum_{i=1}^{n} N_i u_i \qquad (3{:}28)$$

The value of u depends, in general, on all of the u_i's but along an edge it should, if inter-element continuity is to be achieved, depend only on the u_i's at nodes located on the same edge. Consider, for example, the pair of elements shown in Figure 3.10.

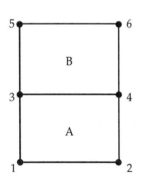

At interior points the displacement field in element A depends on the values of u at nodes 1, 2, 3, and 4, but along the common edge, 3-4, the displacement field in element A must depend only on u_3 and u_4. Otherwise, the parts proportional to u_1 and u_2 could not be matched by element B. Thus the shape functions for nodes 3 and 4 in the two elements must be the same along edge 3-4 and the shape functions for nodes 1, 2, 5, and 6 must be zero along edge 3-4.

Figure 3.10   Adjacent Elements.

The requirement of C^0 continuity can be achieved in general by ensuring that, on a given edge, the shape functions for nodes on or adjacent to the edge are computed identically in adjacent elements and by ensuring that the shape functions for nodes not adjacent to the edge are zero.

We can easily determine whether elements based on polynomial functions of position are conforming. Consider, for a start, the constant strain triangle which has $\lfloor 1, x, y \rfloor$ as basis functions. Thus $u = a_1 + a_2 x + a_3 y$. We can perform a coordinate transformation so that any given edge, say edge 1-2, is the $\bar{x}$-axis and node 1 is the origin, as shown in Figure 3.11. Then $u = a_1' + a_2' \bar{x}$ along the edge and the coefficients a_1' and a_2' can be evaluated in terms of nodal displacements to give $u = \left(1 - \bar{x}/\ell\right)u_1 + \left(\bar{x}/\ell\right)u_2 = N_1 u_1 + N_2 u_2$. Since u depends only on u_1 and u_2, the element is conforming.

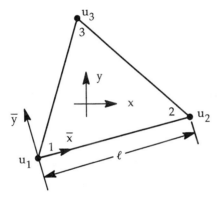

Figure 3.11 The Constant Strain Triangle Is a Conforming Element.

We can apply the same procedure to the four-node quadrilateral shown in Figure 3.12.

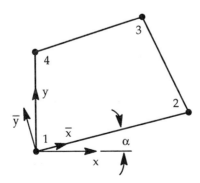

Figure 3.12 Is the Four-Node Quadrilateral Conforming?

Here

$$u = a_1 + a_2 x + a_3 y + a_4 xy \qquad (3:29)$$

so that the coordinate transformation

$$x = \bar{x} \cos \alpha - \bar{y} \sin \alpha$$
$$y = \bar{x} \sin \alpha + \bar{y} \cos \alpha \qquad (3:30)$$

gives

$$u = a_1 + a_2' \bar{x} + a_3' \bar{y} + a_4' \overline{xy} + a_5' \left(\bar{x}^2 - \bar{y}^2 \right) \qquad (3:31)$$

Along the $\bar{x}$-axis, $u = a_1 + a_2' \bar{x} + a_5' \bar{x}^2$ which, since it has three terms, cannot be evaluated in terms of u_1 and u_2 only. As a result, the value of u along side 1-2 depends on u_3 and/or u_4 and the element is not conforming.

We can, of course, make the displacement field conform along edge 1-2 by orienting the x-axis parallel to that edge. But we cannot simultaneously arrange for an axis to be parallel to the other three edges except in the case of a rectangle. With this exception, four-node quadrilateral elements based on power series expansions of the Cartesian position coordinates are nonconforming.[*]

How about a triangle with edge nodes or a tetrahedron with edge nodes? These elements are conforming as long as the edges are straight and all edges have the same number of nodes. The key is that the basis functions are complete polynomials so that, in a coordinate transformation, they transform into linear recombinations of themselves. Suppose that the basis functions are $\lfloor 1, \ x, \ y, \ x^2, \ xy, \ y^2 \rfloor$. Then, if the $\bar{x}$-axis is selected to coincide with edge 1-2, $u = a_1 + a_2' \bar{x} + a_4' \bar{x}^2$ along that edge. Clearly the three coefficients

---

[*]A conforming four-noded parallelogram can be constructed by using skewed Cartesian coordinates.

a_1, a_2', and a_4' can be evaluated in terms of the value of u at three points on the edge, which ensures that u has inter-element continuity.

General quadrilateral elements with edge nodes, on the other hand, do not conform. The reason is that the incomplete basis functions of order n generate an $\bar{x}^n$ term upon coordinate rotation through an arbitrary angle. Thus, if the sides are straight, triangular elements conform for more general shapes than quadrilaterals.

But what if the edges are curved? In that case elements based on power series expansions of position coordinates are non-conforming. To see this consider the curved edge with one edge node shown in Figure 3.13.

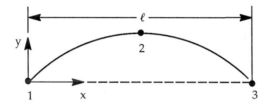

Figure 3.13   A Curved Edge.

Let the x-axis lie along the chord 1-3 with the origin at node 1. Let us assume that the edge has parabolic shape so that

$$y = \frac{cx}{\ell}\left(1 - \frac{x}{\ell}\right) \tag{3:32}$$

Then, if the element is a six-node triangle, we can write, for displacement along the curved edge,

$$u = a_1 + a_2x + a_3y + a_4x^2 + a_5xy + a_6y^2$$

$$= a_1 + a_2x + a_4x^2 + a_3\frac{cx}{\ell}\left(1 - \frac{x}{\ell}\right) + a_5\frac{cx^2}{\ell}\left(1 - \frac{x}{\ell}\right) + a_6\frac{c^2x^2}{\ell^2}\left(1 - \frac{x}{\ell}\right)^2 \tag{3:33}$$

which contains all powers of x up to the fourth power. As a result u contains five independent functions of x along the edge, whose coefficients clearly cannot be evaluated in terms of the values of u at only three nodes. Consequently inter-element displacement continuity does not exist.

The results we have obtained are, on balance, discouraging. They show that, for elements based on power series expansions of Cartesian position coordinates, inter-element displacement continuity is achievable for triangular and tetrahedral elements with straight edges and for rectangles, but not for elements with other shapes. To achieve inter-element continuity for a larger range of shapes, we must look to another formulation of the basis functions. We will see in the next section that parametric mapping provides such a possibility. We will also see that the price is not negligible.

3.5 PARAMETRIC MAPPING

In finite element design, parametric mapping is used to map the interior of a finite element into a standard shape. For example, Figure 3.14 illustrates the mapping of a general quadrilateral element into a square.

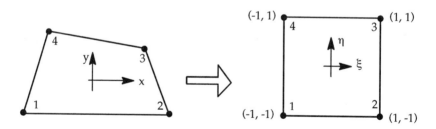

Figure 3.14   Parametric Mapping of a General Quadrilateral into a Square.

Let the mapping be expressed by

$$x = \lfloor x' \rfloor \{b\}$$

$$y = \lfloor x' \rfloor \{c\}$$

(3:34)

where the elements of the position basis vector, $\lfloor X' \rfloor$, are functions of the parametric coordinates (ξ, η). The displacement field, u, may similarly be expressed as

$$u = \lfloor X \rfloor \{a\} \qquad (3{:}35)$$

where the elements of the displacement basis vector $\lfloor X \rfloor$ are also functions of the parametric coordinates. Note that $\lfloor X \rfloor$ and $\lfloor X' \rfloor$ are not necessarily equal to each other or to linear recombinations of each other. If they are, the mapping is said to be *isoparametric*. Finite element theory recognizes two other cases: if $\lfloor X' \rfloor$ is a subset of $\lfloor X \rfloor$, the mapping is called *subparametric*; if $\lfloor X' \rfloor$ includes basis functions that are not in $\lfloor X \rfloor$, the mapping is called *superparametric*.

We may regard the standard shape in the (ξ, η) plane (or the (ξ, η, ζ) space) as a substitute element and use the basis functions previously derived for triangles, rectangles, rectangular bricks, etc., to describe the displacement field. Thus, for the case of the four-node quadrilateral element shown in Figure 3.14, the logical choice for a set of displacement basis functions is $\lfloor X \rfloor = \lfloor 1, \xi, \eta, \xi\eta \rfloor$. Since the substitute element is a square, we know, from the results of Section 3.4, that the displacement field has inter-element continuity in the (ξ, η) plane. Inter-element continuity translates immediately into the (x, y) plane if the mapping is continuous, i.e., if adjacent points in the (ξ, η) plane are mapped into adjacent points in the (x, y) plane.

From this example we see that parametric mapping provides a general solution to the problem of inter-element displacement continuity. All that is required is that the substitute element's shape allow inter-element displacement continuity and that the mapping be continuous. From Section 3.4 we know that straight-sided rectangles and triangles allow inter-element displacement continuity. By extension, straight-sided rectangular parallelepipeds (bricks), tetrahedra, and right triangular prisms also allow inter-element displacement continuity. These, then, must be the standard substitute shapes in the ξ, η plane and the ξ, η, ζ space. They are illustrated in Figure 3.15. Note that isosceles right triangular shapes have been assumed. Any other triangular shapes with straight edges would do as well. For higher order elements, the edge nodes and interior nodes take symmetrical positions in the substitute elements.

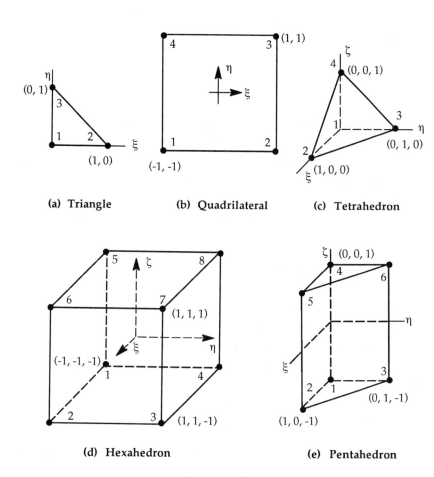

Figure 3.15   Standard Substitute Element Shapes.

The concept of parametric mapping was first introduced into finite element analysis in 1961 by Taig[10] who used it to design his famous four-node isoparametric element. A few years later (1966) Irons[11] extended the concept to membrane and solid elements of all orders. Quite independently, Coons[12] (1967) applied the concept to the generation of curved surfaces of engineering interest. Today extensions of the latter development find application in CAD systems and in the automatic generation of finite element meshes.[13]

The mechanics of generating elements which use parametric mapping will be taken up in the next chapter. Here we concern ourselves only with the relationship between accuracy and the selection of position basis functions, $\lfloor X' \rfloor$, to map position.

We begin by examining the smoothness of the mapping for typical cases. Consider first the four-node quadrilateral element of Figure 3.14 and let the position basis vector be $\lfloor X' \rfloor = \lfloor 1, \xi, \eta, \xi\eta \rfloor$. Then

$$x = b_1 + b_2\xi + b_3\eta + b_4\xi\eta$$

$$y = c_1 + c_2\xi + c_3\eta + c_4\xi\eta$$

(3:36)

For constant η, both x and y are linear functions of ξ and, consequently, y is a linear function of x. Thus lines of constant ξ or η are straight lines in the (x, y) plane. The mapping can be represented by a set of rulings on the element surface, as shown in Figure 3.16.

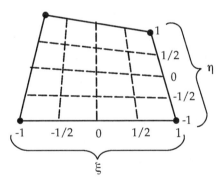

Figure 3.16 Lines of Constant ξ and η for a Four-Node Quadrilateral.

This mapping seems quite reasonable and easy to grasp. Consider, however, what happens when extra nodes are added to an element, such as in the one-dimensional case illustrated in Figure 3.17.

The physical location of the middle node, x_2, may be anywhere between the end points, ± 1, but the parametric coordinate of the middle node is $\xi = 0$

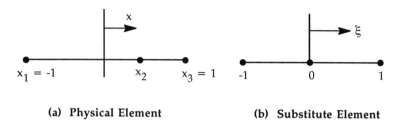

(a) **Physical Element**   (b) **Substitute Element**

Figure 3.17   One-Dimensional Element with Three Nodes.

regardless of its physical position.   Thus the edge node in the substitute
element obeys the symmetry property noted earlier.   The logical choice for a
vector of position basis functions is $\lfloor X' \rfloor = \lfloor 1, \xi, \xi^2 \rfloor$.   The positions selected
for the nodes $\left(-1, x_2, +1\right)$ then give

$$x = \xi + \left(1 - \xi^2\right)x_2 \qquad (3:37)$$

Figure 3.18 shows a plot of Equation 3:37 for three values of x_2.   Observe that
for $x_2 > 1/2$, the value of x can lie outside the range -1 to +1.   This unexpected
result is certainly not what a prudent analyst would have intended.   It carries
over into two and three dimensions where edge node locations that are too near
the corners can cause elements to spill over their boundaries.   Clearly finite
element users must exercise caution in selecting the positions of edge nodes.
But it is an ill wind indeed that blows nobody some good.   It turns out that
placing edge nodes at the quarter points $\left(x_2 = \pm\, 1/2\right)$ allows the element to
simulate a stress singularity at the corner.[14]   For example, in the one-
dimensional case just examined $x_{/\xi} = 0$ at $\xi = 1$ for $x_2 = 0.5$.   Thus
$\varepsilon_x = u_{/x} = u_{/\xi}\, \xi_{/x}$ becomes infinitely large at $\xi = 1$ provided that $u_{/\xi}$ is not
zero at that point.

The selection of the position basis functions in $\lfloor X' \rfloor$ also affects the ability of
the element to represent various strain states.   We can reasonably insist, as a
minimum, that any respectable element should be able to represent rigid body
motion or a constant strain state, say, for example, $u = a + bx + cy$.   Elements

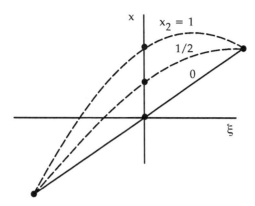

Figure 3.18   Parametric Mapping of a Three-Node Line Element with Various Positions of the Middle Node.

which have this property are said to be *complete* or to satisfy completeness. (We will use this terminology sparingly to avoid confusion with the important issue of polynomial completeness.) We can determine whether an element does, in fact, satisfy completeness by examining how the element forms its displacement states from nodal values. In general, from Equation 3:12,

$$u = \lfloor N \rfloor \{u_i\} = \lfloor X \rfloor \left[X_{ij} \right]^{-1} \{u_i\} \tag{3:38}$$

If we let $\{u_i\} = \{X_{ij}\}$, a column vector representing the jth basis function at nodes, then $u = X_j$. This follows because

$$\{X_{ij}\} = \left[X_{ij} \right] \{\delta_{ij}\} \tag{3:39}$$

where δ_{ij} is the Kroneker delta. Thus we can say that the element will properly evaluate any of its basis functions that is prescribed at nodes. The same cannot be said of any other function. For example, in the case of the four-node quadrilateral, if $\{u_i\} = \{\xi_i^2\}$, then $u = 1$ because ξ_i^2 has the value of unity at all nodes.

This argument has direct bearing on whether or not u can properly represent a linear function of position, i.e., on whether the representation will be the same as that given by the position basis functions. For example, the latter representation of y is

$$y = \lfloor N' \rfloor \{y_i\} = \lfloor X' \rfloor \left[X'_{ij} \right]^{-1} \{y_i\} \tag{3:40}$$

while the former is

$$u = \lfloor N \rfloor \{u_i\} = \lfloor X \rfloor \left[X_{ij} \right]^{-1} \{y_i\} \tag{3:41}$$

The two representations will be the same, i.e., u will equal y and the shape functions will satisfy completeness, if $\lfloor X' \rfloor = \lfloor X \rfloor$ or if the basis functions in $\lfloor X' \rfloor$ are a subset of those in $\lfloor X \rfloor$. In other words, they will be the same if the mapping is *isoparametric* or *subparametric*. In either case, y from Equation 3:40 can be expanded in terms of basis functions contained in $\lfloor X \rfloor$. If, on the other hand, the mapping is *superparametric*, then $\lfloor X' \rfloor$ will contain basis functions not contained in $\lfloor X \rfloor$. In this case Equations 3:40 and 3:41 will give different results.

Subparametric and superparametric mapping are contrasted in Figure 3.19. In Figure 3.19(a), the edges are necessarily straight because x and y are defined only at the corners. The edge nodes are also centered. In this case x and y have four basis functions $\left(1, \xi, \eta, \xi\eta\right)$ and u has eight $\left(1, \xi, \eta, \xi\eta, \xi^2, \eta^2, \xi^2\eta, \xi\eta^2\right)$. In Figure 3.19(b), on the other hand, the edges may be curved because x and y are specified at eight points. In this case x and y have eight basis functions and u has four.

If superparametric mapping is employed, then expressions for displacement like $u = a + bx + cy$ will not, in general, be properly represented within the element. This has two immediate consequences. The first is that the element cannot represent constant strain states exactly. The second is that the element cannot properly represent rigid body rotations. The latter consequence is, if anything, worse than the first. Failure of the rigid body property is the most virulent of all the disorders that can afflict a finite element. As Irons and

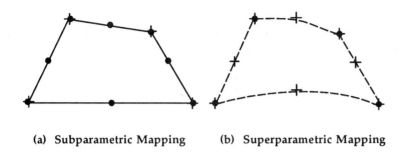

(a) Subparametric Mapping   (b) Superparametric Mapping

+ Points where x and y are given
• Nodes where u is evaluated

Figure 3.19   Subparametric and Superparametric Mapping Contrasted.

Ahmad propose in their list of "Standards and Feelings" about finite elements: [15]

"We shall try to kill any element that fails a test for rigid body motions."

Every structural analyst with gray hair probably remembers his first encounter with rigid body failure.   In the author's case, that first encounter involved a pair of adjacent ribs on a highly swept wing.   Since the angle between the ribs was only one or two degrees, he decided it was unnecessary to account for the angle in the representation of a torsion box between the two ribs.   The result was a dramatic detour of transverse shear from the rear spar to the front spar, contrary to all experience and common sense.

Still, superparametric mapping has powerful attraction.   The unsophisticated user who understands geometry but not the intricacies of finite element design, may insist on an accurate representation of the edges of his structure.   And the element designer may accede to such insistence.   He might, for example, elect to provide exact representation of circular arcs and other curves by special blending functions. [16]

Subparametric and isoparametric mapping appear, so far, to be on an equal footing, at least with respect to the representation of rigid body motion and constant strain states. To distinguish them, we must go further and consider linear strain states such as $u = y^2$. In this case,

$$y^2 = \left(\lfloor X' \rfloor \{c\} \right)^2 = \sum_j \sum_k X'_j X'_k c_j c_k \tag{3:42}$$

so that, to represent $u = y^2$ correctly, $X'_j X'_k$ must be a displacement basis function. We see immediately that this is not generally possible if $\lfloor X' \rfloor = \lfloor X \rfloor$, i.e., if the element is isoparametric.

This is a most important result because it shows that an isoparametric element cannot exactly represent linear strain states when the full geometric capability of the element is employed. The only way in which linear strain states can be properly represented is to use less than the full geometric capability or, in other words, to restrict the element to have subparametric shape. In the case of the six-node triangle for instance, this means keeping the edges straight and the edge nodes centered. Then $y = \lfloor X' \rfloor \{c\} = c_1 + c_2 \xi + c_3 \eta$ so that y^2 will be included within $\left(1, \xi, \eta, \xi^2, \xi\eta, \eta^2 \right)$.

The case of the quadrilateral is even more restrictive. If we keep the edges straight and the edge nodes centered, then the nonzero basis functions for y are $\left(1, \xi, \eta, \xi\eta \right)$. The resulting nine terms for y^2 are $\left(1, \xi, \eta, \xi^2, \xi\eta, \eta^2, \xi^2\eta, \xi\eta^2, \xi^2\eta^2 \right)$. The lowest order element that includes all of these terms is the nine-node Lagrange element. Thus an important advantage of the nine-node Lagrange element over the eight-node serendipity element is its ability to represent linear strain states exactly. Even then the node locations must be effectively subparametric.[*]

---

[*]Edges straight, edge nodes centered, and center node located at the intersection of straight lines joining opposite edge nodes.

A practical example of the importance of keeping the edges of isoparametric elements straight is illustrated in Figure 3.20.

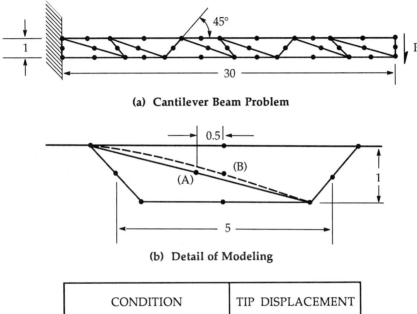

(a) Cantilever Beam Problem

(b) Detail of Modeling

CONDITION	TIP DISPLACEMENT
Edge Node Centered (A)	.953
Edge Node Displaced (B)	.391
Exact Solution	1.000

(c) Results

Figure 3.20   Slender Cantilever Beam Modeled by Six-Node Isoparametric Elements.

In this example, which involves the bending of a slender cantilever beam, the displacement field is strongly quadratic with a little cubic content thrown in. The elements are six-node triangles which are arranged in pairs that form trapezoids. Results are shown for two cases. For the first case, in which all edge nodes are centered, the error in tip deflection is less than 5%. For the second case, in which the node on each trapezoid's diagonal was moved to the center of the trapezoid, the error exceeds 60%. Thus it is seen that, in extreme cases such as this, the small displacement of an edge node can have a disastrous effect.

The same problem has been solved with an element, the MSC/NASTRAN TRIA6, that does not employ parametric displacement mapping (the element uses an assumed strain field). For this element, the error did not increase significantly when the diagonal edge node was displaced (see Table 8.2). The large increase in error experienced with the isoparametric element is directly attributable to parametric mapping.

In spite of the large errors that can occur when edge nodes are not centered, the designs of most six-, eight-, and nine-node elements allow unrestricted placement of edge nodes. The first six-node membrane element in MSC/NASTRAN (called TRIM6 out of respect for an earlier element with the same name and similar design in ASKA) had a subparametric design which centered the edge nodes regardless of the positions specified for them. It was not a popular element with users.

We have seen, in this section, that parametric mapping provides a general solution for the problem of inter-element displacement continuity. At the same time parametric mapping places severe limitations on the accuracy of strain representation. With no limits on element shape, the best that can be achieved, even with an element of arbitrarily high order, is the exact representation of constant strain states. To achieve accurate linear or higher order strains, restrictions must be placed on the geometry.

In contrast, elements which are derived without benefit of parametric mapping yield correct linear strains so long as their displacement basis functions are complete to second degree in the position coordinates. To give a name to such elements we will call them *metric* elements because they employ metric rather than parametric interpolation. The great disadvantage of metric elements, and the reason for the introduction of parametric mapping, is that they exhibit inter-element displacement continuity only when restrictions are placed on *their* geometry.

Another disadvantage of metric elements is that evaluation of their shape functions requires the inversion of a matrix (see Equation 3:12) while parametric shape functions can usually be evaluated by inspection due to the simplicity of element shapes in the parametric domain. Even though matrix inversion causes only a small increase in computer time, it looms large in the minds of element designers who find the attendant loss of numerical control distasteful.

Nearly all finite elements use parametric mapping and we will concentrate on them. We will, however, have occasion to apply metric interpolation as a remedy for certain of the disorders of parametric elements (see Section 8.3).

REFERENCES

3.1   A. G. Peano, "Hierarchies of Conforming Finite Elements for Plane Elasticity and Plate Bending," *Comp. & Maths. with Appls.*, 2, pp 211-24, 1976.

3.2   G. R. Heppler and S. J. Hansen, "Timoshenko Beam Finite Elements Using Trigonometric Basis Functions," *J. AIAA*, 26, pp 1378-86, 1988.

3.3   A. Anandarajah, "Time-Domain Radiation Boundary for Analysis of Plane Love-Wave Propagation Problem," *Intl. J. Numer. Methods Eng.*, 29, pp 1049-63, 1990.

3.4   R. J. Guyan, "Reduction of Stiffness and Mass Matrices," *J. AIAA*, 3, No. 2, 1965.

3.5   B. M. Irons and S. Ahmad, *Techniques of Finite Elements*, Ellis Horwood, Chichester, pp 411-14, 1980.

3.6   M. J. Turner, R. W. Clough, H. C. Martin and L. J. Topp, "Stiffness and Deflection Analysis of Complex Structures," *J. Aeronautical Sci.*, 23, p. 805, 1956.

3.7   R. J. Melosh, "Basis for Derivation of Matrices for Direct Stiffness Method," *J. AIAA*, 1, pp 1631-7, 1963.

3.8   D. J. Allman, "A Compatible Triangular Element Including Vertex Rotations for Plane Elasticity Analysis," *Comput. Struct.*, 19, pp 1-8, 1984.

3.9   W. Ritz, "Über eine neue Methode zur Lösung gewissen Variations-Probleme der mathematischen Physik," J. Reine & Angew. Math., 135, pp 1-61, 1909.

3.10   I. C. Taig, "Structural Analysis by the Matrix Displacement Method," Engl. Electric Aviation Report No. 5017, 1961.

3.11   B. M. Irons, "Engineering Application of Numerical Integration in Stiffness Methods," *J. AIAA*, 14, pp 2035-7, 1966.

3.12   S. A. Coons, "Surfaces for Computer-Aided Design of Space Form," MIT Project MAC, MAC-TR-41, 1967.

3.13   R. B. Haber, M. S. Shephard, J. F. Abel, R. H. Gallagher, and D. P. Greenberg, "A General Two-Dimensional Finite Element Preprocessor Utilizing Discrete Transfinite Mappings," *Intl. J. Numer. Methods Eng.*, 16, pp 1015-44, 1981.

3.14   R. D. Henshell and K. G. Shaw, "Crack Tip Elements Are Unnecessary," *Intl. J. Numer. Methods Eng.*, 9, pp 495-509, 1975.

3.15   B. M. Irons and S. Ahmad, *Techniques of Finite Elements*, Ellis Horwood, Chichester, p. 155, 1980.

3.16   W. J. Gordon and C. A. Hall, "Construction of Curvilinear Coordinate Systems and Applications to Mesh Generation," *Intl. J. Numer. Methods Eng.*, 7, pp 461-77, 1973.

4

Isoparametric Membrane and Solid Elements

This chapter concentrates on the mechanics of isoparametric elements, i.e., on the calculation of their shape functions, stiffness matrices, and load vectors. Later we will examine the disorders of isoparametric elements and ways to improve their performance. Standard isoparametric elements will, thereby, serve us as a base on which to build additional design concepts.

4.1 THE CALCULATION OF SHAPE FUNCTIONS

As we have seen, if boundary displacements are concentrated at nodes, shape functions can be calculated from basis functions according to the formula (see Equation 3:12)

$$\lfloor N_i \rfloor = \lfloor X \rfloor \left[X_{ij} \right]^{-1} \qquad\qquad (4{:}1)$$

where X_{ij} is the value of the jth basis function at node i. Once the basis functions are selected, according to the principles described in Section 3.2, the evaluation of shape functions is an arithmetic exercise which can be performed by brute force using Equation 4:1. On the other hand, because the substitute element shapes of isoparametric elements are so simple, it is usually possible to avoid numerical inversion at a slight cost in intellectual effort. The payoff is algebraic conciseness and assured numerical precision. We will outline some of the techniques.

It is an easy matter to verify that a candidate algebraic expression satisfies the requirements to be a shape function. These conditions are that N_i have unit value at node i, zero value at other nodes, and consist only of a linear combination of the basis functions. They are obviously satisfied, for example, by the shape functions for the three-node triangle listed in Table 4.1(a). Verification of the shape functions for the six-node triangle (Table 4.1(b)) may require more than a quick glance. Here the expressions into which the shape functions are factored provide a clue. Each such expression is the equation of a straight line passing through nodes where the shape function must be zero. This reveals, in fact, the way in which the shape functions were constructed. The leading numerical factor serves to provide unit value of N_i at node i.

Table 4.1(a)

Shape Functions for Triangular Elements (Three-Node Triangle)

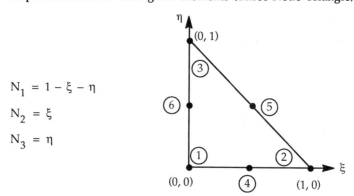

$$N_1 = 1 - \xi - \eta$$
$$N_2 = \xi$$
$$N_3 = \eta$$

Table 4.1(b)

Shape Functions for Triangular Elements (Six-Node Triangle)

$$N_1 = 2\left(1 - \xi - \eta\right)\left(\frac{1}{2} - \xi - \eta\right)$$

$$N_2 = 2\xi\left(\xi - \frac{1}{2}\right)$$

$$N_3 = 2\eta\left(\eta - \frac{1}{2}\right)$$

$$N_4 = 4\xi\left(1 - \xi - \eta\right)$$

$$N_5 = 4\xi\eta$$

$$N_6 = 4\eta\left(1 - \xi - \eta\right)$$

The *line-product* method just described generalizes to "complete" triangles (i.e., triangles with complete polynomials as basis functions) of arbitrary order. Each higher order introduces a new line of nodes in the $\xi\eta$ plane, as shown in Figure 4.1. The number of such lines, e.g., three for the ten-node triangle and four for the fifteen-node triangle, is just sufficient to provide the required polynomial order in the shape functions. In fact, the Pascal triangle (Figure 3.2) maps directly onto Figure 4.1.

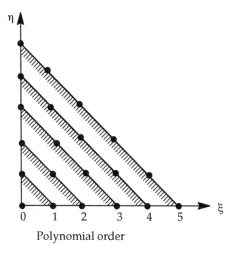

Figure 4.1   Triangular Elements with Complete Basis Functions.

The *line-product* method of deriving shape functions generalizes into a *plane-product* method for tetrahedra, as illustrated in Table 4.2. The shape function for node 1 of the ten-node tetrahedron is, for example, expressed as the product of the equations for two planes passing, respectively, through nodes 2, 3, 4 and nodes 5, 6, 7. Each higher order "complete" tetrahedron adds a triangular array of nodes (parallel, for example, to the $\xi\eta$ plane). Figure 4.1 could represent a view from the ζ axis of a fifth-order "complete" tetrahedron which has 56 nodes.

Table 4.2(a)

Shape Functions for Tetrahedral Elements (Four-Node Tetrahedron)

$N_1 = 1 - \xi - \eta - \zeta$

$N_2 = \xi$

$N_3 = \eta$

$N_4 = \zeta$

Table 4.2(b)

Shape Functions for Tetrahedral Elements (Ten-Node Tetrahedron)

$N_1 = 2(1 - \xi - \eta - \zeta)\left(\dfrac{1}{2} - \xi - \eta - \zeta\right)$

$N_2 = 2\xi\left(\xi - \dfrac{1}{2}\right)$

$N_3 = 2\eta\left(\eta - \dfrac{1}{2}\right)$

$N_4 = 2\zeta\left(\zeta - \dfrac{1}{2}\right)$

$N_5 = 4\xi(1 - \xi - \eta - \zeta)$

$N_6 = 4\eta(1 - \xi - \eta - \zeta)$

$N_7 = 4\zeta(1 - \xi - \eta - \zeta)$

$N_8 = 4\xi\eta$

$N_9 = 4\eta\zeta$

$N_{10} = 4\xi\zeta$

Another way to derive the shape functions of triangles is to use the method of *area coordinates*. This method, which is explained in standard finite element texts,[1] uses three coordinates, one for each vertex. It has beautiful symmetry but it is not computationally superior to the line-product method. It generalizes into a *volume-coordinate* method for tetrahedra.

The line-product method also works for simple quadrilateral elements (up to nine nodes). The resulting shape functions are illustrated in Table 4.3(a), (b), and (c). The method does not work for higher-order elements. For example, the shape function for a corner node of the twelve-node element in Table 4.3(d) is seen to be the product of the equations for two straight lines and a circle which passes through all the edge nodes. The discovery of this fact would appear to have required a degree of ingenuity. Perhaps that is why elements with edge and corner nodes only were dubbed serendipity elements.

Table 4.3(a)

Shape Functions for Quadrilateral Elements (Four-Node Quadrilateral)

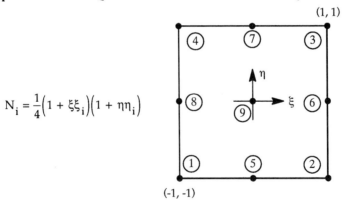

$$N_i = \frac{1}{4}\left(1 + \xi\xi_i\right)\left(1 + \eta\eta_i\right)$$

Table 4.3(b)

Shape Functions for Quadrilateral Elements (Eight-Node Quadrilateral)

$$N_i = \frac{1}{4}\left(1 + \xi\xi_i\right)\left(1 + \eta\eta_i\right)\left(\xi\xi_i + \eta\eta_i - 1\right) \qquad i = 1, 2, 3, 4$$

$$N_i = \frac{1}{2}\left(1 - \xi^2\right)\left(1 + \eta\eta_i\right) \qquad i = 5, 7$$

$$N_i = \frac{1}{2}\left(1 - \eta^2\right)\left(1 + \xi\xi_i\right) \qquad i = 6, 8$$

Table 4.3(c)

Shape Functions for Quadrilateral Elements (Nine-Node Quadrilateral)

$$N_i = \frac{1}{4}\xi_i\eta_i\xi\eta\left(1 + \xi\xi_i\right)\left(1 + \eta\eta_i\right) \qquad i = 1, 2, 3, 4$$

$$N_i = \frac{1}{2}\eta\eta_i\left(1 - \xi^2\right)\left(1 + \eta\eta_i\right) \qquad i = 5, 7$$

$$N_i = \frac{1}{2}\xi\xi_i\left(1 - \eta^2\right)\left(1 + \xi\xi_i\right) \qquad i = 6, 8$$

$$N_i = \left(1 - \xi^2\right)\left(1 - \eta^2\right) \qquad i = 9$$

Table 4.3(d)

Shape Functions for Quadrilateral Elements (Twelve-Node Quadrilateral)

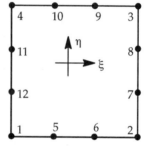

$$N_i = \frac{1}{32}\left(1 + \xi\xi_i\right)\left(1 + \eta\eta_i\right)\left(9\left(\xi^2 + \eta^2\right) - 10\right) \qquad i = 1, 2, 3, 4$$

$$N_i = \frac{9}{32}\left(1 + \eta\eta_i\right)\left(1 - \xi^2\right)\left(1 + 9\xi\xi_i\right) \qquad i = 7, 8, 11, 12$$

$$N_i = \frac{9}{32}\left(1 + \xi\xi_i\right)\left(1 - \eta^2\right)\left(1 + 9\eta\eta_i\right) \qquad i = 5, 6, 9, 10$$

Systematic procedure do, however, exist for generating the shape functions of higher-order quadrilateral elements. They are, to a greater or lesser extent, based on Lagrange interpolation polynomials. Consider the one-dimensional element shown in Figure 4.2.

Figure 4.2   A One-Dimensional Element with n + 1 Nodes.

The shape function for the ith node can be expressed as

$$N_i = \ell_i^n(\xi) = \frac{\Pi_j\left(\xi - \xi_j\right)}{\Pi_j\left(\xi_i - \xi_j\right)} \qquad j = 0, 1, 2, \ldots, i-1, i+1, \ldots, n$$

(4:2)

$$= \frac{\left(\xi - \xi_0\right)\left(\xi - \xi_1\right)\cdots\left(\xi - \xi_{i-1}\right)\left(\xi - \xi_{i+1}\right)\cdots\left(\xi - \xi_n\right)}{\left(\xi_i - \xi_0\right)\left(\xi_i - \xi_1\right)\cdots\left(\xi_i - \xi_{i-1}\right)\left(\xi_i - \xi_{i+1}\right)\cdots\left(\xi_i - \xi_n\right)}$$

The Lagrange interpolation polynomial, ℓ_i^n, clearly satisfies the requirements to be the shape function for the ith node. It has unit value at node i, it vanishes at all other nodes, and it has no more than the maximum polynomial order, n.

The extension to two- and three-dimensional Lagrange elements is straightforward. In two dimensions we express the shape function for the ith node as the product of two Lagrange interpolation polynomials

$$N_i = \ell_j^n(\xi)\ell_k^n(\eta)$$

(4:3)

Here we need a road map such as that shown in Figure 4.3 to identify the values of j and k corresponding to node i. For example, the node identified as node I has j = 1 and k = 2.

For three-dimensional Lagrange elements, each shape function can be written as the product of three Lagrange polynomials. Clearly there is no need, in either two or three dimensions, for the number of nodes in each direction to be equal.

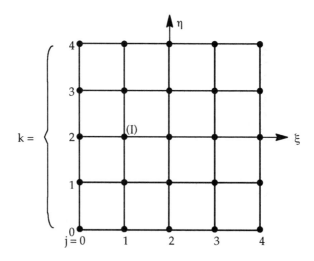

Figure 4.3   A Quartic Lagrangian Element.

A systematic method for deriving the shape functions of higher-order serendipity elements can also be devised.[2] Consider, as an example, the cubic element shown in Figure 4.4.

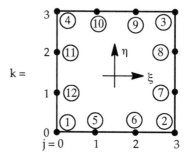

Figure 4.4   A Cubic Serendipity Element.

The shape function for an edge node can be expressed as the product of a Lagrange polynomial and a *blending function*, which makes the shape function vanish on the opposite edge. Thus, for node ⑤ in Figure 4.4,

$$N_5 = \frac{1}{2}\ell_1^3(\xi)(1 - \eta)$$ (4:4)

where

$$\ell_1^3(\xi) = \frac{(\xi + 1)\left(\xi - \frac{1}{3}\right)(\xi - 1)}{\left(\frac{2}{3}\right)\left(-\frac{2}{3}\right)\left(-\frac{4}{3}\right)} = \frac{9}{16}\left(1 - \xi^2\right)(1 - 3\xi)$$ (4:5)

The shape function for a corner node of a higher-order serendipity element can be expressed as the sum of the shape function for the corner of a four-node element, $N_i^{(4)}$, and a linear combination of the shape functions for nodes on the adjacent edges. Figure 4.5 shows a plot along edge ①-② of $N_i^{(4)}$, N_5, and N_6 for the element in Figure 4.4.

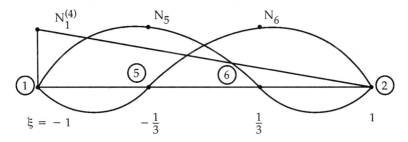

Figure 4.5   **Shape Functions Along the Edge of a Cubic Serendipity Element.**

The linear combination which produces zero value at nodes ⑤ and ⑥ and unit value at node ① is $N_i^{(4)} - \frac{2}{3}N_5 - \frac{1}{3}N_6$. Inclusion of the effect of nodes ⑪ and ⑫ in Figure 4.4 gives the result

$$N_1^{(12)} = N_1^{(4)} - \frac{2}{3}N_5 - \frac{1}{3}N_6 - \frac{2}{3}N_{12} - \frac{1}{3}N_{11}$$ (4:6)

This method can be extended to any number of edge nodes and to three dimensions. Shape functions for eight- and twenty-node serendipity brick elements are listed in Table 4.4. Inspection clearly shows, however, that the expressions given for these shape functions were derived by the plane-product method.

Table 4.4(a)

Shape Functions for Hexahedral Brick Elements (Eight-Node Brick)

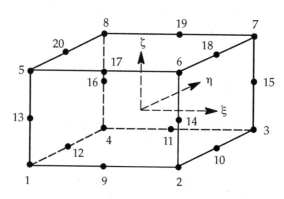

$$N_i = \frac{1}{8}\left(1 + \xi\xi_i\right)\left(1 + \eta\eta_i\right)\left(1 + \zeta\zeta_i\right) \qquad i = 1, 2, \dots, 8$$

Table 4.4(b)

Shape Functions for Hexahedral Brick Elements (Twenty-Node Brick)

$$N_i = \frac{1}{8}\left(1 + \xi\xi_i\right)\left(1 + \eta\eta_i\right)\left(1 + \zeta\zeta_i\right)\left(\xi\xi_1 + \eta\eta_1 + \zeta\zeta_i - 2\right) \quad i = 1, 2, \dots, 8$$

$$N_i = \frac{1}{4}\left(1 - \xi^2\right)\left(1 + \eta\eta_i\right)\left(1 + \zeta\zeta_i\right) \qquad i = 9, 11, 17, 19$$

$$N_i = \frac{1}{4}\left(1 - \eta^2\right)\left(1 + \xi\xi_i\right)\left(1 + \zeta\zeta_i\right) \qquad i = 10, 12, 18, 20$$

$$N_i = \frac{1}{4}\left(1 - \zeta^2\right)\left(1 + \xi\xi_i\right)\left(1 + \eta\eta_i\right) \qquad i = 13, 14, 15, 16$$

Finally, the shape functions for six- and fifteen-node pentahedral elements are listed in Table 4.5. They were derived by the plane-product method.

Table 4.5(a)

Shape Functions for Pentahedral Elements (Six-Node Pentahedron)

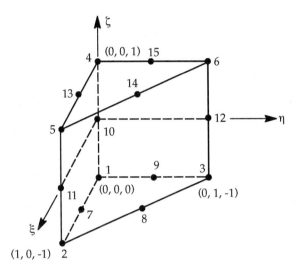

$$N_i = \frac{1}{2}\left(1 - \xi - \eta\right)\left(1 + \zeta\zeta_i\right) \qquad i = 1, 4$$

$$N_i = \frac{1}{2}\xi\left(1 + \zeta\zeta_i\right) \qquad i = 2, 5$$

$$N_i = \frac{1}{2}\eta\left(1 + \zeta\zeta_i\right) \qquad i = 3, 6$$

Table 4.5(b)

Shape Functions for Pentahedral Elements (Fifteen-Node Pentahedron)

$$N_i = \frac{1}{2}(1 - \xi - \eta)\left(1 + \zeta\zeta_i\right)\left(\zeta\zeta_i - 2\xi - 2\eta\right) \qquad i = 1, 4$$

$$N_i = \frac{1}{2}\xi\left(1 + \zeta\zeta_i\right)\left(\zeta\zeta_i + 2\xi - 2\right) \qquad i = 2, 5$$

$$N_i = \frac{1}{2}\eta\left(1 + \zeta\zeta_i\right)\left(\zeta\zeta_i + 2\eta - 2\right) \qquad i = 3, 6$$

$$N_i = 2\xi(1 - \xi - \eta)\left(1 + \zeta\zeta_i\right) \qquad i = 7, 13$$

$$N_i = 2\xi\eta\left(1 + \zeta\zeta_i\right) \qquad i = 8, 14$$

$$N_i = 2\eta(1 - \xi - \eta)\left(1 + \zeta\zeta_i\right) \qquad i = 9, 15$$

$$N_{10} = (1 - \xi - \eta)\left(1 - \zeta^2\right)$$

$$N_{11} = \xi\left(1 - \zeta^2\right)$$

$$N_{12} = \eta\left(1 - \zeta^2\right)$$

The method described above for generating the shape functions of higher-order serendipity elements can also be used to delete edge nodes or, in fact, to allow any number of nodes on any edge. If we wish, for example, to delete nodes ⑤ and ⑥ from the example of Figure 4.4, then we may use in place of Equation 4:6,

$$N_1^{(10)} = N_1^{(4)} - \frac{2}{3}N_{12} - \frac{1}{3}N_{11} \qquad (4:7)$$

The deletion of only a single node, node ⑤ for example, is a little more complicated. In this case N_6 should be recomputed using a Lagrange interpolation function with two factors rather than three. Then, if $\overline{N}_6$ is the recomputed value,

$$N_1^{(11)} = N_1^{(4)} - \frac{1}{3}\overline{N}_6 - \frac{2}{3}N_{12} - \frac{1}{3}N_{11} \qquad (4:8)$$

If, in addition, node ⑥ is centered, the factor for $\overline{N}_6$ becomes 1/2 rather than 1/3. Many finite element programs include provision for automatic calculation of the effects of edge node deletion. The user specifies which nodes are connected and the program does the rest. The shape function logic is, as we have seen, quite straightforward.

An important application of edge node deletion occurs in connection with changes in mesh density. A typical example is shown in Figure 4.6.

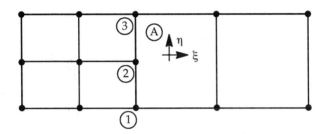

Figure 4.6 A Change in Mesh Density. All Elements Except Element Ⓐ Have Bilinear Shape Functions.

All the elements except element Ⓐ are four-node elements with bilinear shape functions $\left(1+\xi\xi_i\right)\left(1+\eta\eta_i\right)$. There are (at least) three different ways to model the transition. One way is to take the shape function for node ② of element Ⓐ to be $\left(1-\xi\right)\left(1-\eta^2\right)/2$, which is the standard serendipity shape function for a single edge node. Many finite element programs will do this automatically. Note that, with this selection, displacement continuity is not preserved along the interface because the displacement varies linearly in each of the two elements to the left of node ② and quadratically in element Ⓐ. As a result, the junction will have excessive flexibility.

This defect can be remedied, as Hughes has suggested,[23] by selecting the shape function for node ② in element Ⓐ to be $\left(1-\xi\right)\left(1-|\eta|\right)/2$. Note that, with this novel application of a nonanalytic shape function, special care must be taken in the evaluation of strains because $u_{,\eta}$ and $v_{,\eta}$ do not exist along $\eta = 0$.

The third way to model the transition is to apply a rigid (multipoint) constraint so that $u_2 = \left(u_1 + u_3 \right)/2$. Element Ⓐ can then be a standard four-node element (with no connection to node ②) and inter-element displacement continuity will be preserved. While either the second method or the third should be preferred to the first, the second method uses a special element which is not likely to be found in the average commercial finite element program.

The shape functions for low order elements with the five basic shapes (triangle, quadrilateral, tetrahedron, pentahedron, and hexahedron) are given in Tables 4.1 to 4.5 as products of linear and quadratic factors in ξ, η, and ζ. This functional form is as good as any other for computations in which ξ, η, and ζ take on numerical values. We will, however, in analyzing the disorders of finite elements, be interested in determining which basis functions are present when displacement fields are prescribed at the nodes. It is useful, for the purpose, to express shape functions in the factored matrix form

$$\lfloor N_i \rfloor = \lfloor X \rfloor \left[A_{ji} \right] \qquad (4:9)$$

where $\lfloor X \rfloor$ is the row vector of basis functions and $\left[A_{ji} \right]$ is a square matrix of constants. Then, since $u = \lfloor N_i \rfloor \{u_i\} = \lfloor X \rfloor \left[A_{ji} \right] \{u_i\}$, the coefficient of X_j in u is just $\left[A_{ji} \right] \{u_i\}$. Table 4.6 lists $\lfloor X \rfloor$ and $\left[A_{ji} \right]$ for the five simplest two-dimensional elements (TRIA3, QUAD4, TRIA6, QUAD8, and QUAD9).

Table 4.6

**Shape Functions for Two-Dimensional Elements
in Factored Matrix Form,** $\lfloor N_i \rfloor = \lfloor X \rfloor \lfloor A_{ji} \rfloor$

(a) TRIA3:

$$\lfloor X \rfloor = \lfloor 1, \xi, \eta \rfloor$$

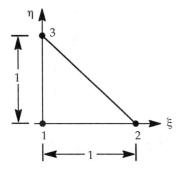

$$\left[A_{ji} \right] = \begin{bmatrix} 1 & 0 & 0 \\ -1 & 1 & 0 \\ -1 & 0 & 1 \end{bmatrix}$$

(b) QUAD4:

$$\lfloor X \rfloor = \lfloor 1, \xi, \eta, \xi\eta \rfloor$$

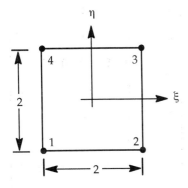

$$\left[A_{ji} \right] = \frac{1}{4} \begin{bmatrix} 1 & 1 & 1 & 1 \\ -1 & 1 & 1 & -1 \\ -1 & -1 & 1 & 1 \\ 1 & -1 & 1 & -1 \end{bmatrix}$$

Table 4.6 (continued)

**Shape Functions for Two-Dimensional Elements
in Factored Matrix Form, $\lfloor N_i \rfloor = \lfloor X \rfloor \lfloor A_{ji} \rfloor$**

(c) TRIA6:

$$\lfloor X \rfloor = \lfloor 1, \xi, \eta, \xi^2, \xi\eta, \eta^2 \rfloor$$

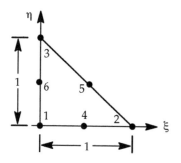

$$\lfloor A_{ji} \rfloor = \begin{bmatrix} 1 & 0 & 0 & 0 & 0 & 0 \\ -3 & -1 & 0 & 4 & 0 & 0 \\ -3 & 0 & -1 & 0 & 0 & 4 \\ 2 & 2 & 0 & -4 & 0 & 0 \\ 4 & 0 & 0 & -4 & 4 & -4 \\ 2 & 0 & 2 & 0 & 0 & -4 \end{bmatrix}$$

(d) QUAD8:

$$\lfloor X \rfloor = \lfloor 1, \xi, \eta, \xi^2, \xi\eta, \eta^2, \xi^2\eta, \xi\eta^2 \rfloor$$

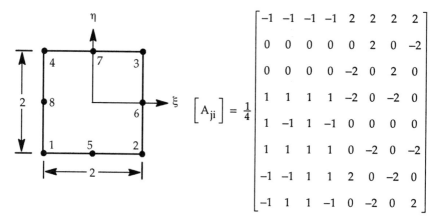

$$\lfloor A_{ji} \rfloor = \frac{1}{4} \begin{bmatrix} -1 & -1 & -1 & -1 & 2 & 2 & 2 & 2 \\ 0 & 0 & 0 & 0 & 0 & 2 & 0 & -2 \\ 0 & 0 & 0 & 0 & -2 & 0 & 2 & 0 \\ 1 & 1 & 1 & 1 & -2 & 0 & -2 & 0 \\ 1 & -1 & 1 & -1 & 0 & 0 & 0 & 0 \\ 1 & 1 & 1 & 1 & 0 & -2 & 0 & -2 \\ -1 & -1 & 1 & 1 & 2 & 0 & -2 & 0 \\ -1 & 1 & 1 & -1 & 0 & -2 & 0 & 2 \end{bmatrix}$$

Table 4.6 (continued)

**Shape Functions for Two-Dimensional Elements
in Factored Matrix Form,** $\lfloor N_i \rfloor = \lfloor X \rfloor \lfloor A_{ji} \rfloor$

(e) QUAD9:

$$\lfloor X \rfloor = \left\lfloor 1, \xi, \eta, \xi^2, \xi\eta, \eta^2, \xi^2\eta, \xi\eta^2, \xi^2\eta^2 \right\rfloor$$

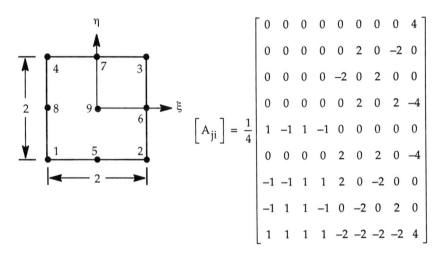

$$\left[A_{ji} \right] = \frac{1}{4} \begin{bmatrix} 0 & 0 & 0 & 0 & 0 & 0 & 0 & 0 & 4 \\ 0 & 0 & 0 & 0 & 0 & 2 & 0 & -2 & 0 \\ 0 & 0 & 0 & 0 & -2 & 0 & 2 & 0 & 0 \\ 0 & 0 & 0 & 0 & 0 & 2 & 0 & 2 & -4 \\ 1 & -1 & 1 & -1 & 0 & 0 & 0 & 0 & 0 \\ 0 & 0 & 0 & 0 & 2 & 0 & 2 & 0 & -4 \\ -1 & -1 & 1 & 1 & 2 & 0 & -2 & 0 & 0 \\ -1 & 1 & 1 & -1 & 0 & -2 & 0 & 2 & 0 \\ 1 & 1 & 1 & 1 & -2 & -2 & -2 & -2 & 4 \end{bmatrix}$$

4.2 THE STRAIN-DISPLACEMENT MATRIX

As summarized in Section 2.3, evaluation of the stiffness matrix by the basic stiffness formulation consists of the following three steps.

1. Evaluate the strain-displacement matrix, [B], in the relationship

$$\{\varepsilon\} = [B]\{u_i\} \tag{4:9}$$

2. Form the triple matrix product $[B]^T[D][B]$.

3. Integrate over the element's volume to find

$$\left[K_{ii}^e\right] = \int_{V_e} [B]^T [D][B] dV \tag{4:10}$$

In the general discussion of Section 2.3, the vector $\left\{u_i\right\}$ included all of the element's boundary displacement variables. It is convenient, for more detailed analysis, to treat $\left\{u_i\right\}$ as the vector of displacement components at node i. Then, in place of Equation 4:9, we have

$$\{\varepsilon\} = \sum_i \left[B_i\right]\left\{u_i\right\} \tag{4:11}$$

and, in place of Equation 4:10, we have for the ij partition of the element's stiffness matrix

$$\left[K_{ij}^e\right] = \int_{V_e} \left[B_i\right]^T [D]\left[B_j\right] dV \tag{4:12}$$

The elastic strain components in $\{\varepsilon\}$ are defined to be, in two dimensions,

$$\{\varepsilon\} = \left\{ \begin{array}{c} \varepsilon_x \\ \varepsilon_y \\ \gamma_{xy} \end{array} \right\} = \left\{ \begin{array}{c} u_{,x} \\ v_{,y} \\ u_{,y} + v_{,x} \end{array} \right\} \tag{4:13}$$

and, in three dimensions,

$$\{\varepsilon\} = \left\{ \begin{array}{c} \varepsilon_x \\ \varepsilon_y \\ \varepsilon_z \\ \gamma_{xy} \\ \gamma_{xz} \\ \gamma_{yz} \end{array} \right\} = \left\{ \begin{array}{c} u_{,x} \\ v_{,y} \\ w_{,z} \\ u_{,y} + v_{,x} \\ u_{,z} + w_{,x} \\ v_{,z} + w_{,y} \end{array} \right\} \tag{4:14}$$

In isoparametric elements, or in parametric elements generally, the assumed displacement field is described in terms of the parameters ξ, η, ζ. As a result, evaluation of the strain components from displacement requires application of the chain rule. For example,

$$\varepsilon_x = u_{,x} = u_{,\xi}\,\xi_{,x} + u_{,\eta}\,\eta_{,x} + u_{,\zeta}\,\zeta_{,x} \tag{4:15}$$

Then, since $u = \sum_i N_i u_i$, where N_i is the ith displacement shape function,

$$\varepsilon_x = \sum_i N_{i,x} u_i = \sum_i \left(N_{i,\xi}\xi_{,x} + N_{i,\eta}\eta_{,x} + N_{i,\zeta}\zeta_{,x} \right) u_i \tag{4:16}$$

Treating all other strain components in similar fashion, we obtain

$$\{\varepsilon\} = \begin{Bmatrix} \varepsilon_x \\ \varepsilon_y \\ \varepsilon_z \\ \gamma_{xy} \\ \gamma_{xz} \\ \gamma_{yz} \end{Bmatrix} = \sum_i \begin{bmatrix} N_{i,x} & 0 & 0 \\ 0 & N_{i,y} & 0 \\ 0 & 0 & N_{i,z} \\ N_{i,y} & N_{i,x} & 0 \\ N_{i,z} & 0 & N_{i,x} \\ 0 & N_{i,z} & N_{i,y} \end{bmatrix} \begin{Bmatrix} u_i \\ v_i \\ w_i \end{Bmatrix} = \sum_i [B_i]\{u_i\} \tag{4:17}$$

where

$$\begin{aligned} N_{i,x} &= N_{i,\xi}\,\xi_{,x} + N_{i,\eta}\,\eta_{,x} + N_{i,\zeta}\,\zeta_{,x} \\ N_{i,y} &= N_{i,\xi}\,\xi_{,y} + N_{i,\eta}\,\eta_{,y} + N_{i,\zeta}\,\zeta_{,y} \\ N_{i,z} &= N_{i,\xi}\,\xi_{,z} + N_{i,\eta}\,\eta_{,z} + N_{i,\zeta}\,\zeta_{,z} \end{aligned} \tag{4:18}$$

Other physical disciplines follow the same procedure with different content for the B matrices. For example, in heat conduction, the gradient of the temperature

$$\{\nabla u\} = \sum_i \{B_i\} u_i \tag{4:19}$$

where

$$\{B_i\} = \begin{Bmatrix} N_{i,x} \\ N_{i,y} \\ N_{i,z} \end{Bmatrix} \qquad (4{:}20)$$

In magnetostatics, the magnetic induction

$$\begin{Bmatrix} B_x \\ B_y \\ B_z \end{Bmatrix} = \sum_i [B_i]\{A_i\} \qquad (4{:}21)$$

where

$$[B_i] = \begin{bmatrix} 0 & -N_{i,z} & N_{i,y} \\ N_{i,z} & 0 & -N_{i,x} \\ -N_{i,y} & N_{i,x} & 0 \end{bmatrix} \qquad (4{:}22)$$

(Unfortunately, electromagnetism and finite element theory use B to represent different things which happen to intersect here.)

To complete the calculation for any physical discipline, we need to know the partial derivatives of the parametric coordinates ξ, η, ζ with respect to the Cartesian metric coordinates x, y, z. The metric coordinates are related to parametric coordinates by

$$\{x\} = \sum_i N_i'\{x_i\} \qquad (4{:}23)$$

where the geometric shape function, N_i', is an assumed function of ξ, η, ζ. In isoparametric elements, N_i' and N_i are identical, but we will keep them distinct to facilitate the investigation of subparametric geometry. The partial derivatives of metric coordinates with respect to parametric coordinates follow from Equation 4:23. For example,

$$x_{,\xi} = \sum_i N'_{i,\xi} x_i \tag{4:24}$$

What we require, however, are the inverse derivatives, $\xi_{,x}$, etc. To form the latter, first arrange $x_{,\xi}$, etc., into a square array called the *Jacobian matrix*

$$[J] = \begin{bmatrix} x_{,\xi} & y_{,\xi} & z_{,\xi} \\ x_{,\eta} & y_{,\eta} & z_{,\eta} \\ x_{,\zeta} & y_{,\zeta} & z_{,\zeta} \end{bmatrix} \tag{4:25}$$

Then the desired derivatives are obtained from

$$\begin{bmatrix} \xi_{,x} & \eta_{,x} & \zeta_{,x} \\ \xi_{,y} & \eta_{,y} & \zeta_{,y} \\ \xi_{,z} & \eta_{,z} & \zeta_{,z} \end{bmatrix} = [J]^{-1} \tag{4:26}$$

To verify that this is true, form the products of the columns of Equation 4:25 and the rows of Equation 4:26. For example,

$$\begin{aligned} x_{,\xi}\, \xi_{,x} + x_{,\eta}\, \eta_{,x} + x_{,\zeta}\, \zeta_{,x} &= x_{,x} = 1 \\ x_{,\xi}\, \xi_{,y} + x_{,\eta}\, \eta_{,y} + x_{,\zeta}\, \zeta_{,y} &= x_{,y} = 0 \end{aligned} \tag{4:27}$$

The steps required to compute $\begin{bmatrix} B_i \end{bmatrix}$ are now complete. They are:

1. Find the parametric derivatives of the shape functions $N_{i,\xi}$, etc., and $N'_{i,\xi}$, etc.

2. Form the partial derivatives of metric coordinates with respect to parametric coordinates by means of Equation 4:24.

3. Form and invert the Jacobian matrix $[J]$ to find the derivatives of ξ, η, ζ with respect to x, y, z.

4. Form the metric derivatives of the displacement shape functions, $N_{i,x}, N_{i,y}, N_{i,z}$, from Equation 4:18 and substitute into Equation 4:17 (or 4:20 or 4:22) to form $\left[B_i\right]$.

The equations that describe the steps have assumed three dimensions. The reduction to two dimensions is self evident.

The calculation is straightforward if values of $\left[B_i\right]$ are required only at specific points. This will be the case if numerical integration is used. If the functional form of $\left[B_i\right]$ is desired, for example, to study accuracy, the only awkward step is the inversion of $[J]$. In two dimensions

$$\begin{bmatrix} \xi_{,x} & \eta_{,x} \\ \xi_{,y} & \eta_{,y} \end{bmatrix} = [J]^{-1} = \frac{1}{J_2}\begin{bmatrix} y_{,\eta} & -y_{,\xi} \\ -x_{,\eta} & x_{,\xi} \end{bmatrix} \qquad (4:28)$$

where J_2, the determinant of the two-dimensional Jacobian matrix, or simply *the Jacobian*, is

$$J_2 = x_{,\xi}\, y_{,\eta} - x_{,\eta}\, y_{,\xi} \qquad (4:29)$$

The corresponding expressions in three dimensions are more cumbersome. It will suffice, for our purposes, to write a typical term of the inverse; for example,

$$\xi_{,x} = \frac{1}{J_3}\left(y_{,\eta}\, z_{,\zeta} - y_{,\zeta}\, z_{,\eta} \right) \qquad (4:30)$$

where J_3 is the determinant of the three-dimensional Jacobian matrix. All other terms may be found by cyclic permutation of x, y, z and ξ, η, ζ.

4.3 NUMERICAL INTEGRATION

The matrices and vectors which express the properties of a finite element at the node point level are evaluated by integration over the element's volume. The most important element properties are the stiffness matrix (expressed here as the partition for nodes i and j),

$$\left[K_{ij}^e \right] = \int\limits_{V_e} \left[B_i \right]^T [D] \left[B_j \right] dV \qquad (4:31)$$

the applied load vector (for the ith node),

$$\left\{ P_i \right\} = \int\limits_{V_e} \left[N_i \right]^T \{p\} dV \qquad (4:32)$$

and the internal force vector (for the ith node),

$$\left\{ F_i \right\} = \int\limits_{V_e} \left[B_i \right]^T \{\sigma\} dV \qquad (4:33)$$

To these properties we should add the ij partition of the element's mass matrix

$$\left[M_{ij}^e \right] = \int \rho \left[N_i \right]^T \left[N_j \right] dV \qquad (4:34)$$

obtained by setting

$$\{p\} = -\rho\{\ddot{u}\} = -\rho \sum_j N_j \left\{ \ddot{u}_j \right\} \qquad (4:35)$$

in Equation 4:32.

4.3.1   Transformation to Parametric Space

Since we are dealing with isoparametric elements, the integrands in the integration formulas are most naturally expressed as functions of ξ, η, ζ. Also, since the elements have simple standard shapes in parametric space, there is considerable numerical advantage to performing the integration in parametric space. To do this, it is first necessary to relate the infinitesimal volume in metric space, dV, to the infinitesimal volume in parametric space $dV' = d\xi \, d\eta \, d\zeta$. Consider the two-dimensional mapping shown in Figure 4.7.

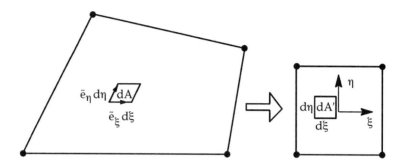

Figure 4.7   Parametric Mapping of an Infinitesimal Area.

In parametric space

$$dV' = t\, d\, A' = t\, d\xi\, d\eta \qquad (4{:}36)$$

where t is the thickness of the element. The corresponding volume in metric space is

$$dV = t\, d\, A = t\left(\tilde{e}_\xi \times \tilde{e}_\eta\right) d\xi\, d\eta \qquad (4{:}37)$$

where $\tilde{e}_\xi$ is the vector in metric space corresponding to a unit vector in the ξ direction in parametric space, i.e.,

$$\tilde{e}_\xi = \tilde{i}\, x_{,\xi} + \tilde{j}\, y_{,\xi} \qquad (4{:}38)$$

where $\tilde{i}$ and $\tilde{j}$ are unit vectors in metric space. Similarly,

$$\tilde{\varepsilon}_\eta = \tilde{i}\, x_{,\eta} + \tilde{j}\, y_{,\eta} \qquad (4{:}39)$$

Carrying out the cross product in Equation 4:37 gives

$$dV = t\left(x_{,\xi}\, y_{,\eta} - y_{,\xi}\, x_{,\eta}\right) d\xi\, d\eta$$
$$= t\, J_2\, d\xi\, d\eta \qquad (4{:}40)$$

where J_2 is the two-dimensional Jacobian determinant (see Equation 4:29).

In three dimensions, the infinitesimal metric volume corresponding to an infinitesimal parametric volume $dV' = d\xi\, d\eta\, d\zeta$ is, from Figure 4.8,

$$dV = \left(\tilde{e}_\xi \times \tilde{e}_\eta \right) \cdot \tilde{e}_\zeta \; d\xi\, d\eta\, d\zeta \qquad (4{:}41)$$

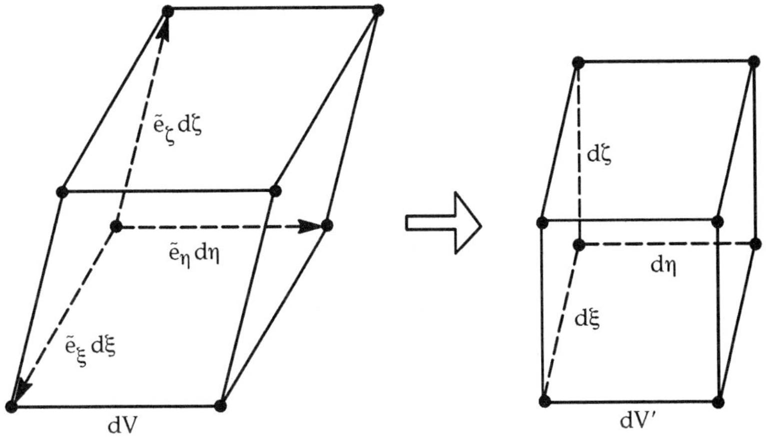

Figure 4.8   Parametric Mapping of an Infinitesimal Volume.

Again, by carrying out the cross product and then the dot product, we can show that

$$dV = J_3 \; d\xi\, d\eta\, d\zeta \qquad (4{:}42)$$

where J_3 is the determinant of the three-dimensional Jacobian matrix. Thus, in both two and three dimensions, the Jacobian determinant is the factor that converts incremental volume in parametric space to incremental volume in metric space. As a result, the volume integrals given by Equations 4:31 to 4:34 can all be written in the general form

$$I = \int_{V_e} f\, dV = \int_{V_e'} f\, J\, dV' = \int_{V_e'} f\, J\, d\xi\, d\eta\, d\zeta \qquad (4{:}43)$$

where f is the integrand, J is the appropriate Jacobian, and for two-dimensional elements, $d\zeta$ is replaced by the thickness, t.

In an element where the relationship between (x, y, z) and (ξ, η, ζ) is linear, the Jacobian is constant over the element's volume. For distorted elements, e.g., four-node quadrilateral elements which are not parallelograms, the Jacobian is not constant and may even be negative at some points if the distortion is severe. This will occur, for example, if edge nodes are placed nearer the corners than the quarter points (see Figure 3.18). Many finite element programs test for negative Jacobians and produce warning messages.

4.3.2 Numerical Integration Rules

Numerical integration evaluates a volume (or surface) integral by a weighted summation of the integrand over a set of points. Except in special cases, the evaluation is approximate. The general form given by Equation 4:43 becomes

$$I = \int_{V_e'} f\, J\, dV' = \sum_g f_g\, J_g\, w_g \qquad (4:44)$$

where w_g is a *weighting factor* and the subscript g identifies the gth *integration point*.

Mathematicians have devised many different numerical integration rules. Well-known rules for one dimension include the trapezoidal rule, Simpson's rule, Newton-Cotes quadrature, and Gauss quadrature. Here we will cover only Gauss quadrature because of its universal preference by finite element designers.

The basic idea behind Gauss quadrature, or Gauss-Legendre quadrature as it is also called, is to use the locations of the integration points and the weighting factors to minimize the error for an integrand that is a general polynomial function of position. Thus, with n integration points and n weighting factors, the error can be nullified for all 2n terms of a polynomial with degree 2n-1 in one dimension. Table 4.7 lists weighting factors and the locations of *Gauss points* in the interval $-1 < \xi < 1$ for the three lowest orders of Gauss integration. Higher order rules are given in standard finite element texts,[3] but most lower order elements use only the ones listed in Table 4.7.

Table 4.7

Gauss Integration Rules

n	ξ_g	w_g	ERROR
1	0	2	$0\left(\xi^2\right)$
2	$\pm \frac{1}{\sqrt{3}}$	1	$0\left(\xi^4\right)$
3	$\left\{\begin{array}{c} \pm\sqrt{0.6} \\ \\ 0 \end{array}\right.$	$\left.\begin{array}{c} \frac{5}{9} \\ \\ \frac{8}{9} \end{array}\right\}$	$0\left(\xi^6\right)$

In quadrilateral isoparametric elements, integration in the η direction is independent of integration in the ξ direction so that Gauss integration may be performed independently in each direction. The same is true for solid hexahedral elements. The integration points form a two- (or three-) dimensional lattice, and the weighting factors are equal to the product of the one-dimensional Gauss weighting factors in each direction. The sum of the weighting factors must equal the area (or volume) of the element in parametric space. Thus for quadrilateral elements $\Sigma \, w_g = 4$, and for hexahedral elements, $\Sigma \, w_g = 8$. Tables 4.8 and 4.9 indicate the weighting factors and the locations of Gauss integration points for quadrilateral and hexahedral elements respectively. The tables also give the order of the lowest error terms. Due to symmetry, functions which are odd in any parametric coordinate, such as $\xi^3\eta^4$, have null integrals. In addition, the error criterion applies to each coordinate independently. As an example, 3 x 3 Gauss integration accurately integrates $\xi^4\eta^4\zeta^4$ in a hexahedral element.

Other integration rules have been proposed for quadrilateral and hexahedral elements. For example, Irons[4] has described a fourteen-point rule for hexahedral elements that is accurate to fifth order, i.e., the rule accurately integrates any term of the form h^5 where h is any linear combination of $\xi, \eta,$ and ζ. If this rule were truly as accurate as 27-point Gauss integration, it

Table 4.8

Gauss Integration Rules for Quadrilateral Elements

(a) 1 x 1 Integration   (b) 2 x 2 Integration   (c) 3 x 3 Integration

$$w_a = \frac{64}{81}$$

$$w_1 = 4 \qquad\qquad w_2 = 1 \qquad\qquad w_b = \frac{40}{81}$$

$$w_c = \frac{25}{81}$$

$$\text{Error} = 0\left(\xi^2\right) \qquad \text{Error} = 0\left(\xi^4\right) \qquad \text{Error} = 0\left(\xi^6\right)$$

Table 4.9

Gauss Integration Rules for Hexahedral Elements

ORDER	ξ_g	η_g	ζ_g	w_g	MULTI-PLICITY	ERROR
1	0	0	0	8	1	$0\left(\xi^2\right)$
2	$\pm\frac{1}{\sqrt{3}}$	$\pm\frac{1}{\sqrt{3}}$	$\pm\frac{1}{\sqrt{3}}$	1	8	$0\left(\xi^4\right)$
3	0	0	0	$\left(\frac{8}{9}\right)^3$	1	
	$\pm\sqrt{.6}$	0	0	$\left(\frac{5}{9}\right)\left(\frac{8}{9}\right)^2$	6	$0\left(\xi^6\right)$
	$\pm\sqrt{.6}$	$\pm\sqrt{.6}$	0	$\left(\frac{8}{9}\right)\left(\frac{5}{9}\right)^2$	12	
	$\pm\sqrt{.6}$	$\pm\sqrt{.6}$	$\pm\sqrt{.6}$	$\left(\frac{5}{9}\right)^3$	8	

would be quite useful because it costs only slightly more than half as much computer time. Unfortunately it does not accurately integrate the important term $\xi^2\eta^2\zeta^2$ which is exactly integrated by even 2 x 2 x 2 Gauss integration.

Gauss quadrature can be extended to triangles by treating them as quadrilaterals with one degenerate edge of zero length. Since such formulas lack isotropy with respect to the directions of the edges, it is better for most applications to use specially devised formulas that possess isotropy. Table 4.10 includes a sampling of successful lower order rules. Rules which include negative weights have been omitted because they cannot be used for mass matrices in dynamic analysis (negative weights produce instant instability). A comprehensive list of higher order rules will be found in Reference 4.5. The error for each rule in Table 4.10 is expressed as a function of h, which is an arbitrary linear combination of ξ and η. The sum of the weights for any rule adds up to 0.5, the area of the standard triangle in parametric space.

Table 4.10

Integration Rules for Triangular Elements

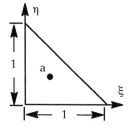

POINT	COORDINATES ξ	COORDINATES η	WEIGHT	ERROR
a	$\frac{1}{3}$	$\frac{1}{3}$	$\frac{1}{2}$	$0\left(h^2\right)$

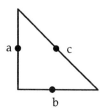

POINT	COORDINATES ξ	COORDINATES η	WEIGHT	ERROR
a	0	$\frac{1}{2}$	$\frac{1}{6}$	
b	$\frac{1}{2}$	0	$\frac{1}{6}$	$0\left(h^3\right)$
c	$\frac{1}{2}$	$\frac{1}{2}$	$\frac{1}{6}$	

Table 4.10 (continued)

Integration Rules for Triangular Elements

POINT	COORDINATES ξ	η	WEIGHT	ERROR
a	$\frac{1}{6}$	$\frac{2}{3}$	$\frac{1}{6}$	
b	$\frac{1}{6}$	$\frac{1}{6}$	$\frac{1}{6}$	$0\left(h^3\right)$
c	$\frac{2}{3}$	$\frac{1}{6}$	$\frac{1}{6}$	

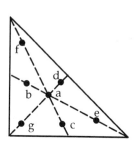

POINT	COORDINATES ξ	η	WEIGHT	ERROR
a	$\frac{1}{3}$	$\frac{1}{3}$	0.1125	
b	α_1	β_1		
c	β_1	α_1	.06619 70763	
d	β_1	β_1		$0\left(h^6\right)$
e	α_2	β_2		
f	β_2	α_2	.06296 95902	
g	β_2	β_2		

where $\alpha_1 = 0.05971\ 58717$

$\beta_1 = 0.47014\ 20641$

$\alpha_2 = 0.79742\ 69853$

$\beta_2 = 0.10128\ 65073$

Similar rules have been devised for tetrahedra. Table 4.11 lists the two lowest order rules. Reference 4.6 describes higher order rules. Again, as in the case of triangles, only rules which are isotropic with respect to the edges and which have all positive weights are acceptable for finite elements with general application. In Table 4.11, the sum of the weights adds up to 1/6, the volume of the standard tetrahedron in parametric space.

Table 4.11

Integration Rules for Tetrahedral Elements

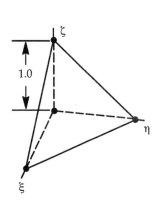

NO. OF POINTS	ξ	η	ζ	WEIGHT	ERROR
1	$\frac{1}{4}$	$\frac{1}{4}$	$\frac{1}{4}$	$\frac{1}{6}$	$0\left(h^2\right)$
4	α	β	β	$\frac{1}{24}$	$0\left(h^3\right)$
	β	α	β	$\frac{1}{24}$	
	β	β	α	$\frac{1}{24}$	
	β	β	β	$\frac{1}{24}$	

where $\alpha = 0.5854\ 1020$

$\beta = 0.1381\ 9660$

Rules for pentahedral elements are constructed by combining triangular rules in the $\xi\eta$ plane with Gauss rules in the ζ direction. Popular pentahedral rules are the six-point (3 x 2) rule which is accurate to $0\left(h^2\zeta^3\right)$ and the 21-point (7 x 3) rule which is accurate to $0\left(h^5\zeta^5\right)$.

4.3.3 How Many Integration Points Are Enough?

Two considerations which always enter into the selection of an integration rule are precision and cost. Other factors which must be considered in finite element analysis are the accuracy of the integrand and the possibility of instability due to an insufficient number of integration points. We will treat these important matters in later chapters. For the present we consider only the effects of precision and cost on the selection of integration rules from those described in Section 4.3.2.

With regard to cost, we need only note that the computer time required to perform the numerical integration of a given integrand is proportional to the number of integration points. Consequently, if a particular integration rule can integrate a given integrand exactly, it makes no sense to consider higher order rules. This very simple logic gives us a conservative way to select integration rules for a broad spectrum of situations: simply select the fewest number of integration points that will exactly compute the integral.

Equations 4:31 to 4:34 list formulas for the (most important) element properties which require numerical integration. It will be assumed that all of the factors which appear in the integrands $(N, B, D, p, \sigma, \rho)$ are polynomial functions of the parametric coordinates. Then it is an easy matter to compute the highest powers of ξ, η, and ζ which occur in the integrand for a particular situation. Comparison with the degree of the leading error term given for each integration rule in Section 4.3.2 will indicate the appropriate rule.

The integrands for the element properties are matrix products with many terms, but for our present purpose it is sufficient to consider only a typical term in each integrand. Thus a typical term in the stiffness matrix is

$$T(K) = \left(T(B)\right)^2 T(D)J \tag{4:45}$$

where $T(B)$ and $T(D)$ are typical terms in $[B]$ and $[D]$ and J is the Jacobian determinant. In like manner, typical terms in the load vector, internal force vector, and mass matrix are, respectively,

$$T(P) = T(N)T(p)J$$

$$T(F) = T(B)T(\sigma)J \tag{4:46}$$

$$T(M) = \left(T(N)\right)^2 T(\rho)J$$

Since every shape function, N, consists of a linear combination of basis functions, we may take

$$T(N) = X \qquad (4:47)$$

where X is one of the basis functions, usually one with the highest degree in ξ.

A typical term in the strain displacement matrix is, therefore,

$$T(B) = X_{,x} = X_{,\xi} \, \xi_{,x} \qquad (4:48)$$

Evaluation of $\xi_{,x}$ requires, it will be recalled, inversion of the Jacobian matrix. In two dimensions

$$\xi_{,x} = y_{,\eta} / J_2 \qquad (4:49)$$

while in three dimensions

$$T(\xi_{,x}) = y_{,\eta} \, z_{,\zeta} / J_3 \qquad (4:50)$$

Substitution into Equation 4:48 gives

$$T(B) = X_{,\xi} \, y_{,\eta} / J_2 \qquad (4:51)$$

in two dimensions, and

$$T(B) = X_{,\xi} \, y_{,\eta} \, z_{,\zeta} / J_3 \qquad (4:52)$$

in three dimensions. It is important to note that, in all of the terms of $[B]$ and not just in $T(B)$, derivatives with respect to each of $\xi, \eta, (\text{and } \zeta)$ occur once and only once. Likewise, typical terms in the Jacobian determinant are

$$T(J_2) = x_{,\xi} \, y_{,\eta}$$
$$\qquad (4:53)$$
$$T(J_3) = x_{,\xi} \, y_{,\eta} \, z_{,\zeta}$$

Finally, we may write the typical terms for each of the integrands, as

$$
\begin{array}{c|c}
\text{In Two Dimensions} & \text{In Three Dimensions} \\[2mm]
\mathrm{T(K)} = \left(x_{,\xi}\, y_{,\eta} \right)^2 D \Big/ J_2 & \left(x_{,\xi}\, y_{,\eta}\, z_{,\zeta} \right)^2 D \Big/ J_3 \\[2mm]
\mathrm{T(P)} = \left(x\, x_{,\xi}\, y_{,\eta} \right) p & \left(x\, x_{,\xi}\, y_{,\eta}\, z_{,\zeta} \right) p \\[2mm]
\mathrm{T(F)} = \left(x_{,\xi}\, y_{,\eta} \right) \sigma & \left(x_{,\xi}\, y_{,\eta}\, z_{,\zeta} \right) \sigma \\[2mm]
\mathrm{T(M)} = \left(x^2\, x_{,\xi}\, y_{,\eta} \right) \rho & \left(x^2\, x_{,\xi}\, y_{,\eta}\, z_{,\zeta} \right) \rho
\end{array}
\qquad (4\!:\!54)
$$

Note that the notation $\mathrm{T(\ )}$ has been dropped from $D, p, \sigma,$ and ρ. These quantities are usually treated as constant or, at most, as linear functions of the parametric coordinates. More importantly, note that the Jacobian determinant appears in the denominator of $\mathrm{T(K)}$ but not in the other element properties. This means that the stiffness matrix can be evaluated exactly only in cases where the determinant is constant, i.e., only in cases where the metric position coordinates are linear functions of the parametric coordinates. One might then ask whether, if the stiffness matrix cannot be evaluated exactly, there are good reasons to evaluate the other element properties exactly. The answer is that there is a very good reason, related to convergence, to evaluate the nodal force vector, $\{F\}$, exactly. This topic will be taken up in the next chapter.

It is also interesting to note that, for constant σ, the typical term in the nodal force vector, $\mathrm{T(F)}$, has the same dependence on $\xi, \eta,$ and ζ as the Jacobian determinant, $\mathrm{T(J)}$, provided only that $\mathrm{T}\!\left(x_{,\xi} \right) = \mathrm{T}\!\left(x_{,\xi} \right)$. The latter property follows from the definition of an isoparametric element. Put succinctly, we assert that any integration rule which can exactly integrate the volume of an element can also exactly integrate its nodal force vector for constant stress.

Tables 4.12 and 4.13 list the minimum number of integration points or the precision required to integrate various element properties exactly for a variety of situations. In the tables "geometry" refers to the relationship between metric and parametric coordinates and "loading" refers to the variation of $D, p, \sigma,$ or ρ with parametric position. To illustrate the process, consider the

Table 4.12

Minimum Number of Gauss Integration Points
To Exactly Integrate Various Properties of Quadrilateral and Brick Elements

ELEMENT	GEOMETRY	LOADING	MINIMUM NUMBER OF INTEGRATION POINTS TO EXACTLY INTEGRATE			
			$[K]$	$\{P\}$	$\{F\}$	$[M]$
QUAD4	Linear	Constant	4	1	1	4
QUAD4	Bilinear	Constant	*	4	1	4
QUAD4	Linear	Linear	4	4	4	4
QUAD4	Bilinear	Linear	*	4	4	9
QUAD8	Linear	Constant	9	4	4	9
QUAD8	Bilinear	Constant	*	4	4	9
QUAD8	Bilinear	Linear	*	9	4	16
QUAD8	General	Constant	*	9	4	16
HEXA8	Linear	Constant	8	1	1	8
HEXA8	Trilinear	Constant	*	8	8	27
HEXA20	Linear	Constant	27	8	8	27
HEXA20	Trilinear	Constant	*	27	8	64
HEXA20	Trilinear	Linear	*	27	27	64
HEXA20	General	Constant	*	64	27	125

*$[K]$ cannot be integrated exactly.

Table 4.13

Minimum Order of Precision to Exactly Integrate
Various Properties of Triangular and Tetrahedral Elements

ELEMENT	GEOMETRY	LOADING	MINIMUM ORDER OF PRECISION TO EXACTLY INTEGRATE			
			$[K]$	$\{P\}$	$\{F\}$	$[M]$
TRIA3	Linear	Constant	1	h	1	h^2
TRIA3	Linear	Linear	h	h^2	h	h^3
TRIA6	Linear	Constant	h^2	h^2	h	h^4
TRIA6	Linear	Linear	h^3	h^3	h^2	h^5
TRIA6	Quadratic	Constant	*	h^4	h^2	h^6
TETRA4	Linear	Constant	1	h	1	h^2
TETRA10	Linear	Constant	h^2	h^2	h	h^4
TETRA10	Linear	Linear	h^3	h^3	h^2	h^5
TETRA10	Quadratic	Constant	*	h^5	h^3	h^7

*$[K]$ cannot be integrated exactly.

QUAD8 element with bilinear "geometry" and linear "loading." A typical term in the internal force vector, $\{F\}$, is, according to Equation 4:54, $T(F) = \left(X_{,\xi} \, y_{,\eta} \right) \sigma$. The basis function, X, of highest degree in ξ is either ξ^2 or $\xi^2 \eta$. Since the geometry is "bilinear," the term of highest degree in y is $\xi\eta$ and we may use $y_{,\eta} = \xi$. Likewise, since the loading is "linear" we may select ξ as the term of highest degree in σ. Thus, adding powers of ξ, we see that ξ^3 is the term of highest degree in $T(F)$. Finally, referring to Table 4.8, we find that 2 x 2 Gauss integration is the lowest order pattern that will evaluate $\{F\}$ exactly.

As another example, consider the case of the twenty-node brick, HEXA20, with "trilinear" geometry and "constant" loading. For this case, a typical term in $\{F\}$ is $T(F) = X_{,\xi} \, y_{,\eta} \, z_{,\zeta}$. The term of highest order in both y and z is $\xi\eta\zeta$ from which we obtain $y_{,\eta} \, z_{,\zeta} = \xi^2\eta\zeta$. Although X can be any of the quartic terms $\xi^2\eta\zeta$, $\xi\eta^2\zeta$, or $\xi\eta\zeta^2$, it is seen that the selection of a quartic term makes the integral of $T(F)$ null because $T(F)$ will be odd in one of the parametric coordinates. Thus we select $X = \xi\eta\zeta$ which gives $T(F) = \xi^2\eta^2\zeta^2$. As noted earlier, this term can be integrated exactly by an eight-point Gauss pattern but not by Irons' fourteen-point formula.[4]

It may be noted that different orders of integration (different patterns of integration points) can be used for $[K]$, $\{P\}$, and $[M]$. On the other hand, $\{F\}$ uses the same integration pattern as $[K]$. As a matter of fact, $\{F\}$ is infrequently computed and then only as an output quantity. Its importance lies in its relationship to node point equilibrium. The selection of an integration pattern for $[K]$, and indirectly for $\{F\}$, depends on more considerations than just the precision of the integration. Designers are, however, at liberty to select the integration patterns for $\{P\}$ and $[M]$ strictly on the basis of precision. As a result, the precision selected for $[K]$ may very well be less than the precision selected for $\{P\}$ and $\{M\}$.

The influence of element shape on the required precision of integration is clearly revealed by the results given in Tables 4.12 and 4.13. For example, we have shown that, in the case of HEXA20, $2 \times 2 \times 2$ integration is sufficient to evaluate $\{F\}$ exactly if the geometry is trilinear. Table 4.12 shows, in addition, that $3 \times 3 \times 3$ integration is required for more general shapes. We will have occasion to refer to these tables, particularly to find the minimum integration pattern that integrates $\{F\}$ exactly. Reference 4.7 contains additional treatment of the issue of required integration precision.

4.4 LOADS, MASSES, AND OUTPUT

The discussion of finite element design tends to concentrate on the elastic stiffness matrix and its troubles. On the other hand, much of the designer's effort and most of the element code are devoted to such other element properties as:

- load vectors
- mass matrices
- damping matrices
- geometric stiffness matrices
- output data recovery

In this section we will take a brief look at some of the considerations which enter the calculation of load vectors and mass matrices, and the recovery of output data. Regarding the other properties, element damping is frequently treated by a complex material modulus which yields a separate imaginary stiffness matrix which can, if required, be converted into a viscous damping matrix.[8] Sometimes, as in MSC/NASTRAN, special viscous damper elements are provided. Geometric stiffness matrices are used to describe geometric nonlinear effects. They are an essential part of the nonlinear analysis of plates and shells.

4.4.1 Loads

Loads can be classified either by their origin or by their type of distribution. With regard to origin, all finite element programs must, as a minimum, allow the *user* to specify the distribution of loads. Most finite element programs also provide a degree of automatic calculation for some loads such as loads due to inertia forces, thermal expansion, fluid-structure interaction, and various electromagnetic forces. We shall treat only the first two of these types.

With regard to type of distribution, loads can be classified as body forces, surface tractions, line loads, or point loads. Even when the user specifies the magnitude and distribution of loads, it is usually advisable to have the computer program handle the transfer of loads to node points. For example, in the case of body forces on an element, the equivalent nodal forces are computed by (see Equation 2:50),

$$\left\{P_i^e\right\} = \int_{V_e}\left[N_i\right]^T\left\{p\right\}dV \qquad (4\text{:}55)$$

The same form may be used for surface tractions or line loads with dS or dℓ substituted for dV. The reason why the computer program rather than the user should handle the distribution of loads is that, except for the very simplest elements, the distribution is non-intuitive. Figure 4.9 shows the node point distribution of surface and line loads for an eight-node rectangular element. The negative corner forces in Figure 4.9(a) are particularly non-intuitive. Not all computer programs provide automatic transfer of such loads. MSC/NASTRAN, for example, provides automatic transfer of surface loads but not line loads.

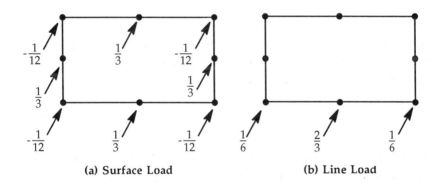

(a) Surface Load   (b) Line Load

Figure 4.9   Distribution of Uniform Surface and Line Loads to Node Points on an Eight-Node Rectangle.

The load distributions shown in Figure 4.9 illustrate one of the differences between corner nodes and edge nodes. There are other differences as well. Consider, for example, a circular ring made up of one-dimensional, three-node elements. The shape of each element will be parabolic rather than circular and, if a uniform radial load is applied according to the distribution of Figure 4.9(b), the edge nodes and corner nodes will displace by slightly different amounts, giving rise to false bending moments. This difficulty does not occur when two-node elements are used.

Point loads are entirely the responsibility of the user as are enforced displacements. Both are applied directly to nodes. It can be argued, correctly, that point loads and enforced displacements have no place in elasticity because they produce singular stresses. This argument overlooks the role of spatial abstraction in continuum mechanics. A Timoshenko beam or a Kirchhoff plate can certainly undergo concentrated loads without experiencing infinite deformations. Engineers are accustomed to the simplification of structural effects and the lumping of distributed loads into point loads. Any complex structure, such as a bridge, a building, or an airplane, must accommodate idealization to be analyzable. The finite element user should, however, be aware that local stresses may rise indefinitely as the elements near a concentrated load are subdivided or refined. This is particularly a problem for higher order elements, such as the hierarchical elements described in the next section.

As has been mentioned, loads due to inertia forces (including gravity) and thermal expansion are computed automatically by most finite element programs. In the case of inertia forces

$$\left\{P_i\right\} = -\left[M_{ii}\right]\left\{\ddot{u}_i\right\} \qquad (4:56)$$

The acceleration vector, $\left\{\ddot{u}_i\right\}$, may be specified by the user as a constant acting in a particular direction (the gravity vector) or, more generally, by the rigid translational and rotational acceleration of the structure. MSC/NASTRAN includes the additional automatic capability to compute the acceleration of a free body subjected to arbitrary time-independent loads, and then to subtract the resulting inertia forces from the load distribution (the so-called *inertia relief* effect).

In the case of thermal expansion, the strains due to temperature rise can be expressed as

$$\left\{\varepsilon^t\right\} = \left\{\alpha\right\}\left(T - T_0\right) \qquad (4:57)$$

where $\{\alpha\}$ is a vector of thermal expansion coefficients. (We use T for temperature here to distinguish it from the symbol used for displacement.)

Uniform thermal expansion produces no stress so that, in linear analysis,

$$\{\sigma\} = [D]\{\varepsilon - \varepsilon^t\} \tag{4:58}$$

where $\{\varepsilon\}$ is the total observed strain. For example, if a uniformly heated, homogeneous body is rigidly restrained, the stress will be $\{\sigma\} = -[D]\{\varepsilon^t\}$.

Thermal expansion is included in finite element analysis by supplying an equivalent nodal load distribution that produces strains equal to $\{\varepsilon^t\}$. To compute the equivalent load distribution, define the fictitious thermal stress $\{\sigma^t\} = [D]\{\varepsilon^t\}$ and then use Equation 2:53 to find

$$\{P_i^{et}\} = \int_{V_e} [B_i]^T \{\sigma^t\} dV = \int_{V_e} [B_i]^T [D]\{\alpha\}(T - T_o) dV \tag{4:59}$$

Equation 4:58 gives the true stresses for use in data recovery.

So far we have said nothing about loads in dynamic analysis. Such loads have spatial distributions like those of static loads and an added dimension which can be expressed as a time history or as a frequency spectrum. The added dimension complicates the preparation of input data to the point where a comprehensive user-friendly treatment of all practical situations is next to impossible. The reader is invited to examine the dynamic load provisions of any general purpose finite element system to observe the compromises that are made.

4.4.2 Mass Matrices

Mass matrices are most often computed in finite element analysis by Equation 4:34, which gives, for the ij partition of an element's mass matrix,

$$[M_{ij}^e] = \int \rho [N_i]^T [N_j] dV \tag{4:60}$$

A mass matrix computed according to Equation 4:60 is called a *consistent* mass matrix[9] because the shape functions $\left[N_i\right]$ are the same as those used to compute stiffness. Note that a consistent mass matrix has the same extent of coupling as the corresponding stiffness matrix. A diagonal mass matrix can be obtained by the simple expedient of equating each diagonal term to the sum of all terms in the same row.* The resulting matrix, called a *lumped* mass matrix, will produce the same inertia forces as the consistent mass matrix if the acceleration is uniform, but not if the motion is more general.

A lumped mass matrix can have a significant computational advantage over a consistent mass matrix in some circumstances. For example, explicit integration of transient dynamics requires that the mass matrix be inverted. This will be less costly, particularly for large problems, if the mass matrix is diagonal.

In general, the use of a lumped mass matrix degrades the accuracy of the solution. There is one exception which occurs when the simplest elements of any type (i.e., rods, triangles, quadrilaterals, bricks, etc., with no edge nodes), sometimes called *simplex* elements, are used to model wave phenomena. In such cases it can be shown that a compromise between the lumped and consistent mass matrices, e.g., one half the sum of the two matrices, minimizes the error in frequency or wave velocity.[10]

Lumped mass matrices for elements with edge nodes have the disadvantage that some of the masses may be negative. For example, the lumped mass distribution for an eight-node rectangle is the same as the node point distribution of the uniform surface load shown in Figure 4.9(a). The reason is that uniform acceleration yields uniform inertia force. Negative masses cannot, of course, be used in transient dynamic analysis because they cause exponential divergence.

---

*It is important for this calculation that all degrees of freedom have the same scale. For example, terms coupling rotations to translations should not be included in the sum.

On balance, while lumped mass matrices have some legitimate applications, particularly for simplex elements, they tend to degrade performance and should not be used uncritically. For general applications, consistent mass matrices are better. Very often they are the only mass matrices provided in a finite element program.

Besides the mass of the elements themselves, two other kinds of mass sometimes appear in finite element programs. The first kind includes the concentrated mass properties of rigid bodies, connected directly to nodes. The second kind includes distributed nonstructural mass, such as the floor loading in a building, which may be spread over a surface element or a line element.

4.4.3 Output Data Recovery

In linear static analysis by the displacement method, displacements are recovered first by solution of the stiffness equation

$$\left[K_{ii} \right] \left\{ u_i \right\} = \left\{ P_i \right\} \tag{4:61}$$

Then, if an assumed displacement formulation is used, strains are recovered by

$$\left\{ \varepsilon \right\} = \left[B \right] \left\{ u_i \right\} \tag{4:62}$$

and stresses are recovered by Equation 4:58.

Various stress resultants are usually provided, such as the maximum shear stress, the Hencky-von Mises stress, or the mean pressure. The strain energy of the element

$$W^e = \tfrac{1}{2} \int_{V_e} \left\{ \varepsilon \right\}^T \left\{ \sigma \right\} dV \tag{4:63}$$

may also be computed by the numerical integration rules described in Section 4.3.2.

Stress and strain components, and their resultants, are most often output at integration points. They can also be extrapolated to other points, by using shape functions of the parametric coordinates. The only difference from the procedure used to interpolate displacements is that ξ, η, and ζ must be scaled to the locations of the integration points rather than to the locations of the nodes.

The points most commonly selected for extrapolated stress and strain output are the center and the corners of the element. In a four-node quadrilateral, for example, stresses are output at the center because the stresses at 2×2 integration points are inaccurate. The incompleteness of the linear terms in the strain field causes the inaccuracy.

Extrapolation of stress and strain output to corner nodes provides the ability to average the results from adjacent elements. This will often improve the accuracy of the data. Extrapolation to corner nodes also furnishes a way to estimate the error in the data. A statistical measure of probable error is easily constructed by treating the values of stress components from adjacent elements like random experimental data.

This approach to à posteriori error estimation has the capital advantages that it is easy to code and inexpensive to use. No information whatever is needed about the formulation of the adjacent elements and no extra runs are required. Although the errors in finite element analysis are more nearly systematic than random, this procedure can provide meaningful error estimates in practical applications. [11,12]

The internal force vector, $\{F_i\}$ (see Equation 4:33), is occasionally provided as output data. In the early days, finite element users liked to check that the internal forces were in equilibrium. More significantly, the availability of $\{F_i\}$ allows the user to construct his own stress distribution for a particular element. This capability is useful for cases where an element is an idealized representation of complex behavior. The ability of modern computers to handle very detailed finite element models makes this capability less important now than formerly.

The calculation of loads and output data for other disciplines generally follows the procedures that have been outlined for structural mechanics. Electromagnetism tends to have more kinds of output than structural mechanics. The important results may include, in addition to the strengths of the fields themselves, the mechanical forces and ohmic heating in physical media. In addition, such overall properties as the resistance, inductance, or capacitance of a device may be needed as output.

Other disciplines also have analogs to the mass matrix. The analogous quantity in heat conduction is the heat capacity matrix, although it comes with the difference that the associated heat flow is proportional to the first rather than to the second time derivative of temperature. In electromagnetism the analog of the mass matrix is a matrix of dielectric permittivities.

4.5 HIERARCHICAL ELEMENTS

The theoretical foundation for hierarchical elements is described in Section 3.1. As will be recalled, interpolation of displacement takes the form

$$u = \lfloor N_j \rfloor \left\{ u_j \right\} + \lfloor N_k \rfloor \left\{ a_k \right\} \qquad (4:64)$$

where $\left\{ u_j \right\}$ is a vector representing the values of u at all connected nodes and $\left\{ a_k \right\}$ are other "generalized" or "hierarchical" displacement variables. The nodal shape function, N_j, satisfies the usual properties that $N_j = 1$ at node j and $N_j = 0$ at all other nodes. The hierarchical shape function, N_k, satisfies only the property that N_k is null at nodes. The designation *hierarchical* derives from the facts that an addition to the number of nodes changes all of the shape functions but that an addition to the number of hierarchical displacement variables changes none of the other shape functions, nodal or hierarchical. As a result, hierarchical degrees of freedom can be added to a finite element without adding to the number of nodes or changing the existing parts of the stiffness matrix, load vector, etc. This feature facilitates *hierarchical analysis* in which a given structure is analyzed by a sequence of finite element assemblies with increasing complexity. The number and shape of the elements remains fixed.

Only the number of hierarchical variables changes. The element stiffness matrix for the nth stage of the sequence has the partitioned form

$$
\left[K^e_{nn} \right] = \left[\begin{array}{c|c} K^e_{mm} & K^e_{md} \\ \hline K^{eT}_{md} & K^e_{dd} \end{array} \right] \tag{4:65}
$$

where $\left[K^e_{mm} \right]$ is the stiffness matrix for the preceding stage.

We can arrange matters so that each stage corresponds to a particular order, p, of complete polynomials in the displacement basis functions. Thus, as p increases, we obtain a sequence of solutions from which we can estimate accuracy and rate of convergence. This approach to finite element analysis is called the p method. [13]

We consider here the particular implementation of the p method that is embodied in the MSC/PROBE computer program. [14] The formulation of the quadrilateral membrane element, which we will call QUADP, will illustrate the general procedure. The QUADP element has nodes only at the corners. Thus the nodal shape functions are just the standard four-node functions, $N_j = \left(1 + \xi \xi_j \right)\left(1 + \eta \eta_j \right) / 4$. The basis functions of the hierarchical variables are selected to make complete polynomials of ascending orders, p, with as few extra terms as possible. The selection process is illustrated with Pascal's triangle in Figure 4.10. As will be recalled, a serendipity element, i.e., an element with edge nodes only, requires basis functions of the form $\xi^n, \eta^n, \xi^n \eta, \xi \eta^n, n = 0, 1, \ldots, p$. These terms are indicated within the outer bands of the Pascal triangle.

The terms within the inner triangle, i.e., $\xi^2 \eta^2, \xi^3 \eta^2, \xi^2 \eta^3$, etc., are added for $p \geq 4$ to make the basis functions complete to order p. If the element were an "ordinary" finite element, these extra terms would require interior nodes (one node for p = 4, three nodes for p = 5, etc.). A hierarchical element, on the other hand, requires no edge nodes or interior nodes which is just as well, because we would not know how to locate the interior nodes for $p \geq 5$ without inducing anisotropy of the stiffness matrix.

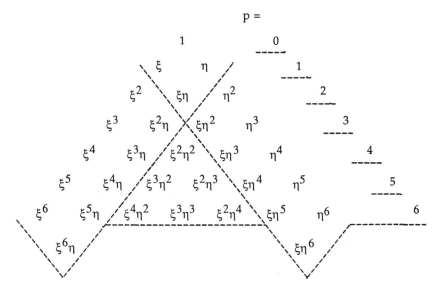

Figure 4.10 Selection of Basis Functions for a Quadrilateral Membrane Element of Order p = 1 to p = 6.

The hierarchical shape functions in MSC/PROBE are taken to be slight modifications of Legendre polynomials which retain some of the orthogonality properties of Legendre polynomials. The use of orthogonal polynomials brings the advantage that it avoids the severe round-off errors usually associated with high degree polynomials. In addition, coupling between hierarchical degrees of freedom is minimized, and even eliminated altogether for some terms in the case of simply geometries with constant Jacobian determinants.

The first five Legendre polynomials are

$$P_0(\xi) = 1$$

$$P_1(\xi) = \xi$$

$$P_2(\xi) = \tfrac{1}{2}\left(3\xi^2 - 1\right) \tag{4:66}$$

$$P_3(\xi) = \tfrac{1}{2}\left(5\xi^3 - 3\xi\right)$$

$$P_4(\xi) = \tfrac{1}{8}\left(35\xi^4 - 30\xi^2 + 3\right)$$

Additional Legendre polynomials can be generated with *Bonnet's recursion formula*

$$(n + 1)P_{n+1}(\xi) = (2n + 1)\xi\, P_n(\xi) - n\, P_{n-1}(\xi) \qquad (4{:}67)$$

Legendre polynomials satisfy the orthogonality property

$$\int_{-1}^{1} P_m(\xi)P_n(\xi)d\xi = \begin{cases} \dfrac{2}{2n + 1} & \text{for } m = n \\[2mm] 0 & \text{for } m \ne n \end{cases} \qquad (4{:}68)$$

It is seen, from Equation 4:66, that Legendre polynomials take on the values ±1 at $\xi = \pm 1$. Thus they cannot qualify as hierarchical shape functions if nodes are located at $\xi = \pm 1$. (The hierarchical shape functions must vanish at node points.) The following functions do, however, qualify

$$\phi_n(\xi) = P_n(\xi) - P_{n-2}(\xi) \qquad n \ge 2 \qquad (4{:}69)$$

as a glance at Equation 4:66 will verify. It can also be shown[15] that the derivatives of $\phi_n(\xi)$ with respect to ξ satisfy the orthogonality property

$$\int_{-1}^{1} \phi'_m(\xi)\phi'_n(\xi)d\xi = 0 \qquad m \ne n \qquad (4{:}70)$$

As a result, if the ϕ's are used as displacement shape functions, the corresponding strain states will be orthogonal, provided that the metric to parametric mapping is uniform (J is constant).

Since the ϕ's are functions of only one variable, they must be used in combination with other functions to form the shape functions for a two-dimensional element. The hierarchical shape functions which replace the shape functions for edge nodes are formed as follows:

$$N_n^{(1)} = \tfrac{1}{2}(1 - \eta)\phi_n(\xi)$$

$$N_n^{(2)} = \tfrac{1}{2}(1 + \xi)\phi_n(\xi)$$

$$N_n^{(3)} = \tfrac{1}{2}(1 + \eta)\phi_n(\xi) \qquad n = 2, 3, \ldots, p \qquad (4{:}71)$$

$$N_n^{(4)} = \tfrac{1}{2}(1 - \xi)\phi_n(\xi)$$

where superscripts $^{(1)}$ etc., refer to the four sides of the quadrilateral domain (see Figure 4.11). The blending functions $(1 - \eta) / 2$, etc., ensure that the shape functions corresponding to a particular edge are null on the opposite edge and thus are null on all other edges.

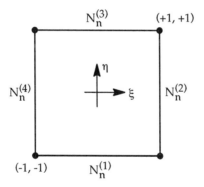

Figure 4.11 Hierarchical Element in Parametric Space.

The shape functions for the interior hierarchical variables can be formed[*] as products of the ϕ's. Thus

---

[*]MSC/PROBE uses a slightly different formulation.

$$N_1^{(0)} = \phi_2(\xi)\phi_2(\eta)$$

$$N_2^{(0)} = \phi_3(\xi)\phi_2(\eta)$$

$$N_3^{(0)} = \phi_2(\xi)\phi_3(\eta)$$

$$N_4^{(0)} = \phi_4(\xi)\phi_2(\eta)$$

etc.

(4:72)

where it is seen that the subscripts of ϕ correspond to the powers of ξ and η within the inner triangle of Pascal's triangle (see Figure 4.10).

The number of displacement variables and shape functions increases rapidly with increasing p, as the following table shows.

p	1	2	3	4	5	6	7	8
No. of Terms	4	8	12	17	23	30	38	47

The MSC/PROBE computer program allows p to range from 1 to 8. Numerical integration of the quadrilateral membrane element is performed with Gauss integration using $(p_{max} + 1)^2$ integration points. Using the same integration rule for all values of p ensures that the stiffness matrix for the pth order will be a partition of the stiffness matrix for the $p + 1^{st}$ order. Thus the stiffness matrix for $p = p_{max}$ is computed first and partitioned down to get results for lower orders.

If adjacent elements have different orders of p, the elements should be modified to ensure displacement continuity along the common boundary. For example, extra hierarchical displacement variables can be added to the lower order element along the common edge so as to provide a unique set of displacement variables on the edge. The shape functions for the additional variables take one of the forms indicated in Equation 4:71 with n ranging from $p_\ell + 1$ to p_h, where p_ℓ and p_h are, respectively, the lower and higher values of p for the adjacent elements. The array of integration points for a lower order element must also be changed from a square array with $\left(p_\ell + 1\right)^2$ points to a rectangular array with $\left(p_\ell + 1\right)\left(p_h + 1\right)$ points.

As we can see, hierarchical elements provide an elegant extension of finite element concepts to higher orders. They avoid proliferation of edge and interior nodes and provide a degree of computational efficiency for nested sets of analyses with different values of p. They also allow the use of orthogonal polynomials to form shape functions, thereby enhancing numerical stability and providing additional computational efficiency for certain simple shapes.

4.6 HIGHER ORDER VERSUS LOWER ORDER ELEMENTS

The material in this chapter has concentrated on the procedures used to construct isoparametric finite elements. We have encountered elements of different shapes and different orders of complexity. We have seen, most clearly in Section 4.5, how to extend the polynomial order of elements indefinitely and have noted the availability of elements with a complete polynomial order, p, equal to eight. The question naturally arises as to how best to select p for a given application. Is large p always better than small p? Or are there some applications where small p is better?

These are not new questions. In the early editions of his classic finite element text,[16] Zienkiewicz noted that *"a dramatic improvement of accuracy arises with the same number of degrees of freedom when complex elements are used."** At the time (ca. 1970) elements with p = 2 or p = 3 were considered to be "complex." Zienkiewicz also pointed to the advantages of lower order elements: lower computer costs to form the elements and greater ability to fit local geometries with the same number of degrees of freedom. The very poor accuracy displayed by quadrilateral elements of the lowest order, p = 1, were due to shear locking, so that when this disorder was diagnosed and corrected[17] the accuracy argument in favor of complex elements became less decisive.[18]

---

*Italics in original text.

Today, elements with $p = 1$ and $p = 2$ exist side by side in many finite element codes and seem to enjoy about equal use.

The issue of higher order versus lower order elements has been rekindled by the availability of high-order hierarchic elements. The issue has two aspects: economics and accuracy. Without doubt, a single low-order element is less accurate and more economic, in terms of computer time, than a single high-order element. It is more instructive, however, to compare accuracy and economy for fields of elements with the same number of degrees of freedom. Ultimately we are interested in the cost to achieve desired accuracy in real life applications.

4.6.1   Economic Arguments

As an introduction to the economic aspect of the issue, consider the rectangular region shown in Figure 4.12(a). We shall assume either that it contains an array of simple, $p = 1$, elements or that it contains a smaller number of higher order elements such that the number of displacement variables on an outside edge remains the same. As a result, the number of elements is $N_e = N_1 / p^2$ where N_1 is the number of elements for $p = 1$. Unless the elements are Lagrange elements, the total number of degrees of freedom for the array declines as p is increased. As shown in Figure 4.12(b), each element adds N_c displacement variables so that the total number of degrees of freedom[*] is equal to $N_e N_c$, not counting the number on the bottom and left edges. For the $p = 8$ version of the QUADP element described in Section 4.5, $N_c = 30$ compared to $N_c = 64$ for the $p = 8$ version of a Lagrange element. Thus an array of QUADPs with $p = 8$ has approximately 30/64 times the number of degrees of freedom that exist for $p = 1$.

---

[*]Here we assume one component of motion per node point or hierarchical displacement vector. The number of components is immaterial to the discussion.

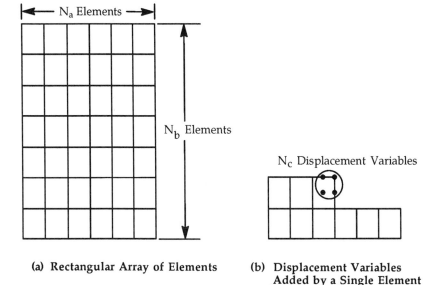

(a) Rectangular Array of Elements   (b) Displacement Variables
Added by a Single Element

Figure 4.12   Finite Element Model of a Rectangle.

Two time-consuming parts of the finite element solution are formation of the
element stiffness matrices and decomposition of the combined overall stiffness
matrix. For large problems we can obtain a reasonably accurate estimate of
computer time by counting the number of arithmetic operations. For example,
assume that formation of the ij partition of an element's stiffness matrix requires
computer time, T_o, to perform the arithmetic operations at one integration
point. The time required to form the *stiffness* matrices for all of the elements in
the array is then no less than

$$T_s = T_o N_g N_v^2 N_e \qquad (4.73)$$

where N_g is the number of integration points and N_v is the number of
displacement variables assigned to one element. For the QUADP element,
$N_g = (p+1)^2$ and $N_v = 4p + (p-2)(p-3)/2$. (In the latter expression 4p is
the number of exterior displacement variables and $(p-2)(p-3)/2$ is the

number of interior displacement variables.) Consequently, for the QUADP
element,

$$T_{s_p} = T_o N_1 \frac{(p+1)^2 \left(4p + \frac{(p-2)(p-3)}{2}\right)^2}{p^2} \qquad (4:74)$$

where $N_e = N_1 / p^2$ has been used. The ratio T_{s_p} / T_{s_1} is tabulated below for
$p = 2, 3, \dots 8.$

p	2	3	4	5	6	7	8
T_{s_p} / T_{s_1}	2.25	4.00	7.06	11.90	19.14	29.46	43.68

The time required to form the stiffness matrices of $p = 1$ elements is frequently
considered to be an insignificant part of the total solution time. The time for an
equivalent array of $p = 8$ element, which is seen to be higher by a large factor, is
not likely to be insignificant. In this calculation, no credit has been given for
possible orthogonality of the shape functions, which would, for simple shapes,
result in many null partitions of an element's stiffness matrix.

Once the stiffness matrix is formed it must be solved. For very large problems,
the controlling time is the matrix decomposition time which can be
approximated by

$$T_d = \tfrac{1}{2} T_o N_n B^2 \qquad (4:75)$$

where N_n is the total number of equivalent nodes (displacement variables) and
B is the semi-bandwidth, i.e., the number of terms, including embedded zeroes,
in a typical row of the stiffness matrix between the diagonal and the last
nonzero term to the right. T_o is just the time required to multiply two
numbers and accumulate the product. For a regular array, such as that shown
in Figure 4.12, degrees of freedom can be ordered so that B is equal to the
number of elements in the shorter direction (N_a in Figure 4.12(a)) times the
number of degrees of freedom added by each element (N_c in Figure 4.12(b)).

Thus $B = N_a N_c$. Also, $N_n = N_e N_c$ so that the time to decompose the stiffness matrix is

$$T_d = \tfrac{1}{2} T_o N_e N_a^2 N_c^3 \tag{4:76}$$

If we keep the ratio of the number of elements in the horizontal and vertical directions fixed, then $N_a = cN_b$ and $N_a^2 = cN_a N_b = cN_e$. As a result

$$T_d = \tfrac{1}{2} cT_o N_e^2 N_c^3 \tag{4:77}$$

Equation 4:77 is an important result which we will use in other contexts. It shows that, in two-dimensional arrays, stiffness matrix decomposition time, which is the controlling efficiency factor for very large problems, is proportional to the square of the number of elements and to the cube of the number of degrees of freedom added by each element. For the QUADP element, $N_c = 2p - 1 + (p - 2)(p - 3)/2$ so that, using $N_e = N_1 / p^2$

$$T_{d_p} = \tfrac{1}{2} cT_o N_1^2 \frac{\left(2p - 1 + \frac{(p-2)(p-3)}{2}\right)^3}{p^4} \tag{4:78}$$

The ratio T_{d_p} / T_{d_1} is tabulated below for $p = 2, 3, \ldots 8$.

p	2	3	4	5	6	7	8
T_{d_p} / T_{d_1}	1.69	1.54	2.0	2.76	3.79	5.07	6.59

Again we see that the higher order elements take more computer time. The disadvantage can be reduced by partitioning each element's stiffness matrix according to interior and exterior displacement variables and solving out the interior variables in a preliminary operation. For a very large number of elements, the time required to eliminate the interior variables becomes a negligible part of the total time, so that the total decomposition time may be estimated by Equation 4:77 with N_c equal to the number of exterior variables added by each element. In that case, $N_c = 2p - 1$ and the asymptotic value of T_{d_p} for large N_e is

$$T^a_{d_p} = \tfrac{1}{2} c T_o N_1^2 \frac{(2p - 1)^3}{p^4} \qquad (4:79)$$

The ratio $T^a_{d_p} / T_{d_1}$ is tabulated below for $p = 2, 3, \ldots 8$.

p	2	3	4	5	6	7	8
$T^a_{d_p} / T_{d_1}$	1.69	1.54	1.34	1.17	1.03	.92	.82

Here we see that the computational disadvantage has been reduced, and even reversed for large enough p. The reversal should not be taken too seriously because the required number of elements is very large and because a similar strategy (staged decomposition) can be employed for arrays of simple elements.

On balance, considering both stiffness matrix formulation and stiffness matrix decomposition, higher order element are, for equal numbers of degrees of freedom, more costly than lower order elements. To be effective, higher order elements must offset their efficiency disadvantage with greater accuracy.

4.6.2 Accuracy Arguments

The dependence of the accuracy of finite elements on the order of polynomial representation has been studied extensively.[19,20] We consider here only the one-dimensional case. Let the error in a finite element solution be defined as

$$e = u_{ex} - u_{fe} \qquad (4:80)$$

where u_{ex} is the exact solution and u_{fe} is the finite element solution. The *energy norm* of the error is then defined as

$$\|e\| = \sqrt{W(e)} \tag{4:81}$$

where W(e) is the expression for strain energy with e substituted for u.

Consider an interval spanned by N one-dimensional finite elements. Then, if the element's shape functions are polynomials of order p and if u_{ex} is a smooth function with derivatives to order p defined everywhere, it can be shown that[19]

$$\|e\| \leq \frac{k}{N^{p+1-m}} \tag{4:82}$$

where m is the order of differentiation required to obtain strains from displacements and k is a constant. For most applications, including three-dimensional elasticity, heat transfer, and magnetostatics, m = 1. For plate bending, m = 2. In the discussion which follows we will assume m = 1. Equation 4:82 indicates that the error declines very rapidly with increasing p but more slowly with increasing N. Again we should compare examples with equal numbers of degrees of freedom, i.e., we should let $N = N_1 / p$ in Equation 4:82. Then

$$\|e\| \leq k\left(\frac{p}{N_1}\right)^p \tag{4:83}$$

and we see that $\|e\|$ declines with increasing p for all feasible values ($p = N_1$ gives one element). The rate of convergence also depends on the number of elements, N_1, selected for p = 1. To provide a concrete example, the ratio $\|e\|_p / \|e\|_1$ is tabulated below for N = 24 and p = 2, 3, 4, 6, 8 (note that p = 5, 7 give fractional numbers of elements).

p	2	3	4	6	8
$\|e\|_p / \|e\|_1$	.167	.047	.019	.0059	.0037

Quite clearly, raising the polynomial order greatly improves the accuracy as long as the solution is smooth with finite derivatives to order p. But what if the solution is not smooth? Consider, for example, a membrane plate which has an abrupt change in thickness or an abrupt change in Young's modulus, and let the change occur along the y-axis of a rectangular element. Since equilibrium requires that the stress normal to any surface be continuous, the strain must be discontinuous. As a result, $u_{,xx} = \varepsilon_{x,x}$ does not exist along the y-axis. The finite element will approximate ε_x by a polynomial fit in such a way that the energy norm of the error, $\|e\|$, is minimized. For our example that amounts to a simple least square fit.

Figure 4.13 shows how the step change in ε_x is approximated by polynomials of order 1, 3, 5, 7 corresponding to p = 2, 4, 6, 8. The observed 22% overshoot persists to all higher orders of p. This overshoot is known, at least to electrical engineers, as the Gibbs phenomenon.

Generalizing from this example, we can say that finite elements based on polynomial shape functions do not cope well with discontinuities that occur within their boundaries. On the other hand they cope exceedingly well with certain types of discontinuities that occur on their boundaries. For example, if the previously described change in thickness or material property had occurred on the boundary between two elements, then the levels of strain would have been constant or slowly varying within each element and could, very likely, have been accurately represented by simple p = 1 elements.

The point of this discussion is that discontinuities should, if possible, be placed on the boundaries of finite elements, not within them. This observation leads to the mesh generation strategy followed instinctively by all finite element analysts and also, somewhat surprisingly, to a practical guide for choosing the polynomial order, p.

4.6.3 Finite Element Modeling

All objects can be observed on different scales. Take the Eiffel Tower for example. On the largest scale, as seen from far away, it appears as a gigantic column. From somewhat closer, the four support legs and the three levels

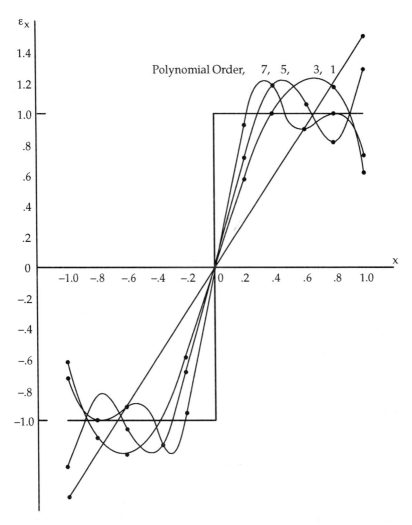

Figure 4.13   Finite Element Approximation to a Step Change in Strain.

come into view, and then we begin to see that the entire structure consists of trusses, and of smaller trusses within trusses. We come finally to the smallest truss members which are pieces of iron no more than two or three feet long. These are connected to plates, and to each other, with rivets. Finally, if we are structural engineers, we worry about the effects of corrosion, or possibly about the effects of fatigue cracks near the joints. Each of these levels of detail corresponds to a different scale, from the largest to the smallest. Eiffel and his assistants probably thought about problems involved with all of these levels— one level at a time. The great advantage of modern computing machinery is that we are able to consider two or more levels of detail, and their interactions, in a single analysis.

The number of levels of detail and their relative size, or scale, sets the minimum number of elements in a finite element analysis. Suppose that three levels of detail are to be included (for example: a whole airplane, its major parts (wings, fuselage, nacelles, etc.), and major structural elements (ribs, spars, frames, etc.)). Consider further that the relative scales of the separate levels are about 5 to 1. Then the size of the whole relative to the smallest level of detail is about 25 to 1. If the smallest detail, such as a panel between two ribs and two spars, is represented by a single finite element and if the analysis is two-dimensional, we can expect the number of finite elements to be about 25 x 25 or 625. Of course this is a very rough estimate. There may be large smooth areas between details that can be represented more coarsely, or perhaps some of the details require more than one finite element. But the order of magnitude of the number of elements has at least been set.

At this point we must try to characterize what is meant by a structural detail. From a geometric viewpoint, it represents a bump, or discontinuity, in what would otherwise be a smooth landscape. Mandelbrot has used this viewpoint to develop what he calls the "Fractal Geometry of Nature."[21] In his world, "bumps" are repeated at an infinite regression of smaller scales, leading to curves of infinite length and surfaces of infinite area. To describe this world, Mandelbrot introduces the idea of fractional dimension. A fairly smooth curve would have a dimension slightly above one, while a very irregular one would have a dimension closer to two (a curve of dimension 2.0 can completely fill a surface). Engineers sometimes talk in this way as, for example, when they refer to a very complex $2 \text{-} \frac{1}{2} \text{ D}$ structure.

Mandelbrot's concept of nature has applicability to man-made structures, at least to the more complex ones. While we need not agree to consider an infinite regression of scales (even physicists would not agree to this), it is sufficient to agree that very often there are some important details too small to be included in a particular finite element analysis. Such details are either ignored or "smeared." For example, joint flexibility is often ignored (sometimes to the analyst's later regret) while fibers or closely-spaced stiffeners are often smeared to produce a composite orthotropic material property.

The details included in a structural analysis are usually characterized by some kind of discontinuity, for example: edges, corners, holes, stiffeners, fasteners, changes in load intensity, changes in material properties. Sometimes these changes are smooth rather than discontinuous as, for example, when fillets are used in corners. Such smoothing is a smaller scale effect which is frequently ignored. The concept that structural detail is accompanied by discontinuities is, as we have seen, an important one for finite element analysis.

Since we should avoid putting discontinuities within elements, the construction of a practical finite element model begins by identifying the discontinuities and arranging the finite elements so that the discontinuities occur on the boundaries of elements. The minimum number of elements is set by the number of levels of detail that the analyst chooses to include. That choice is often, and particularly for large problems, set by economic considerations. The analyst, in such cases, includes as many structural details as his budget can afford. The elements that he selects will most probably have the minimum polynomial order, $p = 1$. To select more costly, higher order elements might, with rigid budgetary constraints, require the elimination of important structural details. The best that the element designer can do for the analyst in this case is to provide him with the best possible $p = 1$ elements.

Suppose that the analyst's budget were suddenly increased. What would he do? He could elect either to include additional structural details, such as small holes, fillets, stabilizing flanges, etc., or to improve overall accuracy by subdividing elements and/or by increasing their polynomial order. The choice would depend on the significance of the suppressed structural details.

In some structures there exists a large enough difference in scale between two succeeding levels of structural detail that their interaction effects can safely be ignored. Such is the case, for example, in the analysis of machined parts, where the difference in scale between the smallest design features (flanges, holes, fillets, threads, etc.) and the largest flaws (cracks, tool marks, surface roughness, voids, etc.) may be very large. For such an example, modeling for structural detail might stop well short of a budgetary limit and the analyst might feel free to refine the model further, simply to improve accuracy.

In general, finite element modeling is a two-step process. The analyst first lays out the minimum number of lowest order finite elements that are required to model the significant structural details. Then, if an economic limit has not been reached, he can further refine the model by subdividing the elements or by increasing their polynomial order. The latter course gives, as we have seen, greater accuracy if the solution is sufficiently smooth.

A simple example, which represents one quarter of a doubly symmetric structure, is shown in Figure 4.14. The model has 18 solid elements which capture all of the design details (intersections of planes, fillets, and a hole). It was analyzed[22] with the hierarchic elements in MSC/PROBE for $p = 1, 2, ..., 8$.

The number of degrees of freedom, an estimate of the energy norm of the error, and the maximum stress in the model, are listed in Table 4.14 for each value of p. It is noted that, while this example has only a few structural details, it is not trivial in terms of the number of degrees of freedom needed to achieve acceptable accuracy.

In summary, we can be satisfied that both higher and lower order elements have useful applications in linear finite element analysis. In nonlinear analysis, on the other hand, lower order elements are generally preferred. One reason is that it is easier to treat large deformations with simple elements. Another

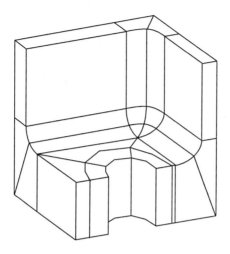

Figure 4.14 Finite Element Model of a Bathtub Fitting (one-quarter model).

Table 4.14

Results for Structural Model of Figure 4.14

p	DEGREES OF FREEDOM	ESTIMATED ENERGY NORM OF ERROR, %	MAXIMUM STRESS
1	124	54.83	31260
2	420	25.98	46620
3	728	14.89	49800
4	1261	8.99	49990
5	2025	5.38	52200
6	3074	3.48	51030
7	4462	2.33	51340
8	6237	1.63	51260

reason is that, with plastic materials, rapid changes in strain can occur anywhere. As we have seen, the attempt to represent such behavior by continuous polynomial functions converges very slowly. It is best to localize the discontinuities in elements which are as small as possible, such as those provided when using a large number of very simple elements.

REFERENCES

4.1   O. C. Zienkiewicz and R. L. Taylor, *The Finite Element Method*, 4th Ed., McGraw-Hill, pp 128-32, 1989.

4.2   R. L. Taylor, "On Completeness of Shape Functions for Finite Element Analysis," *Intl. J. Numer. Methods Eng.*, 4, pp 17-22, 1972.

4.3   Ref. 4.1, p. 173.

4.4   B. M. Irons, "Quadrature Rules for Brick-Based Finite Elements," *Intl. J. Numer. Methods Eng.*, 3, pp 239-94, 1971.

4.5   G. Strang and G. J. Fix, *An Analysis of the Finite Element Method*, Prentice-Hall, pp 183-4, 1973.

4.6   D. C. Hammer, O. P. Marlowe, and A. H. Stroud, "Numerical Integration Over Simplexes and Cones," *Math. Tables Aids Comp.*, 10, pp 130-7, 1956.

4.7   Ref. 4.5, pp 181-92.

4.8   R. H. MacNeal (Ed.), The NASTRAN Theoretical Manual, Level 15.5, NASA SP-222(01), pp 9.3-7, -8, 1972.

4.9   J. S. Archer, "Consistent Mass Matrix for Distributed Mass Systems," *J. Struct. Div. ASCE*, 89, No. ST4, p. 161, 1963.

4.10   Ref. 4.8, pp 5.5-3, -4.

4.11   O. C. Zienkiewicz and J. Z. Zhu, "A Simple Error Estimator and Adaptive Procedure for Practical Engineering Analysis," *Intl. J. Numer. Methods Eng.*, 24, pp 337-57, 1987.

4.12   M. J. Wheeler and S. M. Yunus, "An Efficient Error Approximation Technique for Use with Adaptive Meshing," Proc. 2nd Int. Conf. on Quality Assurance and Standards in Finite Elem. Analysis, NAFEMS, May 1989.

4.13   B. A. Szabó and A. K. Mehta, "p-Convergent Finite Element Approximations in Fracture Mechanics," *Intl. J. Numer. Methods Eng.*, 12, pp 551-60, 1978.

4.14   B. A. Szabó, *PROBE Theoretical Manual*, Release 1.0, Noetic Technologies Corp., St. Louis, MO, 1985.

4.15   B. A. Szabó and I. Babuška, *Finite Element Analysis*, John Wiley & Sons, p. 39, 1991.

4.16   O. C. Zienkiewicz, *The Finite Element Method*, 3rd Ed., McGraw-Hill, p. 211, 1977.

4.17   W. P. Doherty, E. L. Wilson, and R. L. Taylor, "Stress Analysis of Axisymmetric Solids Using Higher Order Quadrilateral Finite Elements," U. of Calif. Berkeley, Struct. Eng. Lab. Report SESM 69-3, 1969.

4.18   R. H. MacNeal, "Higher Order vs. Lower Order Elements—Economics and Accuracy," Proc. 2nd World Conf. on Finite Element Methods, Robinson and Associates, pp 201-15, 1978.

4.19   Ref. 4.5, pp 39-51, 105-16, 165-71.

4.20   I. Babuška, "The p- and hp-Versions of the Finite Element Method: The State of the Art," *Finite Elements: Theory and Applications*, D. L. Dwoyer (Ed.), M. Y. Hussaini, and R. G. Voigt, Springer-Verlag, New York, 1988.

4.21   B. Mandelbrot, *The Fractal Geometry of Nature*, W. H. Freeman, New York, 1983.

4.22   J. E. Schiermeier, "The p-Version of the Finite Element Method in MSC/PROBE," Proc. 1990 MSC World Users' Conf., The MacNeal-Schwendler Corp., Los Angeles, 1990.

4.23   T. J. R. Hughes, *The Finite Element Method*, Prentice-Hall, p. 159, 1987.

5

The Patch Test

The general idea behind the patch test is that the results of a finite element analysis should be exactly correct under sufficiently simple conditions. The genius in this concept lies in its precision. There is no quibbling over degrees of accuracy. An element either passes the test to something like six significant figures or it fails. The most commonly applied test condition is a uniform state of stress imposed through boundary conditions on a small "patch" of irregularly-shaped elements. It is reasoned that, since the element shapes and the load orientation have no special symmetries in the test, an element type which passes the test will produce correct results in any other test or practical application where the state of stress is uniform. This result has wide implications. In effect, it validates the element for use in structures with arbitrary shapes, provided only that the stresses vary slowly enough.

The patch test was first introduced by Irons[1,2] in 1965. At the time the testing of new elements was anything but adequate. The author recalls, for example, that the only set of tests applied to the original NASTRAN plate elements involved a rectangular plate modeled with rectangular elements.[3] There was no attempt to test performance for other shapes.

Recognition of the importance of the patch test did not occur immediately and later developers were sometimes obliged to recall their elements. For example, Wilson's original (1973) incompatible membrane element[4] was modified to pass the constant strain patch test and re-released in 1976.[5] Today it is expected that all commercial element designers will routinely apply the patch test before releasing their elements for public use. As we shall see, there is no need for surprises to occur at patch test time because the passing or failing of a patch test is generally predictable on theoretical grounds.

5.1 PATCH TESTS AS NUMERICAL EXPERIMENTS

Figure 5.1 illustrates a patch of elements which has frequently been used in benchmark tests[6] of two-dimensional elements. The only essential features of the test are the irregular shapes of the elements and the existence of at least one interior node. Usually, but not necessarily, the elements belong to the same type. Also, the exterior shape of the patch is chosen to be rectangular in Figure 5.1 to facilitate the calculation of boundary conditions. The patch shown in Figure 5.2 would, in principle, do as well. Generally speaking, fairly irregular element shapes should be used to emphasize errors in the solution should they occur. Pathological shapes, e.g., quadrilaterals with re-entrant corners or edge node placements that yield negative Jacobians (see Figure 3.18), should be avoided unless there is a special reason.

The patches shown in Figures 5.1 and 5.2 can be applied to triangles by subdividing the elements, and to higher order elements by adding edge nodes. In the latter case, the interior edges can be left straight or made curved. As will be seen, the location of edge nodes has a bearing on the test results.

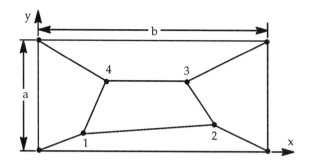

Figure 5.1   A Two-Dimensional Patch of Elements.

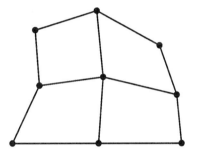

Figure 5.2   Another Two-Dimensional Patch of Elements.

Figure 5.3 illustrates a standard patch[7] for testing three-dimensional elements. Again, it includes the essential requirements—irregular element shapes and at least one interior node—and has a rectangular exterior shape to facilitate boundary conditions. Pentahedra and hexahedra can be tested by subdividing the elements and higher order elements can be tested by adding edge nodes, etc.

The patch test has different versions corresponding to the way in which boundary conditions are applied. In what we will call version 1, the displacements are specified at all exterior nodes. For example, suppose we wish to impose the constant strain condition $\varepsilon_x = B$. The corresponding displacements are $u = a + Bx + cy$ and $v = d - cx$. The integration constants (a, c, d) may be selected arbitrarily since they produce no strains. The values of u and v are rigidly imposed at all exterior nodes. The patch test is passed if the

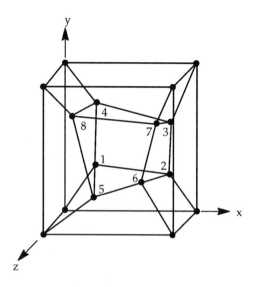

Figure 5.3   A Three-Dimensional Patch of Elements.

values of u and v at interior nodes and the strains in all elements agree with the selected algebraic forms.

We can easily construct patch tests of any polynomial order. For example, if we select $u = xy$, $v = -\left(x^2 + \upsilon y^2\right)/2$ at exterior nodes, then the expected linear stress field is $\sigma_x = y/E$, $\sigma_y = 0$, and $\tau = 0$. Since the patch of elements in Figure 5.1 has no special symmetries, it can be presumed that, if the elements in Figure 5.1 pass this test, they will pass any other linear strain patch test.[*]

In version 2 of the patch test we impose only the minimum number of exterior constraints that will restrain rigid body motions and apply loads to the remaining exterior displacement components to achieve the desired state of stress. The procedure is illustrated in Figure 5.4 for $\sigma_x = 1$.

---

[*]We wish to avoid the complications introduced by test conditions with distributed loads. Thus, for linear strain and higher order test conditions, we will assume that the homogeneous form of the equilibrium equations, Equation 2.13, is satisfied by the desired state of stress.

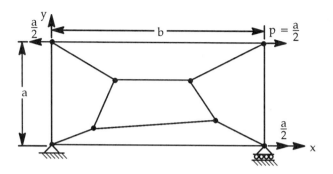

Figure 5.4 Patch Test with Applied Loads (Method 2).

Version 2 has the advantage that it can detect *spurious mechanisms*, i.e., deformations that have zero strain energy. If such are present the stiffness matrix will be singular within round-off error, even with constraints that remove the rigid body modes. The displacements at nodes may or may not be extremely large, depending on whether or not the loads excite the spurious mechanisms.[*] If version 1 were used, the detection of spurious mechanisms would require a separate test such as, for example, an eigenvalue analysis to find the number of zero frequency modes for the patch.

Version 2 requires calculation of the node point loads that correspond to the desired deformation state. This was particularly easy for the patch in Figure 5.4 due to the symmetry of the exterior shape and the absence of edge nodes. For a more general case, such as the patch of Figure 5.2 with higher order elements, resort must be made to the equations which define nodal forces in terms of stresses. Thus we must use either Equation 2:53,

$$\{F_i\} = \sum_e \int_{V_e} [B]^T \{\sigma\} dV_e \qquad (5:1)$$

---

[*]Singular stiffness matrices can occur in practical applications without necessarily causing bad effects. For example, displacements normal to a plane of load application may inadvertently be left unconstrained.

or Equation 2:80,

$$\{F_i\} = \sum_e \int_{S_e} [N]^T \{t\} dS_e \qquad (5:2)$$

where the edge tractions {t} are related to stress components by Equation 2:15. The calculation is not especially difficult but it raises the disturbing thought that the application of boundary conditions is dependent on element properties $([B] \text{ or } [N])$. As a practical matter this means that different element types may require different loads for the same test. It also seems to violate the impartiality of the test. Note, however, that to pass the test the displacements must be consistent with the desired deformation state. Thus if the boundary displacements are the same as those obtained with version 1, the boundary reaction forces in version 1 will equal the applied loads in version 2, showing that the two versions of the patch test are equivalent. The results will differ only if the test is failed.

There exists a third version of the patch test which is less practical for testing elements but more useful for proving theorems. With version 3, displacement constraints are applied to all nodes, interior as well as exterior. The test is passed if the strains are correct in all elements and if the forces of constraint are null at the interior nodes. Version 3 is equivalent to version 1 because, if version 1 is applied and the displacements at interior nodes are correct, the forces of constraint at interior nodes will be zero when interior constraints are added to convert version 1 into version 3.

A patch test is sometimes applied to a single element. The so-called single element test [8] lacks interior nodes and is, as a result, less powerful than a multi-element patch test. The single element test can determine whether strains are correctly evaluated from nodal displacements but it cannot determine whether adjacent elements are compatible. Stated differently, the single element test cannot determine whether correct nodal displacements and element strains imply nodal equilibrium. The latter is a most important point for element design, as we shall see.

5.2 THE PATCH TEST AS PROOF OF CONVERGENCE

Satisfaction of the constant strain patch test would appear to be, on physical grounds alone, both a necessary and a sufficient condition for convergence. Consider, for example, that the finite elements in an idealized mathematical model of a physical structure are repeatedly subdivided into smaller elements. Then, if the strain distribution satisfies certain smoothness criteria everywhere within the structure, the strains within each element will become more nearly uniform as the subdivision continues. In the limit, the nonuniform part of the strain distribution within each element will become vanishingly small. Then, since the patch test assures us that the elements will give correct results for any constant state of strain and any set of element shapes, the error in the solution due to the nonuniform part of the strain distribution will also become vanishingly small.

Mathematical respectability has been added to the above line of reasoning by Strang.[9,10] The difficulty with regard to the sufficiency of the patch test for convergence lies in the smoothness criteria taken for the strains. We know that even in practical problems the stresses and strains may not be defined everywhere. In Section 4.6.2, for example, we examined a case where the strain experienced a step change due to a change in material property. It was shown that a polynomial approximation to the step converges slowly and, in fact, exhibits a finite overshoot regardless of the number of terms used. We could choose to rule out such cases, which would be unfortunate because they occur in practice, or we could accept them and then claim that the finite element solution converges "almost everywhere"; in the example cited it converges everywhere except within a vanishingly small distance of the step. In like manner, we must except the vanishingly small neighborhoods of point loads and other singularities. We can, conveniently, claim convergence in most such cases by using an error "norm," such as the energy norm of the error described by Equation 4:81. Finally, we must reject Mandelbrot's concept of the physical world as an infinite regression of discontinuities at smaller and smaller scales.[11] With these or similar criteria put in suitable mathematical form it is possible to construct a proof of the sufficiency of the patch test for convergence.

The necessity of the patch test for convergence must also be hedged. If a regular rule is adopted for the subdivision of elements, then the elements may approach a shape which is less than perfectly general. For example, Figure 5.5 shows that equal division of the edges of a general quadrilateral leads, in the limit, to elements of parallelogram shape.

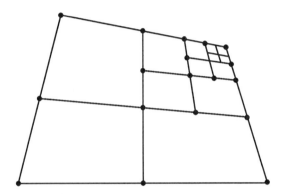

Figure 5.5   **Progressive Halving of the Edges of a General Quadrilateral Leads to Elements of Parallelogram Shape.**

As a result, a quadrilateral element need only pass the constant strain patch test for parallelogram shapes in order to converge for problems in which the size of elements is progressively decreased by equal subdivision of the edges. We will return to this question in Section 5.4.

Ultimately we must leave the question of convergence to mathematicians. Engineers are concerned with the behavior of finite collections of finite elements and in accuracies down to a percent or so. Anything beyond that is academic.

The patch test can also provide a lower bound for the rate of convergence as elements are subdivided. Assume for example that an element type passes the patch test for strains of the form $\varepsilon = x^{p_a}$ (for example, $p_a = 0$ for constant strain and $p_a = 1$ for linear strain). The corresponding correct displacements have the form $u = x^{p_a + m}$, where m is the order of differentiation required to compute strains from displacements (m = 1 for general elasticity and m = 2 for plate bending). Let a polynomial displacement field, $u = \sum_{q=1}^{Q} a_q x^q, q > p_a + m$

be imposed at external nodes. The lead term in the strain error can then be no greater than cx^{p_a+1}. Within any element, let $x = x_o + \bar{x}$, where x_o is the coordinate of the "center" of the element. Thus the lead term in the strain error is no larger than

$$c\left(x_o + \bar{x}\right)^{p_a+1} = c\left(x_o^{p_a+1} + \left(p_a + 1\right)\bar{x}\, x_o^{p_a} + \cdots + \bar{x}^{p_a+1}\right)$$

Since the element properly represents strains to order $\bar{x}^{p_a}$, the constant term, $x_o^{p_a+1}$, cannot be a part of the error and neither can any of the terms except the last. Thus the order of magnitude of the error in strain is

$$E(\varepsilon) = 0\left(\bar{x}^{p_a+1}\right) \tag{5:3}$$

If h is the maximum dimension of the element, the order of magnitude of the maximum error in strain is

$$E(\varepsilon) = 0\left(h^{p_a+1}\right) \tag{5:4}$$

The derivation of an estimate for the error in displacement is trickier. From Reference 5.10,

$$E(u) = 0\left(h^{p_a+m+1}\right) + 0\left(h^{2(p_a+1)}\right) \tag{5:5}$$

The first term is controlling (has a smaller exponent) if $p_a > m - 1$. It is seen that the second term has the same exponent as the error in strain energy. Thus it correctly predicts the rate of convergence of displacement at the point of application of a single concentrated load.

If the structure is uniformly subdivided into elements, then $h = L / N$, where L is a characteristic length of the structure and N is the number of elements along L. If the elements used in the analysis only satisfy a constant strain patch test and if general two- or three-dimensional elasticity is assumed, then $E(\varepsilon) = 0(1 / N)$. This estimate is quite pessimistic. In Section 4.6.2, for example, we quoted a result[12] which stated that, if displacement derivatives to order p exist everywhere, the order of magnitude of the error in the energy "norm" (proportional to the error in strain) is $0(1 / N^p)$ where p was the polynomial order of the element. Our present result contends that all elements converge as if p = 1 unless they can be shown to pass a higher order patch test. The highest order of patch test that can be passed is $p_a = p_c - m$ or possibly $p_a = p_q - m$ where p_c is the order of complete polynomials in the element's displacement basis and p_q is the order of quasi-complete polynomials in the displacement basis (see Section 3.3). It will be shown in Section 5.4 that satisfaction of patch tests at these higher orders depends on element shape. Thus the manner in which elements are subdivided influences the convergence rate.

The characteristic length, L, also influences the convergence rate through the number of its associated elements, N. It is rarely possible to take L to be the maximum dimension of the structure. If it were possible, the number of elements in today's largest analyses, which may exceed 100,000, would produce insanely high levels of accuracy. Such large numbers of elements are, instead, needed to accommodate multiple levels of structural detail in the same analysis.

The scale of the smallest structural detail determines the characteristic length, L, because polynomials cannot usefully be continued across the discontinuities which characterize structural details. In Figure 5.6, for example, the diameter of the small holes determines L. As a result, the number of layers of elements near each small hole establishes the accuracy of the analysis.

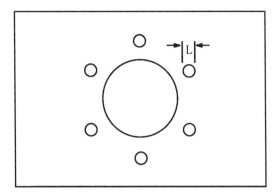

Figure 5.6   **The Smallest Structural Details Determine the Characteristic Length, L, and the Convergence Rate.**

5.3 HOW TO PASS THE PATCH TEST

As we have seen, satisfaction of the constant strain patch test guarantees convergence to the correct solution of well-posed problems as elements are subdivided. It also appeals to users as a reasonable requirement. Who can say that users are not justified in expecting finite element models to give correct answers for constant stress conditions?

Given the practical requirement of satisfying constant strain patch tests, element designers need to know how this can be done. It is not difficult, in fact, to lay out a set of rules, or sufficient conditions, that will guarantee the satisfaction of patch tests. The starting point is version 3 of the patch test in which, it will be recalled, the desired displacement field is imposed at all nodes, interior as well as exterior. The patch test is satisfied if the strains (and stresses) are correct and if the resultant forces at interior nodes are null. The latter requirement, or necessary condition, is satisfied if the forces imposed on the interior nodes by the elements are in equilibrium, i.e., if

$$\sum_e \left\{ F_i^e \right\} = 0 \qquad\qquad (5:6)$$

where $\left\{F_i^e\right\}$ is the vector of generalized forces on node i due to element e. (We assume here that body forces $\{p\}$ and loads applied directly to interior nodes, $\left\{P_i^d\right\}$ are null in the test.)

The generalized nodal forces $\left\{F_i^e\right\}$ are related to the tractions $\{t\}$ on element boundaries by

$$\left\{F_i^e\right\} = \int_{S_e} N_i^e \{t\} dS \qquad (5:7)$$

where N_i^e is the shape function for node i in element e and the integral is taken over the surface of the element S_e. (Equation 5:7 is obtained from Equation 2:80 by requiring the shape functions to be the same for all components of displacement.) Substitution into Equation 5:6 then gives

$$\sum_e \int_{S_e} N_i^e \{t\} dS = 0 \qquad (5:8)$$

as the requirement for equilibrium. Figure 5.7 illustrates, in two dimensions, the situation represented by Equation 5:8. Only the elements immediately adjacent to node i are shown because N_i^e is null in all other elements. The element boundaries can be separated into two sets. In set j, two elements share the same boundary; in set k, there is only one element per boundary. All element boundaries belong to one of these two sets in both two and three dimensions.

If the stresses are correct in the elements, the tractions along boundaries in set j, such as along H-i in Figure 5.7, will be equal and opposite in the two elements. As a result, we may write the equilibrium condition for node i as

$$\sum_j \int_{S_j} \left(N_i^{(+)} - N_i^{(-)} \right) \left\{ t^{(+)} \right\} dS + \sum_k \int_{S_k} N_i^{(k)} \{t\} dS = 0 \qquad (5:9)$$

where $N_i^{(+)}$ and $N_i^{(-)}$ are the shape functions for the elements on opposing sides of a boundary in set j.

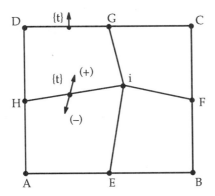

Figure 5.7   An Element Patch Showing Boundary Tractions.

We should recall, at this point, the requirements that shape functions must satisfy to ensure inter-element displacement continuity (see Section 3.4). Those requirements are that, on a given edge, the shape functions for nodes on or adjacent to the edge must be computed identically in adjacent elements and that the shape functions for nodes not adjacent to the edge must be zero. Thus, in Equation 5:9, $N_i^{(+)} = N_i^{(-)}$ and $N_i^{(k)} = 0$ for conforming elements. This leads immediately to the conclusion that the equilibrium condition is automatically satisfied by conforming elements. At last we have discovered a concrete justification for requiring inter-element displacement continuity. It ensures satisfaction of the equilibrium condition and hence, if all other requirements are met, satisfaction of the patch test.

We also see that nonconforming elements have a chance of satisfying the equilibrium condition since Equation 5:9 only requires that the sum of certain integrals involving the shape functions be null. Equation 5:9 may be simplified in certain cases. Consider, for example, that node i is interior to a particular element. Then Equation 5:9 can be written as

$$\int_{S_e} N_i \{t\} dS = 0 \qquad (5:10)$$

where the integration is taken over the entire boundary of the element. In a constant strain patch test, $\{t\}$ is constant and the condition becomes simply

$$\int_{S_e} N_i dS = 0 \qquad (5{:}11)$$

This condition can be used to design nonconforming elements which pass the patch test.[13]

In summary, the equilibrium condition can be satisfied either by using conforming elements or by requiring some weaker condition that allows satisfaction of Equation 5:9. It should also be noted that, in assumed displacement elements, the generalized forces on nodes are computed from volume integrals of stresses rather than from surface integrals of tractions. Thus, from Equation 2:53,

$$\left\{ F_i^e \right\} = \int_{V_e} \left[B_i \right]^T \{\sigma\} dV \qquad (5{:}12)$$

Satisfaction of Equation 5:6 clearly requires that the integral in Equation 5:12 be computed exactly. We have described, in Section 4.3.3, the integration order required to achieve precision in various circumstances.

Equation 5:12 suggests another way in which satisfaction of the equilibrium condition can be assured. Let the strain displacement matrix $\left[B_i \right]$ be written as the sum of the matrix for a conforming element (or for an element which is otherwise known to satisfy equilibrium), $\left[B_i^c \right]$, and a nonconforming part, $\left[B_i^n \right]$, so that

$$\left\{ F_i^e \right\} = \int_{V_e} \left[B_i^c + B_i^n \right]^T \{\sigma\} dV \qquad (5{:}13)$$

Now, since

$$\int_{V_e} \left[B_i^c \right]^T \{\sigma\} dV = \int_{S_e} N_i^c \{t\} dV \tag{5:14}$$

satisfies the equilibrium condition, the condition will continue to be satisfied if

$$\int_{V_e} \left[B_i^n \right]^T \{\sigma\} dV = 0 \tag{5:15}$$

For a constant stress field Equation 5:15 becomes

$$\int_{V_e} B_i^n dV = 0 \tag{5:16}$$

where B_i^n is any component of $\left[B_i^n \right]$. It is relatively easy, in any particular case, to determine whether this condition is satisfied. For example, Taylor, et al.,[5] used it to establish a modified version of Wilson's original nonconforming element[4] that would pass the patch test.

We are now prepared to state conditions which will guarantee satisfaction of a patch test to any desired order. To repeat, the general required condition is that the known theoretical solution be reproduced exactly within the patch. This is equivalent, in version 3 of the patch test, to the requirement that strains and stresses be correctly computed from nodal displacements and that the forces exerted by elements on interior nodes be in equilibrium, i.e., that the forces of constraint due to the displacements imposed on interior nodes be null. Examination of the equilibrium condition has revealed a variety of ways in which it can be satisfied.

Put formally, we may say that the following is a set of sufficient conditions for satisfaction of patch tests by elements derived from an assumed displacement field:

1. Strains and stresses computed from nodal displacements are correct to the desired order.

2. The integral relating the generalized forces on grid points to stresses,

$$\{F_i\} = \int_{V_e} [B_i]^T \{\sigma\} dV \tag{5:17}$$

is exact for the desired order of $\{\sigma\}$.

3. Equilibrium at interior nodes is satisfied because either

   (a) the elements are conforming, or

   (b) the nonconforming part of the strain-displacement matrix of each element produces zero grid point forces, i.e., for a general vector $\{\sigma\}$ of desired order,

$$\int_{V_e} [B_i^n]^T \{\sigma\} dV = 0 \tag{5:18}$$

   or,

   (c) the shape functions associated with interior nodes can be shown to satisfy Equation 5:9 for the desired order of $\{\sigma\}$.

Failure to satisfy these conditions leads to patch test failure. Thus we may say that the cause of patch test failure is either *interpolation failure* (failure to compute strains correctly from nodal displacements), *integration failure*, or *equilibrium failure*.

These conditions apply without qualification if all of the elements are of the same type. They also apply if elements of different types are used in the same patch test, provided that inter-element displacement continuity (required in condition 3(a) and in the implied conforming part of condition 3(b)) holds for adjacent dissimilar elements. It will not, for example, hold for an interface between linear and quadratic isoparametric elements (see Figure 4.6).

5.4 ISOPARAMETRIC ELEMENTS AND THE PATCH TEST

Since isoparametric elements are conforming, they have no trouble with the equilibrium requirement for passing patch tests. Also, since isoparametric elements satisfy completeness, they can correctly interpolate linear displacement fields regardless of element shape.

The patch test troubles of isoparametric elements begin with quadratic displacement states. From the discussion of Section 3.5 we know that isoparametric elements cannot accurately represent quadratic and higher order displacement states when the full geometric capability of the element is employed. This has important implications for plate and shell elements because a state of constant bending curvature corresponds to a quadratic variation of displacement normal to the neutral surface. As a result, isoparametric plate and shell elements cannot pass a constant curvature patch test unless restrictions are placed on element geometry. The same restrictions also apply to solid elements when they are used to model relatively thin shells. Patch tests which employ quadratic or higher-order displacement states have less significance for membrane elements and for general solid elements. Still we would like some assurance that higher accuracy is achievable as we increase the complexity of elements.

If passing a patch test requires that restrictions be placed on element geometry, then we should ascertain whether the restrictions are reasonable. The greatest selling point of finite elements is that they can fill spaces of arbitrary complexity and we should not toss this advantage away lightly. For example, a restriction to rectangular elements would be unacceptable.

The elements which can most easily fill two- and three-dimensional spaces, respectively, are the general (scalene) triangle and the tetrahedron. Designers of automatic mesh generators frequently prefer these two elements. It matters only a little to users whether the edges of these elements are straight or curved or whether edge (or face) nodes are evenly spaced.[*] Thus we can readily accept a restriction to triangles and tetrahedra with linear position basis functions $\lfloor X' \rfloor = \lfloor 1, \xi, \eta \rfloor$ and $\lfloor X' \rfloor = \lfloor 1, \xi, \eta, \zeta \rfloor$ respectively if that will ensure passage of a patch test. A six-node isoparametric triangle or a ten-node isoparametric tetrahedron restricted to linear shape can correctly interpolate quadratic displacement states because their displacement basis functions are complete through the quadratic terms in the parametric coordinates.

---

[*]This is clearly the case for interior edges. It is less likely to be true for exterior edges, particularly where edge displacements are constrained.

The possible restrictions on quadrilaterals and hexahedra are a little more complicated. The analog of the scalene triangle is the general four-node quadrilateral. A restriction to this shape, i.e., a requirement that edges be straight and that edge nodes be evenly spaced, is readily acceptable, at least for lower order elements. The geometry for this case may be called "bilinear" because the position basis vector, $\lfloor X' \rfloor = \lfloor 1, \xi, \eta, \xi\eta \rfloor$, includes the bilinear term. The three-dimensional equivalent is the general eight-node hexahedron with "trilinear" shape, $\lfloor X' \rfloor = \lfloor 1, \xi, \eta, \zeta, \xi\eta, \xi\zeta, \eta\zeta, \xi\eta\zeta \rfloor$.

As explained in Section 3.5, the nine-node Lagrange quadrilateral with bilinear shape can correctly interpolate a quadratic displacement field but the eight-node serendipity element cannot. The reason is that $u = y^2 = \left(c_1 + c_2\xi + c_3\eta + c_4\xi\eta \right)^2$ includes the term $c_4^2 \xi^2\eta^2$ which is in the displacement basis of the nine-node element but not in that of the eight-node element. As a consequence, a nine-node plate element can pass a constant curvature patch test with bilinear element shapes but an eight-node element cannot.

To find a patch test which the eight-node plate element can pass, we must restrict the element geometry to be linear in the parametric coordinates, i.e., to $\lfloor X' \rfloor = \lfloor 1, \xi, \eta \rfloor$, which corresponds to parallelogram shape. A restriction this severe is not useful for filling arbitrary spaces with coarse elements. We should recall, however, from Figure 5.5, that the parallelogram is the shape obtained in the limit when an arbitrary quadrilateral is repeatedly subdivided by equal division of the edges. Thus we can say, with a little chutzpah, that an eight-node isoparametric shell element can pass the constant curvature patch test *in the limit* as elements are subdivided in a regular way. This has also been called a *weak form* of the patch test. Other examples of weak patch tests are discussed in Reference 5.14.

When we come to consider three- and four-node isoparametric bending elements we find that no restrictions on element shape will allow them to pass a constant curvature patch test. The reason is that the displacement basis vectors of these elements do not contain the complete quadratic terms which are the minimum requirement. In these cases satisfaction of the patch test can be achieved only by partial abandonment of parametric interpolation. We will return to this subject in Section 9.4.

The final requirement for the satisfaction of patch tests is that the integral needed to form the vector of generalized forces on nodes, $\left\{ F_i \right\}$, be exact. We considered, in Section 4.3.3, the requirements on integration order that would allow the computation of $\left\{ F_i \right\}$ and other integrals to be exact. Results were presented in Tables 4.12 and 4.13 for various combinations of element type, element geometry, and the polynomial degree of the "loading."

The integration rules which enable isoparametric elements to pass low order patch tests are identified in Table 5.1. From the discussion in Section 4.3.3, we know that any rule which integrates the volume of an isoparametric element exactly will enable it to pass linear displacement patch tests. It is seen that only low integration orders are required and that the integration order does not increase when passing from linear to quadratic displacement states. The reason is that the element shapes which allow linear strains to be computed exactly are necessarily simpler. It should also be noted that the simplest element in each class (QUAD4, HEXA8, TRIA3, and TETRA4) cannot pass quadratic displacement patch tests for any geometry. Finally, it is seen that the QUAD8 and HEXA20 pass the quadratic displacement patch test for less general shapes that the QUAD9 and HEXA27.

It is not difficult to extend the results in Table 5.1 to higher order elements and higher order patch tests. Shape restrictions rather than integration rules are the issue because a sufficiently accurate integration scheme can always be constructed. The superiority of Lagrange elements over serendipity elements extends to higher orders. For example, the sixteen-node cubic Lagrange quadrilateral can pass a cubic displacement patch test with bilinear element geometry while the twelve-node cubic serendipity element can only pass the cubic test with linear geometry. The same conclusion applies to n^{th} order patch tests and n^{th} order elements of regular or hierarchic form. It should also be noted that quartic and higher order serendipity elements do not provide basis functions which are complete to n^{th} order.

Table 5.1

**Integration Rules Required to Pass
the Patch Test with Isoparametric Elements**

TEST TYPE:	LINEAR DISPLACEMENTS		QUADRATIC DISPLACEMENTS			
Element Type	Element Geometry	Integration Rule	Element Geometry	Integration Rule	Element Geometry	Integration Rule
QUAD4	Most General	1 pt.	Linear	*	—	—
QUAD8	"	2 x 2	"	2 x 2	Bilinear	*
QUAD9	"	2 x 2	"	2 x 2	Bilinear	2 x 2
HEXA8	"	2 x 2 x 2	"	*	—	—
HEXA20	"	3 x 3 x 3	"	2 x 2 x 2	Trilinear	*
HEXA27	"	3 x 3 x 3	"	2 x 2 x 2	Trilinear	3 x 3 x 3
TRIA3	"	1 pt.	"	*	—	—
TRIA6	"	3 pts.	"	3 pts.	—	—
TETRA4	"	1 pt.	"	*	—	—
TETRA10	"	5 pts.**	"	4 pts.	—	—

*Strains not exact.

**See, for example, Reference 5.20.

In practical analysis the shapes of higher order elements will rarely be simple enough to allow them to pass higher order patch tests. The reason is that the smaller number of higher order elements must have more complex shapes in order to represent the same amount of structural detail. If, for example, we permit the edges of a plane quadrilateral element to be general cubic curves, the nonzero terms in the position basis vector must include all terms of cubic or lower degree plus $\xi^3\eta$ and $\xi\eta^3$. Satisfaction of even a quadratic patch test then requires that the terms $\xi^6\eta^2$, $\xi^4\eta^4$, and $\xi^2\eta^6$ be in the displacement basis vector. This is achieved by the p = 8 hierarchical element described in Section 4.5, but not by any lower order element.

5.5 A NEW ORDER OF RESPECTABILITY

Since isoparametric elements are both complete and conforming, they can accurately interpolate linear displacement states and satisfy nodal equilibrium. They can also be derived from variational principles, as was demonstrated in Section 2.3. Mathematicians are, as a result, quite contented with isoparametric elements, particularly since convergence proofs can be constructed for methods based on variational principles.[15]

We have confirmed the convergence property of isoparametric elements by noting that they pass constant strain patch tests which, as discussed in Section 5.2, guarantees convergence. We have emphasized that isoparametric elements can pass patch tests with quadratic displacement states only when restrictions are placed on element shape. Indeed, no three- and four-node isoparametric plate element of any shape can pass a constant curvature patch test. Herein lies the beginning of a quarrel with mathematicians because element designers, being pragmatic, have tended to cope with this problem, and with the other disorders of finite elements, on an ad hoc basis without necessarily appealing to variational principles. In so doing they may have used nonconforming shape functions or inexact integration, thereby committing "variational crimes."[9]

A remedial effort has been mounted by mathematicians and by more mathematically-inclined element designers to relate such "tricks" as *reduced and selective order integration*, *incompatible modes*, and *assumed strain fields* to variational principles.[10,16,17] Thus they have sought to wrap these devices, which have proven their utility, in the mantle of respectability. While this respectability derives from the use of accepted mathematical principles, it does not of itself confer either utility or accuracy. In fact, elements without a variational pedigree often exhibit superior performance.

We propose, instead, a new order of respectability based on the patch test. It will be an order based on merit because satisfaction of a patch test is no mean accomplishment. It requires that displacements and strains be correctly interpolated from nodal displacements, that nodal forces be integrated exactly from stresses, and that nodal equilibrium be satisfied. Each of these requirements defines, in a negative sense, one of the principle disorders of finite elements: interpolation failure, which is the cause of locking; integration failure; and equilibrium failure. We have already discussed how integration failure can be avoided (see Section 4.3.3). Interpolation failure is the sole subject of the next chapter. We will encounter equilibrium failure as a side effect of remedies for interpolation failure. One of those remedies, reduced order integration, can also create spurious mechanisms as a side effect. Version 2 of the patch test can, as discussed in Section 5.1, also detect this disorder.

While we assert that satisfaction of a patch test confers respectability, we do not claim that it necessarily implies superior performance. In Section 11.4 we will encounter cases where elements can only pass patch tests if we allow them to lock. We will find that other important benefits can result from relaxing patch test requirements. Still, it is best for an element developer to keep the patch test in mind and to deviate from it only when there are clear advantages to be gained.

Our new order will have degrees of respectability which depend on the complexity of element shape and the polynomial degree of the strain (or bending curvature) for which the patch test is passed. At the bottom will be elements which cannot pass a constant strain (or constant curvature) test for any element shape. They will be classed as *unusable*. Next will come quadrilateral elements which pass constant strain tests for rectangular shapes (the unacceptables) followed by elements which pass for parallelogram shapes (the weaklings), bilinear shapes, and general shapes. Above these will be elements which pass patch tests for linear strain states, quadratic strain states, etc. In all of this we note that bending elements are at a disadvantage because they require patch tests with displacements one degree higher than membrane elements.

Elements can have their degree of respectability raised through design changes. For example, the original (1976) release of the MSC/NASTRAN QUAD4 bending element[18] could satisfy a constant curvature patch test for rectangular shapes only. (Even this was an improvement over four-node bending elements based on a strict isoparametric formulation.) The current version of the QUAD4, reported[19] in 1982, satisfies constant curvature patch tests for general (bilinear) shapes. There was also an unreported intermediate configuration which only passed the patch test for parallelogram shapes.

REFERENCES

5.1   B. M. Irons, "Numerical Integration Applied to Finite Element Methods," Conf. on Use of Digital Computers in Structural Engineering, Univ. of Newcastle, 1966.

5.2   G. P. Bazely, Y. K. Chueng, B. M. Irons, and O. C. Zienkiewicz, "Triangular Elements in Plate Bending. Conforming and Nonconforming Solutions," Proc. 1st Conf. on Matrix Methods in Struct. Mech., pp 547-76, AFFDLTR-CC-80, Wright-Patterson AFB, OH, 1966.

5.3   R. H. MacNeal (Ed.), The NASTRAN Theoretical Manual, Level 15.5, NASA SP-221(01), pp 15.2-1—11.

5.4   E. L. Wilson, R. L. Taylor, W. P. Doherty, and J. Ghaboussi, "Incompatible Displacement Models," *Numerical and Computer Meth. in Struc. Mech.*, S. T. Fenves, et al. (Eds.), Academic Press, pp 43-57, 1973.

5.5   R. L. Taylor, P. J. Beresford, and E. L. Wilson, "A Nonconforming Element for Stress Analysis," *Intl. J. Numer. Methods Eng.*, 10, pp 1211-9, 1976.

5.6   J. Robinson and S. Blackham, An Evaluation of Lower Order Membranes as Contained in the MSC/NASTRAN, ASAS, and PAFEC FEM Systems, Robinson and Associates, Dorset, England, 1979.

5.7   R. H. MacNeal and R. L. Harder, "A Proposed Standard Set of Problems to Test Finite Element Accuracy," *Finite Elem. Analysis & Design*, 1, pp 3-20, 1985.

5.8   R. L. Taylor, O. C. Zienkiewicz, J. C. Simo, and A. H. C. Chan, "The Patch Test—A Condition for Assessing F.E.M. Convergence," *Intl. J. Numer. Methods Eng.*, 22, pp 39-62, 1986.

5.9   G. Strang, "Variational Crimes and the Finite Element Method," Proc. Foundations of the Finite Element Method, A. K. Aziz (Ed.), Academic Press, pp 689-710, 1972.

5.10   G. Strang and G. J. Fix, *An Analysis of the Finite Element Method*, Prentice-Hall, 1973.

5.11   B. Mandelbrot, *The Fractal Geometry of Nature*, W. H. Freeman, New York, 1983.

5.12   W. Gui and I. Babuška, "The h-, p-, and hp-Versions of the Finite Element Method in One Dimension. Part I: The Error Analysis of the p-Version; Part II: The Error Analysis of the h- and hp-Versions; Part III: The Adaptive hp-Version;" to appear in *Numerishe Mathematik*.

5.13   B. Spect, "Modified Shape Functions for the Three-Node Plate Bending Element Passing the Patch Test," *Intl. J. Numer. Methods Eng.*, 26, pp 705-15, 1988.

5.14   O. C. Zienkiewicz and R. L. Taylor, *The Finite Element Method*, 4th Ed., McGraw-Hill, p. 295, pp 311-2, 1989.

5.15   S. C. Mikhlin, *The Problem of the Minimum of a Quadratic Functional*, Holden-Day, 1966.

5.16   D. S. Malkus and T. J. R. Hughes, "Mixed Finite Element Methods— Reduced and Selective Integration Techniques," *Comput. Methods Appl. Mech. Engrg.*, 15, pp 63-81, 1978.

5.17   H. C. Huang and E. Hinton, "A New Nine-Node Degenerated Shell Element with Enhanced Membrane and Shear Interpolation," *Intl. J. Numer. Methods Eng.*, 22, pp 73-92, 1986.

5.18   R. H. MacNeal, "A Simple Quadrilateral Shell Element," *Comput. Struct.*, 8, pp 175-83, 1978.

5.19   R. H. MacNeal, "Derivation of Element Stiffness Matrices by Assumed Strain Distributions," *Nucl. Eng. Design*, 70, pp 3-12, 1982.

5.20   H. Kardestuncer (Ed.), *Finite Element Handbook*, McGraw-Hill, p. 2.105, 1987.

6
Interpolation Failure: Locking and Shape Sensitivity

The term *interpolation failure* was introduced in Section 5.3 to describe one of the reasons for an element's inability to pass a patch test—namely the incorrect interpolation of a displacement field and its derivatives from its values at nodes. Since a finite element has only a finite number of basis functions, interpolation failure occurs naturally for higher-order functions beyond the range of the basis functions. Interpolation failure can also occur prematurely, i.e., for lower-order functions than might be expected. We have already encountered premature interpolation failure as one of the consequences of parametric mapping. For example, it was shown in Section 3.5 that, for elements of any order, superparametric mapping does not guarantee correct interpolation of linear displacements and that isoparametric mapping does not guarantee correct interpolation of quadratic displacements. This argument was extended in Section 5.4 to determine the restrictions on element shapes which allow displacement fields of various orders to be correctly interpolated by isoparametric elements.

While the failure to pass patch tests is an important disorder, it is not the only consequence, nor even the most serious consequence of interpolation failure. That distinction belongs to the phenomenon of *locking*, which is a condition of excessive stiffness for a particular deformation state or even, as inferred by the lay meaning of the term, a condition of grossly excessive stiffness bordering on rigidity.

We encountered the locking phenomenon in Chapter 1 where it was used to demonstrate the need for studying element design. It was shown in Figure 1.4, for example, that the bending deflection of the finite element model of a cantilever beam varied from 1.4% to 99.5% of the correct value, depending on the elements used and their shapes. Figures 1.6 and 3.20 illustrated other, less severe examples of locking.

It may not be an exaggeration to say that locking has driven the design of finite elements for the last twenty-five years or perhaps longer. By the mid-1960s, developers were aware of and began to understand [1] the poor performance of the early three-node triangles and four-node quadrilaterals for particular applications. The first result of this awareness was the push to develop higher-order elements. [2] Fixes for the locking problems of the lower-order elements appeared more slowly, the first in 1969, [3] and then throughout the 1970s and 1980s. The term locking, itself, did not appear in the technical literature until the mid-1970s, and as late as 1979 it was possible to discover [4] that many commercial finite elements had easily corrected locking flaws which produced enormous errors. The revelation of these errors was largely responsible for the formation of NAFEMS, the British finite elements standards organization, and the publication of problem sets [5,6] to detect locking and other flaws in finite elements.

Even today it cannot be said that the locking phenomenon is well understood by finite element users and developers. Standard texts have little to say on the subject and rarely, if ever, analyze the locking disorders of specific elements.

Element designers are more interested in finding fixes for locking than in understanding why locking occurs. As a result, experts tend to disagree on the causes. The recent work of Babuška and Suri[7] on locking is perhaps the most illuminating. Their focus is on the parameters which characterize locking and on the influence which locking has on convergence rates.

The emphasis in this chapter, derived from the author's experience, is on the underlying causes of locking and their manifestation in particular elements. The author's conviction that interpolation failure must be present for locking to occur is recent and as yet inadequately tested by peer review. The many examples that will be presented may convince the reader of the plausibility of this thesis.

We also know that some elements, such as the early NASTRAN QDMEM element, lock for all element shapes (see Figure 1.4), while others, such as the MSC/NASTRAN QUAD4, give much better results for some shapes than for others. Thus we are led to consider the *shape sensitivity* of elements with respect to locking. Shape sensitivity has also been encountered in Chapter 5 as an important consideration for the satisfaction of patch tests.

In order to study locking disorders, we will, for the first time in this book, make detailed calculations of strain states for both regular and distorted element shapes. We will, therefore, begin the study by introducing an important tool which can simplify the calculations.

6.1 ALIASING

The term *aliasing* is borrowed from sample data theory where it is used to describe the misinterpretation of a time signal by a sampling device. If, for example, the time interval between samples is too long, the frequency of a sine wave may be assigned a value which is much too low. The sine wave is then represented in the output of the sampling device by a sine wave of lower frequency, which passes as an *alias* for the true signal.

In finite element analysis, we can usefully extend the concept of aliasing to spatial discretization.[8] In this case the sample data points are the values of displacements at nodes and the alias is the function which interpolates the nodal displacements within an element.[*]

We have shown that any element will correctly interpolate its basis functions or, indeed, any function that is a linear combination of its basis functions (see discussion following Equation 3:38). The interpolation of any other function will not be correct because the interpolate can only contain a linear combination of basis functions. As a result all functions except basis functions and their linear combinations are represented within an element by aliases.

It is an easy matter to compute the alias for any particular function. Let the function be u and let $\{u_i\}$ be the vector representing the values of u at nodes. Then the interpolated value, or alias, is

$$u^a = \lfloor N \rfloor \{u_i\} = \lfloor X \rfloor \left[A_{ji} \right] \{u_i\}$$ (6:1)

The product $\left[A_{ji} \right]\{u_i\} = \{a_j\}$ is a vector of constant coefficients for the basis functions in $\lfloor X \rfloor$. We can, therefore, easily express u^a in terms of elementary basis functions, which is convenient for the derivative operations needed to compute strains.

Let us begin by computing the alias for $u = \xi^2$ in a three-node isoparametric triangle. Table 4.6(a), reproduced below as Figure 6.1, expresses the shape function for the three-node triangle in factored form $\lfloor N \rfloor = \lfloor X \rfloor \left[A_{ji} \right]$. The nodal displacement vector is

$$\{u_i\} = \{\xi_i^2\} = \begin{Bmatrix} 0 \\ 1 \\ 0 \end{Bmatrix}$$ (6:2)

---

[*]Barlow[10] has developed the same concept. He uses the term *substitute function* rather than alias.

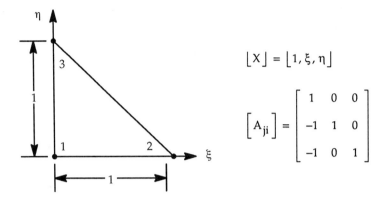

Figure 6.1 Shape Function Factors for the Three-Node Isoparametric Triangle.

so that

$$\left[A_{ji}\right]\left\{u_i\right\} = \begin{Bmatrix} 0 \\ 1 \\ 0 \end{Bmatrix} \qquad (6{:}3)$$

and

$$u^a = \lfloor 1, \xi, \eta \rfloor \begin{Bmatrix} 0 \\ 1 \\ 0 \end{Bmatrix} = \xi \qquad (6{:}4)$$

Thus we can say that within the element the function $u = \xi^2$ is represented by its alias $u^a = \xi$ or, in more compact notation,

$$u = \xi^2 \Rightarrow \xi \qquad (6{:}5)$$

We can, in like manner, work out the alias for any function in any element, even functions like $u = \sin x$. The aliases for some simple functions are evident by inspection, as in the case of the four-node quadrilateral shown in Figure 6.2. There, for example, the value of $u = \xi^2$ is 1.0 at all four nodes which clearly gives the alias $u^a = 1$ within the element. Similarly, the other functions shown

in Figure 6.2 have aliases which mimic each of the element's other elementary
basis functions.

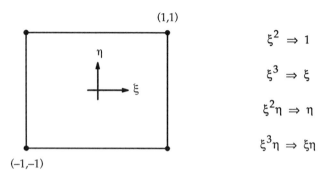

(1,1)

$\xi^2 \Rightarrow 1$

$\xi^3 \Rightarrow \xi$

$\xi^2\eta \Rightarrow \eta$

$\xi^3\eta \Rightarrow \xi\eta$

(−1,−1)

Figure 6.2   Some Aliases for the Four-Node Isoparametric Rectangle.

Table 6.1 records the aliases for the functions $\xi^m\eta^n$, $m+n \leq 4$, in the four
simplest isoparametric membrane elements. The aliases for the quadrilateral
elements, QUAD4 and QUAD8, are evident by inspection except for the alias
$\xi^2\eta^2 \Rightarrow \xi^2 + \eta^2 - 1$ in the QUAD8 element. The aliases for the triangular
elements are not as self evident.

In evaluating the performance of an element we are more interested in its strain
field than in its displacement field. In the formal calculations used to construct
an element's stiffness matrix, the strains are computed from

$$\{\varepsilon\} = \sum_i [B_i]\{u_i\} \tag{6:6}$$

The steps used to compute $[B_i]$ are outlined in Section 4.2. Fortunately we do
not need to follow these rather cumbersome procedures to compute strains for
a given displacement state if we already know the displacement state's alias.
All that is required is to evaluate

$$\{\varepsilon\} = [L]\{u^a\} \tag{6:7}$$

Table 6.1

Aliases in 2-D Isoparametric Elements

ELEMENT	TRIA3	QUAD4	TRIA6	QUAD8
FUNCTION	←		ALIAS	→
ξ^2	ξ	1	C	C
$\xi\eta$	0	C	C	C
η^2	η	1	C	C
ξ^3	ξ	ξ	$\frac{1}{2}\left(3\xi^2 - \xi\right)$	ξ
$\xi^2\eta$	0	η	$\frac{1}{2}\xi\eta$	C
$\xi\eta^2$	0	ξ	$\frac{1}{2}\xi\eta$	C
η^3	η	η	$\frac{1}{2}\left(3\eta^2 - \eta\right)$	η
ξ^4	ξ	1	$\frac{1}{4}\left(7\xi^2 - 3\xi\right)$	ξ^2
$\xi^3\eta$	0	$\xi\eta$	$\xi\eta / 4$	$\xi\eta$
$\xi^2\eta^2$	0	1	$\xi\eta / 4$	$\xi^2 + \eta^2 - 1$
$\xi\eta^3$	0	$\xi\eta$	$\xi\eta / 4$	$\xi\eta$
η^4	η	1	$\frac{1}{4}\left(7\eta^2 - 3\eta\right)$	η^2

C = correctly interpolated function

where $[L]$ is a linear derivative operator. For example, in two-dimensional elasticity

$$\{\varepsilon\} = \begin{Bmatrix} \varepsilon_x \\ \varepsilon_y \\ \gamma \end{Bmatrix} = \begin{Bmatrix} u^a{}_{,x} \\ v^a{}_{,y} \\ u^a{}_{,y} + v^a{}_{,x} \end{Bmatrix} \tag{6:8}$$

For isoparametric elements, the displacement field's alias will, in the first instance, be expressed as a set of polynomials in ξ and η. Thus the strains are computed, in two-dimensional elasticity, from

$$\left. \begin{aligned} \varepsilon_x &= u^a{}_{,x} = u^a{}_{,\xi}\,\xi_{,x} + u^a{}_{,\eta}\,\eta_{,x} \\ \varepsilon_y &= v^a{}_{,y} = v^a{}_{,\xi}\,\xi_{,y} + v^a{}_{,\eta}\,\eta_{,y} \\ \gamma = u^a{}_{,y} + v^a{}_{,x} &= u^a{}_{,\xi}\,\xi_{,y} + u^a{}_{,\eta}\,\eta_{,y} + v^a{}_{,\xi}\,\xi_{,x} + v^a{}_{,\eta}\,\eta_{,x} \end{aligned} \right\} \tag{6:9}$$

The three-node, one-dimensional isoparametric element shown in Figure 6.3 provides a simple illustrative example. The center node is deliberately offset and the exercise is to determine the effect of the offset on the accuracy of the axial strain, $\varepsilon_x = u_{,x}$.

The elementary basis functions of the element are $1, \xi, \xi^2$. The expression for position, x, in terms of the basis functions is

$$x = \xi + \alpha\left(1 - \xi^2\right) \tag{6:10}$$

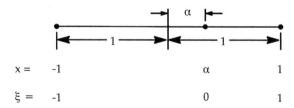

Figure 6.3   Three-Node Isoparametric Element with Offset Center Node.

We know, since the element is isoparametric, that constant strain is correctly evaluated. There may also be a presumption, on the part of a user, that linear strain is correct because the displacement basis includes ξ^2. Let us see whether this is true by assuming a displacement field

$$u = x^2 = \left(\xi + \alpha\left(1 - \xi^2\right)\right)^2 = \alpha^2 + 2\alpha\xi + \left(1 - 2\alpha^2\right)\xi^2 - 2\alpha\xi^3 + \alpha^2\xi^4$$

(6:11)

The higher degree terms ξ^3 and ξ^4 have the aliases

$$\xi^3 \Rightarrow \xi$$
$$\xi^4 \Rightarrow \xi^2$$

(6:12)

which is evident by inspection of Figure 6.3. Substitution into Equation 6:11 gives the displacement alias

$$u^a = \alpha^2 + \left(1 - \alpha^2\right)\xi^2$$

(6:13)

The resulting strain is

$$\varepsilon_x = u^a{}_{,x} = u^a{}_{,\xi}\,\xi_{,x} = u^a{}_{,\xi}\,/\,x_{,\xi} = \frac{2\left(1 - \alpha^2\right)\xi}{1 - 2\alpha\xi}$$

(6:14)

This is to be compared to the correct strain

$$\varepsilon_x^c = 2x = 2\xi + 2\alpha\left(1 - \xi^2\right)$$

(6:15)

We observe that $\varepsilon_x = \varepsilon_x^c$ if $\alpha = 0$. The error in strain for all other cases is

$$E\left(\varepsilon_x\right) = \varepsilon_x - \varepsilon_x^c = \frac{-2\alpha\left(1 - 3\xi^2\right) + 2\alpha^2\xi\left(1 - 2\xi^2\right)}{1 - 2\alpha\xi}$$

(6:16)

Observe that the error in strain becomes infinite for $2\alpha\xi = 1$ so that the quarter points, $\alpha = \pm 0.5$, are the offset limits for finite strains. Note also that the first term in the numerator vanishes for $\xi^2 = 1/3$ and that the second term vanishes for $\xi^2 = 1/2$. The positions $\xi = \pm 1/\sqrt{3}$ just happen to be the locations of integration points for two-point Gauss quadrature. Thus two-point Gauss quadrature will eliminate the first term in the error, which is the more important term if $\alpha \ll 0.5$. We will, in later and more important examples, find many cases where particular Gauss rules minimize errors.

The effect of center node offset can be mitigated by offsetting the node by a proportionate amount in parametric space.[9] Thus, if the center node is placed at $\xi = \alpha$ in parametric space, we will have $x = \xi$ in the example. The modified shape functions for the three-node line element are

$$N_1 = \frac{1}{2}\left(1 - \xi - (1-\alpha)N_2\right) \;\; ; \;\; N_2 = \frac{1 - \xi^2}{1 - \alpha^2} \;\; ; \;\; N_3 = \frac{1}{2}\left(1 + \xi - (1+\alpha)N_2\right)$$

The extension to two and three dimensions is straightforward. Note that this tactic avoids aliasing for edge node offset toward one of the adjacent corner nodes but not for offset in a perpendicular direction.

6.2 LOCKING OF THE FOUR-NODE RECTANGLE

We have defined locking as a condition of [grossly] excessive stiffness for a particular deformation state and have stated that interpolation failure is the primary cause of locking. The word *grossly* is within brackets because the degree of excess stiffness depends on the parameters of the application. Very often we will find that the degree of excess stiffness is strongly dependent on a single parameter which we can use to characterize the severity of the locking phenomenon.

The four-node rectangular membrane element is a good first example because of its simplicity and because it clearly exhibits two of the most common types of locking. Both types involve the same quadratic displacement state, i.e., in-plane bending parallel to one of the rectangle's principal directions.

Given the incompleteness of the QUAD4's quadratic displacement states (two terms out of six), interpolation failure is not surprising. Still, users and developers expect, in fact insist, that the four-node membrane element perform well for in-plane bending. We shall see that locking is a big roadblock in the way of this requirement.

6.2.1 Shear Locking

Consider the four-node rectangular membrane element shown in Figure 6.4.

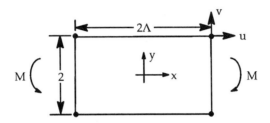

Figure 6.4 **Four-Node Rectangular Membrane Element with In-Plane Bending Load.**

The shape of the element is characterized by a single parameter, its aspect ratio Λ. We wish to determine the element's strain distribution when an in-plane bending load is applied. The desired deformation state is described by the displacement field

$$u = xy$$
$$v = -\frac{1}{2}x^2 \tag{6:17}$$

The corresponding strains and stresses for the case of plane stress (see Equation 2:16) are

$$\varepsilon_x = u,_x = y \qquad \sigma_x = \frac{Ey}{1 - \upsilon^2}$$

$$\varepsilon_y = v,_y = 0 \qquad \sigma_y = \frac{\upsilon Ey}{1 - \upsilon^2} \tag{6:18}$$

$$\gamma_{xy} = u,_y + v,_x = 0 \qquad \tau_{xy} = 0$$

Note that the stresses correspond to pure bending moment only if Poisson's ratio, υ, is zero. For the moment we will ignore this effect of Poisson's ratio.

Due to the simplicity of the element's geometry, we can dispense with parametric mapping and work directly with the metric coordinates, x and y. Since the value of x^2 is Λ^2 at all four nodes, the displacement field's alias is

$$u^a = xy$$
$$v^a = -\tfrac{1}{2}\Lambda^2$$

(6:19)

The corresponding strains and stresses are

$$\varepsilon_x = y \qquad\qquad \sigma_x = \frac{Ey}{1 - \upsilon^2}$$

$$\varepsilon_y = 0 \qquad\qquad \sigma_y = \frac{\upsilon Ey}{1 - \upsilon^2}$$

(6:20)

$$\gamma_{xy} = x \qquad\qquad \tau_{xy} = \frac{Ex}{2(1 + \upsilon)}$$

The shear strain, γ_{xy}, is clearly incorrect. Its effect on stiffness is obtained by comparing strain energies. The strain energy of the element is, if we include the spurious shear strain

$$W_s = \tfrac{1}{2} \int_{V_e} \{\sigma\}^T \{\varepsilon\} dV = \frac{tE}{2\left(1 - \upsilon^2\right)} \int_{-\Lambda}^{\Lambda} \int_{-1}^{1} \left(\varepsilon_x^2 + \frac{1 - \upsilon}{2}\gamma_{xy}^2\right) dxdy$$

$$= \frac{tE}{2\left(1 - \upsilon^2\right)} \int_{-\Lambda}^{\Lambda} \int_{-1}^{1} \left(y^2 + \frac{1 - \upsilon}{2}x^2\right) dxdy$$

$$= \frac{2tE\Lambda}{3\left(1 - \upsilon^2\right)} \left(1 + \frac{1 - \upsilon}{2}\Lambda^2\right)$$

(6:21)

The ratio of this strain energy to the correct strain energy is just the factor $1 + (1 - \upsilon)\Lambda^2 / 2$. This factor must then also be the ratio of the element's bending stiffness to the correct bending stiffness. It is seen that, for reasonably slender elements, say $\Lambda = 10$, the increase in stiffness is large enough to be aptly described as gross. The term *shear locking* has been applied to this phenomenon because its proximate cause is the spurious shear strain.

The spatial distribution of the spurious shear strain, $\gamma_{xy} = x$, provides a clue to an effective remedy. All we need do to get the correct strain energy for bending states and also for constant strain states is to substitute the value of γ_{xy} at $x = y = 0$ for the values at all integration points. This technique is called *selective underintegration* because, in effect, we are using a single point to evaluate the strain energy due to γ_{xy} and a 2 x 2 or perhaps higher order array of Gauss points to evaluate the strain energy due to ε_x and ε_y. The use of selective underintegration as a remedy for locking dates from 1969. [3] Selective underintegration is not an unmixed blessing because it can cause a variety of other disorders as side effects. These matters will be discussed at length in Chapter 7.

A clue to another effective remedy is provided by the deformed shape of the element. Within the element the interpolated value of v is constant while the correct value is proportional to x^2. The remedy is to add a mode of deformation in the y direction proportional to $x^2 - \Lambda^2$. The amplitude of the mode is determined by strain energy minimization. Such a mode is necessarily nonconforming because the two end points of an element's edge cannot determine a quadratic curve. This remedy for shear locking, which dates from 1973, [11] will be developed in Section 8.1.

6.2.2 Dilatation Locking

Dilatation locking[*] is a phenomenon that is most commonly associated with nearly incompressible materials, i.e., with those for which υ approaches one-half very closely. To study the effect we consider the in-plane bending of a

---

[*]This terminology is new. The locking of nearly incompressible materials is called Poisson's ratio locking in Reference 6.23. It has also been called volumetric locking.

four-node rectangular element which is in a state of plane strain. The assumed displacement state, which is a slight modification of that just used to examine shear locking, is

$$u = xy$$

$$v = -\frac{1}{2}x^2 - \frac{v}{2(1-v)}y^2 \qquad (6:22)$$

The associated strains and in-plane stresses* (see Equation 2:17) are

$$\varepsilon_x = y \qquad \sigma_x = \frac{Ey}{1-v^2}$$

$$\varepsilon_y = -\frac{v}{1-v}y \qquad \sigma_y = 0 \qquad (6:23)$$

$$\gamma_{xy} = 0 \qquad \tau_{xy} = 0$$

which shows that the applied loading is pure in-plane bending. It is also useful, for our purpose, to display the volumetric expansion or *dilatation*

$$e = \varepsilon_x + \varepsilon_y + \varepsilon_z = \left(\frac{1-2v}{1-v}\right)y \qquad (6:24)$$

which, it is seen, tends to zero as v approaches 0.5. The mean pressure, $p = -\left(\sigma_x + \sigma_y + \sigma_z\right)/3$, is equal to the dilatation multiplied by the bulk modulus, $K = E/3(1-2v)$. It is noted that, for nearly incompressible materials, the bulk modulus is very much larger than the elastic modulus, E. For our example

$$p = -Ke = \frac{-Ey}{3(1-v)} \qquad (6:25)$$

which remains finite as v approaches 0.5.

Within the four-node rectangular element the displacement field is represented by its alias

---

*Note that σ_z is not zero in this application.

$$u^a = xy$$

$$v^a = -\frac{1}{2}\Lambda^2 - \frac{\upsilon}{2(1-\upsilon)}$$ (6:26)

The associated strains and stresses are

$$\varepsilon_x = y \qquad \sigma_x = \frac{E(1-\upsilon)y}{(1+\upsilon)(1-2\upsilon)}$$

$$\varepsilon_y = 0 \qquad \sigma_y = \frac{E\upsilon y}{(1+\upsilon)(1-2\upsilon)}$$ (6:27)

$$\gamma_{xy} = x \qquad \tau_{xy} = \frac{Ex}{2(1+\upsilon)}$$

The dilatation, $e = \varepsilon_x + \varepsilon_y + \varepsilon_z = y$, is seen to remain finite instead of tending to zero as υ approaches 0.5. The element's strain energy density is

$$W'_s = \frac{1}{2}\{\sigma\}^T\{\varepsilon\} = \frac{E}{2(1+\upsilon)}\left[\left(\frac{1-\upsilon}{1-2\upsilon}\right)y^2 + \frac{x^2}{2}\right]$$ (6:28)

while the correct strain energy density is

$$W_s^{c'} = \frac{E}{2(1-\upsilon^2)}y^2$$ (6:29)

It is seen that there are now two sources of error: the x^2 term due to the spurious shear strain and the magnification of the y^2 term by the factor $(1-\upsilon)^2/(1-2\upsilon)$. The magnification, which becomes very large as υ approaches 0.5, is aptly described as *dilatation locking*. The cause of the locking is the inability of the element to interpolate the y^2 term in v, i.e., the inability to produce the linear strain component $\varepsilon_y = y$. Dilatation locking remains as a source of error even if the x^2 term in W'_s is eliminated by evaluating γ_{xy} at $x = y = 0$. A remedy can, however be provided by adding a nonconforming displacement mode, $v = y^2 - 1$. Wilson's 1973 element[11] included both the $x^2 - \Lambda^2$ and $y^2 - 1$ nonconforming, or incompatible, modes and hence solved the shear locking and dilatation locking problems at the same time.

A selective integration remedy of sorts can be achieved by first subtracting one-third of the dilatation from the direct strains to obtain the so-called *deviatoric strains*, $\bar{\varepsilon}_x = \varepsilon_x - \frac{1}{3}e = \frac{2}{3}\varepsilon_x - \frac{1}{3}\varepsilon_y - \frac{1}{3}\varepsilon_z$, etc. The strain energy due to deviatoric strains is then evaluated at a normal (2 x 2) set of Gauss points while the strain energy due to dilatation is evaluated only at $x = y = 0$.[12] The value of dilatation at this point is correct both for bending and for constant strain states. Note, however, that while locking is avoided, the deviatoric strains at the (2 x 2) Gauss points remain in error for bending states.

If Poisson's ratio equals exactly 0.5 the material becomes *incompressible* and the equation for dilatation becomes a constraint, $e = \varepsilon_x + \varepsilon_y + \varepsilon_z = 0$. Finite elements have been designed to model this limiting condition.[13] The pressure becomes a *Lagrange multiplier* which is treated as a separate degree of freedom. The form of the element's equations becomes

$$\begin{bmatrix} K & \vdots & G^T \\ \cdots & \vdots & \cdots \\ G & \vdots & 0 \end{bmatrix} \begin{Bmatrix} u \\ -- \\ P \end{Bmatrix} = \begin{Bmatrix} P \\ -- \\ 0 \end{Bmatrix} \qquad (6{:}30)$$

where the lower partition, $[G]\{u\} = 0$, expresses the incompressibility constraint. The number of pressure degrees of freedom per element is equal to the number of points at which the incompressibility constraint is enforced. For the rectangular four-node element, dilatation locking will occur if there is more than one such point.

As a practical matter, the ratio of bulk modulus to elastic modulus, B/E, does not exceed 10^4 for most rubbery materials. Since stiffness ratios of this size do not cause serious numerical stability difficulties, the treatment of nearly incompressible materials as incompressible has no computational advantage in modern computers.

The interpolation failure which is responsible for dilatation locking in the plane strain case produces significant but non-catastrophic excess stiffness in the plane stress case. Here we assume a displacement state

$$u = xy \qquad v = -\frac{1}{2}x^2 - \frac{1}{2}vy^2 \qquad (6{:}31)$$

which produces strains and stresses (see Equation 2:16)

$$\varepsilon_x = y \qquad \sigma_x = Ey$$
$$\varepsilon_y = -\upsilon y \qquad \sigma_y = 0 \qquad (6:32)$$
$$\gamma_{xy} = 0 \qquad \tau_{xy} = 0$$

Within the four-node rectangular element of Figure 6.4, the alias of the displacement field is

$$u^a = xy \qquad v^a = -\frac{1}{2}\Lambda^2 - \frac{1}{2}\upsilon \qquad (6:33)$$

and the associated strains and stresses are

$$\varepsilon_x = y \qquad \sigma_x = \frac{Ey}{1 - \upsilon^2}$$

$$\varepsilon_y = 0 \qquad \sigma_y = \frac{\upsilon Ey}{1 - \upsilon^2} \qquad (6:34)$$

$$\gamma_{xy} = x \qquad \tau_{xy} = \frac{Ex}{2(1 + \upsilon)}$$

The element's strain energy density is

$$W'_s = \frac{1}{2}\{\sigma\}^T\{\varepsilon\} = \frac{E}{2}\left(\frac{y^2}{1 - \upsilon^2} + \frac{x^2}{2(1 + \upsilon)}\right) \qquad (6:35)$$

As before, selective underintegration of the shear strain will remove the spurious x^2 term. The y^2 term is magnified over its correct value by a factor $1/(1 - \upsilon^2)$, which has a maximum value of $\frac{4}{3}$ at $\upsilon = \frac{1}{2}$. For metals, which have Poisson's ratios near 0.3, the error in strain energy is approximately ten percent. The MSC/NASTRAN QUAD4 element displays this magnitude of error in Figure 1.4 where the tip deflection of a cantilever beam made up of rectangular elements is seen to be too small by 9.6%.

The errors for plane strain can be much larger, even when bending does not dominate the displacement field. Consider, for example, the standard thick-walled cylinder test shown in Figure 6.5.[5] The radial displacement varies

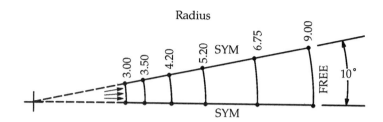

Inner Radius = 3.0; Outer Radius = 9.0; Thickness = 1.0;
E = 1,000; υ = 0.49, 0.499, 0.4999; Plane Strain Condition;
Mesh: 5 x 1 (as shown above); Loading: Unit Pressure at Inner Radius

Figure 6.5   Thick-Walled Cylinder.

approximately as 1 / r for υ near one-half. Within each element, the alias for the radial displacement has the form a + br.   Five elements with graded dimensions are used to minimize the error.

Results for four different MSC/NASTRAN elements and three values of Poisson's ratio are shown in Table 6.2. It is seen that dilatation locking devours the accuracy of the QUAD2 and QUAD4 elements and begins to nibble at the accuracy of the QUAD8 as υ approaches 0.5. Only the QUADR escapes with a small error which is seen to be independent of Poisson's ratio. The observed differences in accuracy are due to differences in element design. The QUAD2 is a four-node element made up of three-node triangles. The QUAD4 is a four-node isoparametric element with selective underintegration for shear. The QUAD8 is an eight-node isoparametric element with selective underintegration. The QUADR is a four-node element with drilling freedoms and nonconforming modes (see Section 8.2).

Table 6.2

Results for Thick-Walled Cylinder

Exact Solution = 1.000

υ	NORMALIZED RADIAL DISPLACEMENT AT INNER BOUNDARY			
	QUAD2	QUAD4	QUAD8	QUADR
.49	.643	.846	1.000	.985
.499	.156	.359	.997	.985
.4999	.018	.053	.967	.985

6.2.3 Locking in Fields of Four-Node Rectangular Elements

So far we have considered only the locking of a single element and have shown
that the source of locking is incorrect interpolation of a higher-order
displacement state.   In the case of four-node rectangles, the assumed
displacement state was quadratic for both shear locking and dilatation locking.
Supplied with such elements, a finite element user may attempt to mitigate the
locking problem by subdividing the elements. We examine here the extent to
which this strategy will succeed.

Figure 6.6 shows three ways in which a rectangular element can be subdivided
into smaller rectangular elements. For the bending states treated in the two
prior subsections, the vertical subdivision would appear to be the most logical
one because the bending strain, ε_x, is proportional to y.   Thus, since
$y = y_0 + \overline{y}$, where $\overline{y}$ is measured from a sub-element's center, the vertical
subdivision will increase the constant part of the bending strain in each element
relative to the linear part.

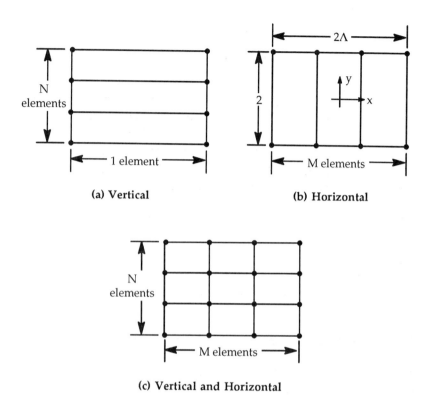

(a) Vertical   (b) Horizontal

(c) Vertical and Horizontal

Figure 6.6   Three Ways to Subdivide a Rectangular Element Into Rectangular Elements.

This line of reasoning is valid for dilatation locking where, for any sub-element, the error in strain energy density is proportional to $\bar{y}^2$. Since the integration limits for $\bar{y}$ are $\pm 1 / N$, the error in strain energy for the field of elements declines as $1 / N^2$.

Vertical subdivision does not, on the other hand, reduce the shear locking problem. In this case the error in strain energy density is proportional to x^2 (see Equation 6:21), so that the proper remedy is to subdivide the element in the horizontal direction as shown in Figure 6.6(b). This remedy, while effective, is also counter-intuitive, showing that errors due to locking do not necessarily behave like discretization errors.

Subdivision in both directions will clearly lessen the effects of both types of locking. If equal numbers of subdivisions, M = N, are used the error in strain energy will decline as $1/N^2$, as predicted by discretization theory (see Section 5.2). The number of elements used will, however, be larger than necessary if only one of the two types of locking is present.

Although the magnitude of the errors due to locking declines as the number of elements increases, there is no guarantee that acceptable accuracy will result from a reasonably small number of elements. Suppose, for example, that we wish to reduce the error in strain energy due to shear locking to one percent. Then, from Equation 6:21,

$$\left(\frac{1-\upsilon}{2}\right)\frac{\Lambda^2}{M^2} <. 01$$

or

$$M > \Lambda\sqrt{50(1-\upsilon)} \qquad (6:36)$$

If the original element's aspect ratio is fairly large, say $\Lambda = 20$, then more than one hundred substitute elements will be required to achieve one percent accuracy.

In the cases of dilatation locking, suppose that $\upsilon = .4999$, which gives a ratio of bulk modulus to Young's modulus, $K / E = 1667$. We can obtain the ratio of the error in strain energy to the correct strain energy by integrating Equations 6:28 and 6:29 over each vertical sub-element and summing. The result is

$$\frac{W_s - W_s^c}{W_s^c} = \frac{\upsilon^2}{(1-2\upsilon)N^2} \qquad (6:37)$$

so that, to achieve one percent accuracy with the prescribed value of υ, $N > 10\upsilon/\sqrt{1-2\upsilon} = 353$. Clearly this result, and also the prior result for shear locking, cannot reasonably be achieved in practical finite element analysis. We conclude, therefore, that living with locking problems is a poor substitute for fixing them.

6.3   LOCKING OF THE CONSTANT STRAIN TRIANGLE

As its name implies, the constant strain triangle experiences interpolation errors for every quadratic or higher-order displacement field. As a result, it is not unreasonable to expect that the constant strain triangle will lock for in-plane bending loads. Consider, for example, the pair of right triangles joined to form a rectangle in Figure 6.7.

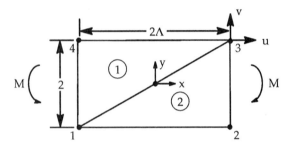

Figure 6.7   Two Constant Strain Right Triangles Subjected to In-Plane Bending.

Also assume a condition of plane strain and let the pair of elements be subjected to in-plane bending with the displacement state

$$u = xy$$

$$v = -\frac{1}{2}x^2 - \frac{v}{2(1-v)}y^2 \qquad (6{:}38)$$

This displacement state is exactly the same as that used to study dilatation locking of the four-node rectangle. We could determine the corresponding strains within the element by finding the displacement state's alias and differentiating. It is, however, more convenient in this case to evaluate the strains by inspection of the nodal displacements shown in Figure 6.8.

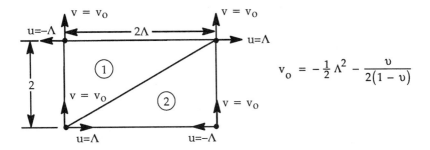

$$v_0 = -\frac{1}{2}\Lambda^2 - \frac{v}{2(1-v)}$$

Figure 6.8 **Nodal Displacements for the Pair of Constant Strain Right Triangles.**

The strains, including dilatation, are compared with their correct values in Table 6.3. We see that the bending strain, ε_x, is correct only at the top and bottom fibers and that the other two strain components are completely incorrect. Shear locking is evident because the ratio of γ_{xy} to ε_x is equal to the aspect ratio of the element pair, Λ. Dilatation locking is also evident because the element's dilatation does not tend to zero as v approaches 0.5; in fact, it is not even a function of v.

Table 6.3

Strains for the Elements in Figure 6.8

STRAIN COMPONENT	CORRECT VALUE	TRIANGLE ①	TRIANGLE ②
ε_x	y	1	-1
ε_y	$-\dfrac{v}{1-v}y$	0	0
γ_{xy}	0	$-\Lambda$	Λ
Dilatation	$\left(\dfrac{1-2v}{1-v}\right)y$	1	-1

While no one (nowadays) would think of using constant strain triangles in the manner illustrated, the implications of the example for *fields* of constant strain triangles apply with a validity equal to those derived in Section 6.2.3 for fields of four-node rectangles. In particular, the error in strain energy due to locking declines in proportion to the square of the number of elements in the appropriate direction, but remains large for reasonable numbers of elements if the locking parameter, Λ or $1/(1-2\upsilon)$, is large. The locking effects are, in fact, stronger for the triangle than for the rectangle because the spurious strain energies are larger by a factor of three due to the constant, as opposed to linearly varying, values of strains over the surfaces of the triangles.

The incompressible case provides particularly instructive insights. For this case, the element cannot experience a change in volume or, if plane strain is assumed, a change in area. In the field of the constant strain triangles shown in Figure 6.9, each new interior node adds two degrees of freedom and two elements. Note, however, that if the nodes below and to the left of node A are rigidly restrained, there can be no motion at point A which does not change the area of triangle ① or triangle ②. In effect, the number of internal constraints added by the two elements equals the number of displacement components added by the node so that there is no *net* increase in the number of independent degrees of freedom when node A is added to the field. As a result, if boundary conditions constrain the bottom and left edges of the field, an ultimate state of locking is reached—total rigidity everywhere.

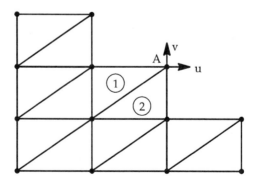

Figure 6.9   A Field of Incompressible Constant Strain Triangles.

The ratio of the number of degrees of freedom added by an element or, in this case, by a pair of elements, to the number of internal constraints added has been called the *constraint ratio*.[14] This ratio gives an indication of the tendency of a field of incompressible or nearly incompressible elements to lock. The constraint ratio for the constant strain triangle, for example, is $2/2 = 1$. The constraint ratio for the four-node rectangle with single-point evaluation of the dilatation is $2/1 = 2$. Clearly, the tendency to lock decreases as the constraint ratio is raised. Hughes[15] argues that the ideal value of the constraint ratio is two for planar elements.

6.4 LOCKING IN OTHER RECTANGULAR ELEMENTS

Before plunging ahead we should perhaps stop and think over what we have learned so far about locking. We started with the proposition that interpolation failure is the cause of locking and we have indeed seen that the failure to properly interpolate a quadratic displacement state of in-plane bending leads to two types of locking—shear locking and dilatation locking—in four-node rectangles and in three-node right triangles. We have also seen that, to produce grossly excessive stiffness, a large value of some parameter is also required.

It can easily be demonstrated that interpolation failure does not inevitably lead to excessive strain energy. Consider the displacement state $u = x^2$ which, for the four-node rectangular element of Figure 6.4, is interpreted as $u = \Lambda^2$ so that the element has no strain energy at all. The difficulty is, of course, that the element cannot distinguish the deformation state $u = x^2$ and the rigid body motion $u = \Lambda^2$. Which displacement state is correct? The expectation that $u = x^2$ exists only in our minds and perhaps also in the real world problem we are trying to model. The finite element model knows nothing about our expectations.

This example illustrates the point that interpolation failure is a denial of expectation or, as a cynic might say, a denial of unreasonable expectation. For example we expect that a rectangular element subjected to in-plane bending moment should deform as shown in Figure 6.10(a), but if the element is a standard isoparametric with only four nodes, it will actually deform as shown

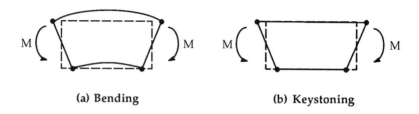

(a) Bending   (b) Keystoning

Figure 6.10   Response of a Four-Node Element to In-Plane Bending Moment.

in Figure 6.10(b).   On the other hand, if incompatible modes are added to the element, or if selective underintegration is used, the element will, in fact, respond in the manner shown in Figure 6.10(a).

We can make up examples, such as the one shown in Figure 6.11, where the "keystoning" illustrated in Figure 6.10(b) is the expected behavior.   Here the response will be too soft if the elements have been "fixed" to eliminate shear locking.   A gap will also open up between the two elements.   You can't win them all, but since bending occurs more frequently than keystoning in practical applications, the developer who has fixed his elements for shear locking should sleep better than the one who hasn't.

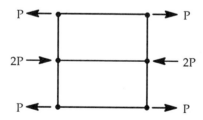

Figure 6.11   Example Where Keystoning is the Expected Behavior.

In the investigation of higher-order rectangles and bricks, we will restrict our attention to cases where at least one of the components of displacement is correctly interpolated.   To do otherwise would provide no logical limit to the number of cases and would strain the bounds of reasonable expectation.   In the case of the three-node triangle, on the other hand, we had to breach this limit to find a meaningful example of locking.

Since interpolation failure occurs only if one or more of the components of displacement is incorrectly interpolated, we can further restrict our attention to polynomial degrees where the basis functions are incomplete. Still, there will be plenty to do.

6.4.1 Locking of the Eight-Node Rectangle

Consider the eight-node rectangular membrane element shown in Figure 6.12. Its displacement basis includes the terms $1, \xi, \eta, \xi^2, \xi\eta, \eta^2, \xi^2\eta, \xi\eta^2$. Since the element is rectangular we can, for convenience, substitute x and y for ξ and η. Only cubic displacement fields need be considered because the quadratic and lower order fields are complete. The aspect ratio of the element, Λ, is assumed to be greater than one.

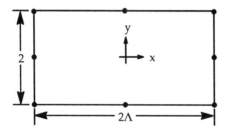

Figure 6.12 An Eight-Node Rectangular Membrane Element.

Let us begin by assuming $u = x^2y$, $v = 0$, which are correctly interpolated, and see if anything interesting develops. For this case

$$\varepsilon_x = 2xy \; ; \; \varepsilon_y = 0 \; ; \; \gamma_{xy} = x^2 \tag{6:39}$$

Observe that with $\Lambda > 1$ the strain energy due to γ_{xy} is larger than the strain energy due to ε_x by a factor proportional to Λ^2. If we expect ε_x to be the only nonzero strain component, as in the case of pure bending, we need to add a displacement component $v = -\frac{1}{3}x^3$. This term will, however, be represented in the element by its alias, $v^a = -\frac{1}{3}\Lambda^2x$. The resulting strain components are

$$\varepsilon_x = 2xy \ , \ \varepsilon_y = 0 \ , \ \gamma_{xy} = x^2 - \tfrac{1}{3}\Lambda^2 \qquad (6{:}40)$$

The shear strain is incorrect, $\gamma_{xy} \neq 0$, at all points except where $x = \pm\Lambda / \sqrt{3}$. Shear locking can therefore be avoided by evaluating γ_{xy} at $x = \pm\Lambda / \sqrt{3}$, $y = \pm 1 / \sqrt{3}$ which happen to be the 2 x 2 integration points for Gauss quadrature. This solution for locking dates from 1971.[16] We also know, from Table 5.1, that constant strain patch tests will be satisfied even if we evaluate *all* strain components at the 2 x 2 Gauss points. Reduced-order (2 x 2) Gauss integration appears to be a fine choice. The only potential drawback is that there may not be enough integration points to provide stiffness for all of the element's strain states. This possibility is examined in Chapter 7.

The next question we ask is whether dilatation locking can result for the bending state under examination. Zero dilatation is achieved by adding a strain $\varepsilon_y = -2xy$ which corresponds to adding the term $-xy^2$ to v. Note that xy^2 is in the displacement basis. The resulting displacement and strain components are

$$u^a = x^2 y \ , \ v^a = -\tfrac{1}{3}\Lambda^2 x - xy^2$$

$$\varepsilon_x = 2xy \ , \ \varepsilon_y = -2xy \ , \ \gamma_{xy} = x^2 - \tfrac{1}{3}\Lambda^2 - y^2 \qquad (6{:}41)$$

Dilatation locking does not occur because $e = \varepsilon_x + \varepsilon_y + \varepsilon_z = 0$. Neither does shear locking if γ_{xy} is evaluated at 2 x 2 Gauss points.

Consider next the displacement state $u = xy^2$, $v = 0$, illustrated in Figure 6.13.

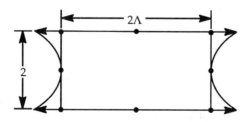

Figure 6.13   The Displacement State $u = xy^2, v = 0$.

The corresponding strains are

$$\varepsilon_x = y^2 \ , \ \varepsilon_y = 0 \ , \ \gamma_{xy} = 2xy \qquad (6{:}42)$$

Although the strain energy for γ_{xy} is larger than that for ε_x if $\Lambda > 1$, shear locking cannot be said to occur because, since xy^2 is in the displacement basis, the deformation state depicted in Figure 6.13 agrees exactly with the desired displacement state. Dilatation locking can, however, occur because to achieve zero dilatation we must set $\varepsilon_y = -y^2$ which requires the unavailable displacement term $v = -y^3 / 3$. The resulting alias for v gives, for the displacement and strain fields,

$$u^a = xy^2 \ , \ v^a = -\frac{y}{3}$$

$$\varepsilon_x = y^2 \ , \ \varepsilon_y = -\frac{1}{3} \ , \ \gamma_{xy} = 2xy \ , \ e = y^2 - \frac{1}{3} \qquad (6{:}43)$$

Note that dilatation locking is avoided if strains are evaluated at $y = \pm 1 / \sqrt{3}$, i.e., if they are evaluated at 2×2 Gauss quadrature points.

The remaining two cubic displacement fields of the eight-node rectangle, $u = 0, v = x^2 y$ and $u = 0, v = xy^2$, add nothing new. They are just x-y permutations of cases already examined.

Locking of the eight-node rectangle is a much less serious problem than locking of the four-node rectangle. When the four-node rectangle locks it locks in its primary bending mode and cannot bend at all. The eight-node rectangle, on the other hand, locks only in its higher modes and can still bend with the help of its lower modes. The result, for most applications, is mild rather than severe loss of accuracy. The cantilever beam problem of Figure 1.4 is a typical application. The shear load at the tip provides a modest linear variation of bending moment within each element and, while the linear variation of curvature may be lost through locking, the constant part remains. The result for this problem is an approximately 3% error in tip displacement for elements which lock compared to approximately 1.5% error for elements which use reduced-order integration. These conclusions cannot be extended to elements with nonrectangular shapes.

6.4.2 Locking of the Eight-Node Brick

The eight-node rectangular brick is the three-dimensional analog of the four-node rectangle. Its displacement basis is $\lfloor 1, x, y, z, xy, xz, yz, xyz \rfloor$. We see that there are four incomplete quadratic and cubic terms, so that with three components of displacement (u, v, w) there are a total of twelve independent displacement states to examine for locking. The dimension of the brick in the z direction is assumed smaller by a factor of $1 / \Lambda$ than the dimensions in the other two directions (see Figure 6.14).

Several of the displacement states are truly two dimensional so that their locking behavior is identical to that for the four-node quadrilateral. The remaining states are u = yz, u = xyz, and their cyclic permutations, v = zx, w = xy, v = xyz, w = xyz. The quadratic displacement states, u = yz, etc., are linearly varying shear states for which $\varepsilon_x = \varepsilon_y = \varepsilon_z = 0$. They can be combined to form pure twists about each of the three axes and they do not lock because none of the missing quadratic terms $\left(x^2, y^2, z^2 \right)$ are needed to describe them.

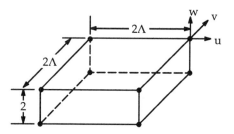

Figure 6.14   Eight-Node Rectangular Brick.

The cubic states do lock. For example, u = xyz represents bending in the xz plane that varies in the y direction. To complete it we need to add the displacement components $v = x^2 z / 2$ and $w = - x^2 y / 2$ which then give $\varepsilon_x = yz$ and $\gamma_{xy} = 2xz$ as the only nonzero strains. Within the element, v and w are represented by aliases. The complete displacement and strain fields within the element are

$$u^a = xyz \;,\; v^a = \frac{z\Lambda^2}{2} \;,\; w^a = -\frac{y\Lambda^2}{2}$$

(6:44)

$$\varepsilon_x = yz \;,\; \varepsilon_y = 0 \;,\; \varepsilon_z = 0 \;,\; \gamma_{xy} = xz \;,\; \gamma_{xz} = xy \;,\; \gamma_{yz} = 0$$

It is seen that, since the x and y dimensions are larger by a factor Λ than the z dimension, the spurious transverse shear strain γ_{xz} has strain energy that is larger by a factor proportional to Λ^2 than the intended strain energy. Shear locking can be avoided by evaluating γ_{xz} in the plane $x = 0$. Note, in addition, that γ_{xy} is too small by a factor of two and that it too would be zero if evaluated at $x = 0$. The error in strain energy due to the inaccuracy of this term hardly matters in comparison with the potentially large strain energy due to locking of γ_{xz}.

Dilatation locking is also a problem for $u = xyz$. To avoid the locking we could add a term $\varepsilon_y = -yz$ which would require the non-available displacement component $v = -y^2 z / 2$. One remedy would be to add this term internally as a nonconforming mode.

In summary the eight-node rectangular brick has the same locking problems as the four-node rectangle plus additional locking difficulties associated with the trilinear displacement states. Selective underintegration of shear terms appears to be a promising remedy for shear locking. The removal of dilatation locking requires a more advanced remedy, such as nonconforming internal modes.

6.4.3 Locking of the Twenty-Node Brick

We come finally to the twenty-node rectangular brick. Its basis functions, which are recorded in Table 3.2(b), include all cubic terms (except x^3, y^3, and z^3) and the quartic functions x^2yz, xy^2z, and xyz^2.

Improper interpolation of the missing cubic terms causes planar locking modes identical to those of the QUAD8 element. The quartic basis functions have locking modes caused by improper interpolation of terms like x^3z and x^2y^2.

The quartic locking modes of the twenty-node brick, HEXA20, are recorded in Table 6.4 along with the locking modes of the other rectangular elements we have studied. Shear locking modes, with $\upsilon = 0$, and dilatation locking modes, with $\upsilon = .5$, are tabulated separately.

The locking modes of the solid elements also include the locking modes of the planar elements with obvious permutations. The degree of multiplicity, i.e., the number of possible permutations of position and displacement coordinates is listed for the locking modes of each element. The relative sizes of the strain components determine which component will cause locking. For this purpose, it is assumed that $\Lambda > 1$ so that x is larger than y for 2-D elements and both x and y are larger than z for 3-D elements.

Examination of Table 6.4 shows that reduced-order integration eliminates locking for nearly all cases. Thus evaluation of shear strain and dilatation at $x = y = z = 0$ eliminates locking for the QUAD4 and HEXA8, and evaluation of strains at two Gauss quadrature points in each direction eliminates all locking modes for QUAD8 and all but one of the locking modes for HEXA20. The exception is a shear locking mode of the HEXA20 in which $w = x^2 y^2 / 2$. This term has the alias $w^a = \Lambda^2 \left(x^2 + y^2 - \Lambda^2 \right) / 2$ for the assumed element geometry. The spurious shear strains for this mode only vanish at $x^2 = y^2 = \Lambda^2$. The 27-node isoparametric brick, HEXA27, does not have this locking mode because $x^2 y^2$ is included in its basis (see Table 3.2(c)).

Most of the locking modes of the solid brick elements are associated with bending deformations in which the u and v components of displacement vary linearly in the thin dimension, z. (The dilatation locking mode $u = xyz^2$, $w = - yz^3 / 3$ is an exception.) Thus a plate bending problem (Figure 6.15) makes a good diagnostic tool to study the locking of solid brick elements. Note that the Λ value for this problem is 100 and that Poisson's ratio is 0.3 so that shear locking predominates. Note also that symmetry permits us to model only one-quarter of the plate.

Table 6.4

Locking Modes of Rectangular Elements

(a) Shear Locking Modes, $v = 0$

ELEMENT	u	v	w	MULTI-PLICITY	ε_x	ε_y	ε_z	γ_{xy}	γ_{xz}	γ_{yz}
QUAD4	xy	$-\frac{1}{2}\underline{x^2}$	—	2	y	0	—	x	—	—
QUAD8	x^2y	$-\frac{1}{3}\underline{x^3}$	—	2	$2xy$	0	—	$x^2 - \frac{1}{3}\Lambda^2$	—	—
HEXA8	xyz	$\frac{1}{2}x^2z$	$-\frac{1}{2}\underline{x^2y}$	3	yz	0	0	xz	xy	0
HEXA20	x^2yz	$\frac{1}{3}x^3z$	$-\frac{1}{3}\underline{x^3y}$	3	$2xyz$	0	0	$z\left(x^2+\frac{1}{3}\Lambda^2\right)$	$y\left(x^2-\frac{1}{3}\Lambda^2\right)$	0
HEXA20	xy^2z	x^2yz	$-\frac{1}{2}\underline{x^2y^2}$	3	y^2z	x^2z	0	0	$x\left(y^2-\Lambda^2\right)$	$y\left(x^2-\Lambda^2\right)$

— Incorrectly interpolated displacement term

☐ Locking strain component

Table 6.4
Locking Modes of Rectangular Elements (continued)

(b) Dilatation Locking Modes, $\upsilon = 0.5$

ELEMENT	u	v	w	MULTI-PLICITY	ε_x	ε_y	ε_z	e
QUAD4	xy	$-\frac{1}{2}y^2$	—	2	y	0	—	y
QUAD8	xy^2	$-\frac{1}{3}y^3$	—	2	y^2	$-\frac{1}{3}\Lambda^2$	—	$y^2 - \frac{1}{3}\Lambda^2$
HEXA8	xyz	$-\frac{1}{2}y^2z$	0	3	yz	0	0	yz
HEXA20	xy^2z	$-\frac{1}{3}y^3z$	0	6	y^2z	$-\frac{1}{3}z\Lambda^2$	0	$z\left(y^2 - \frac{1}{3}\Lambda^2\right)$

— Incorrectly interpolated displacement term

□ Locking strain component

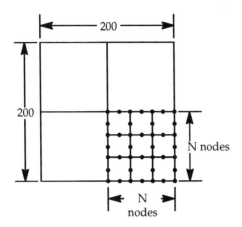

Figure 6.15   Square Plate.   Thickness = 1.0, Poisson's Ratio = 0.3.

Results are recorded in Table 6.5 for three elements and two different combinations of loading and boundary conditions (uniform pressure with simply supported edges, and a concentrated center load with clamped edges). The three elements are: the MSC/NASTRAN HEXA(8) element, which uses selective underintegration for shear strains and nonconforming modes to eliminate dilatation locking; a standard twenty-node solid isoparametric element, HEX20, with full (3 x 3 x 3) Gauss integration; and a twenty-node solid isoparametric element HEX20R with reduced-order (2 x 2 x 2) Gauss integration.   No results were obtained for an eight-node isoparametric with full integration.   Its results are so poor that it has practically vanished from the inventory of commercial finite element programs.

In order to be fair with respect to computational effort, results are compared for equal numbers of nodes, N, along an edge of the quarter plate model rather than for equal numbers of elements.   On this basis the heavily modified HEXA(8) is the clear winner, while the standard isoparametric, HEX20, is the clear loser.   The latter element converges very slowly, indicating a severe case of locking.   The response of the twenty-node isoparametric element with reduced integration, HEX20R, is more subtle.   With only a single element (N = 3), the displacements are larger than the displacements with more

Table 6.5

Normalized Lateral Deflection
at Center of Square Plate (Figure 6.15)
Exact Solution = 1.000

(a) Uniform Load, Simply-Supported Edges

NUMBER OF NODES, N, PER EDGE OF MODEL	HEXA(8)	HEX20	HEX20R
3	.989	.023	1.073
5	.998	.738	.993
7	.999	.967	1.011
9	1.000	.991	1.008

(b) Concentrated Load, Clamped Edges

NUMBER OF NODES, N, PER EDGE OF MODEL	HEXA(8)	HEX20	HEX20R
3	.885	.002	.983
5	.972	.072	.433
7	.988	.552	.813
9	.994	.821	.942

elements, which shows the softening effect of underintegration. For the case
with simply supported edges, the HEX20R converges rapidly even if not
monotonically. With clamped edges, on the other hand, the presence of
locking is indicated by the low value of central displacement for $N = 5$.

A glance at Table 6.4 shows that the only locking mode of the twenty-node brick not removed by $2 \times 2 \times 2$ Gauss quadrature is the mode in which lateral deflection, w, is proportional to $x^2 y^2$. This then must be the mode which causes locking of the HEX20R for the case with clamped boundary conditions. To see this, consider that w varies as $\left(x - x_0\right)^2$ near a clamped edge at $x = x_0$ and as $\left(y - y_0\right)^2$ near a clamped edge at $y = y_0$. Thus, near a clamped corner, w must vary as $\left(x - x_0\right)^2 \left(y - y_0\right)^2$ plus higher-order terms. We conclude that, in Figure 6.15, the twenty-node element adjacent to corner must lock because $x^2 y^2$ is an essential part of the displacement field which the element is called upon to represent. As more elements are added, the corner element continues to lock but its effect on the total strain energy declines.

6.4.4 Locking of Higher-Order Elements

Locking is a serious disorder for the lowest order, $p = 1$, elements and also, as we have just seen, for $p = 2$ elements. The question then arises as to whether locking remains a serious concern as p is increased. The answer is that it does not except in increasingly specialized circumstances. Babuška[7] has provided evidence that increasing p is a more "rugged" solution to locking problems than increasing the number of lower-order elements. The examples developed in this chapter also tend to bear this out, as long as special "fixes" for locking are not used.

The specialized circumstances where locking might be serious for higher-order elements are ones in which the correct displacement fields include important terms not present in the element's basis. Consider the example of a slender beam with displacements $u = x^p y$, $v = - x^{p+1} / (p + 1)$. If the beam is represented by a single, standard isoparametric element of order p, the element will lock because x^{p+1} is not included in the element's basis. A model for the beam made up of a number of lower-order elements, say $p = 1$, may not be accurate but it will not lock if the locking modes have been corrected by reduced-order integration or otherwise.

It is not current practice to pay much attention to the locking modes of hierarchical elements because it is reasoned that the locking of higher-order displacement states can be removed simply by increasing p. Increasing p is not, however, without cost. The remedy of reduced-order integration, at least, should be considered, particularly since it reduces computational cost and has other advantages, as we shall see in the next chapter.

6.5 LOCKING IN OTHER DISCIPLINES

The examples of locking considered so far have all been taken from the theory of elasticity. They have been further characterized by extreme values of parameters and, with rare exceptions, by bending deformations. We will comment here on the extent to which locking can occur in heat conduction and in magnetostatics, the two other disciplines given nominal treatment in this book.

Interpolation failure occurs with certainty in heat conduction and in magnetostatics, as indeed it must in all applications of finite elements. The real issue is whether parameters exist which can intensify interpolation failure to an extent which justifies the use of "locking" as a descriptive term.

In the case of heat conduction it seems improbable that such parameters exist. In elasticity the presence of locking seems to require leverage through the coupling of motions in different directions. Since heat conduction uses only a single scalar potential function (and a single, second-order partial differential equation) it is not clear where the analog of such leverage might come from. Indeed, the concept of amplification through leverage seems to contradict the spirit, if not the letter, of the second law of thermodynamics.

Locking occurs in magnetostatics, although not to the extent that it occurs in elasticity. The three components of vector potential provide the possibility of leverage but since the magnetic induction has only three components (the components are the curl of the vector potential) as opposed to six components of strain for elasticity, the possible number of interactions is fewer. The author is aware of two locking problems in magnetostatics which bear a resemblance to those treated in this chapter. The first problem involves HEXA8 elements arranged in a *nonplanar* loop of magnetic material. [17] (The elements work perfectly well when simulating a planar magnetic loop.) Analysis of the locking problem[18] requires a minimum of three elements arranged as shown in Figure 6.16. Reduced-order (single point) integration relieves the problem at the expense of potential difficulty with singular modes. The use of twenty-node brick elements also relieves the problem.

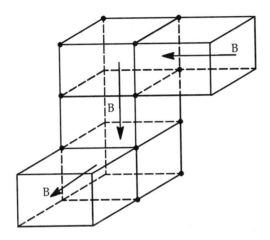

Figure 6.16 Locking Problem in Nonplanar Magnetic Loop.

Another type of locking problem occurs in electromagnetism in connection with the divergence of the vector potential $\nabla \bullet \tilde{A}$. This term is analogous to the dilatation $e = \nabla \bullet \tilde{u}$ in elasticity with the difference that $\nabla \bullet \tilde{A}$ carries no energy. It is, however, usually necessary to assign some "stiffness" to $\nabla \bullet \tilde{A}$ to avoid singular modes, and locking may result if the level of stiffness is too high. Such locking is analogous to dilatation locking. This matter will be taken up in Section 7.5.

6.6 SHAPE SENSITIVITY

So far we have analyzed the locking of rectangular two- and three-dimensional elements and of a pair of three-node triangles which form a rectangle. The locking parameter in these cases was either the aspect ratio of the rectangle or a function of Poisson's ratio. In several cases the analytical results suggested easy fixes to locking problems, such as the evaluation of strain components at particular points. Experience shows, however, that when such fixes are applied to isoparametric elements, the locking phenomenon will often return when the element's shape is altered from rectangular. This is illustrated in Figure 1.4 where it is seen that the MSC/NASTRAN QUAD4 element behaves fairly well for rectangular shape but locks badly for parallelogram or trapezoidal shapes. Figure 3.20 illustrates another example in which the displacement of the edge node of a six-node triangle causes severe stiffening. Such examples lead to the conclusion that the locking phenomenon can depend on more element shape parameters than just the element's aspect ratio, Λ.

As a preliminary to the study of the shape sensitivity of locking, we need to identify the parameters which determine an element's shape. The *number* of such parameters can be easily determined from the number of nodes, N_n, connected to the element. Since six position coordinates can define the position and orientation of an element in three-dimensional space and since one dimension can define its size, it follows that the number of parameters needed to define an element's shape is $N_s = 3N_n - 7$. For a four-node quadrilateral $N_s = 5$, and for a twenty-node brick $N_s = 53$. While it may seem hopeless to attempt to characterize all of the shape parameters of a twenty-node brick,[*] it is quite useful to do so for the four-node quadrilateral. We begin by taking the *standard shape* of the quadrilateral to be a square. We can then see, in Figure 6.17, five independent ways in which the shape can be altered. These include: (1) elongation by increasing the aspect ratio, Λ; (2) skewing the corner angles by an amount, δ; (3) and (4) tapering the element in either of two directions; and (5) warping the element out of its plane. All of these distortion modes are important for element performance in general and for locking in particular. They are easy to identify if the amount of distortion is small. It is also possible to decide how much distortion of each type is present for a general quadrilateral.[19] This is unnecessary for our purposes because we intend to examine the different types of distortion independently or in combination with aspect ratio only.

---

[*]But it has been done (see Reference 6.10).

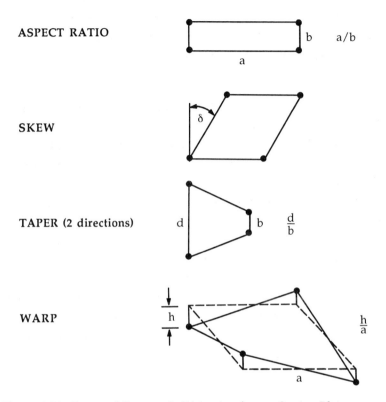

ASPECT RATIO

SKEW

TAPER (2 directions)

WARP

Figure 6.17 Types of Geometric Distortion from a Square Plate.

If we go on to consider the possible distortion modes of an eight-node quadrilateral we must add twelve distortion modes corresponding to three components of displacement at each of the four edge nodes. Such distortions can be classified as either in or out of the mean plane, determined by the locations of corner nodes.[*] The out-of-plane offset of edge nodes converts flat elements into curved shell elements, a subject which will be treated in Chapter 10. In-plane offsets in the directions of the edges cause severe distortions of internal geometry which usually serve no practical purpose. We

---

[*]For warped quadrilaterals, the mean plane may be taken parallel to and midway between the diagonals (see Figure 6.17).

have examined such a case in Section 6.1. In-plane offsets of edge nodes in directions normal to the edges are frequently needed to conform to the exterior shape of a field of elements. All of these movements cause interpolation failure for quadratic displacement fields and are, therefore, potential locking parameters.

Shape sensitivity has been an important consideration since the earliest days of finite element analysis. The need to certify element performance for nonstandard shapes led to the invention of the constant strain patch test.[20] The effect of shape sensitivity on locking has not been recognized nearly so well. Through the years designers have produced elements which avoid locking for types of shape distortion beyond simple elongation but rarely have they provided an adequate description of the locking problem which was avoided. Test results are, for the most part, relied upon to demonstrate shape sensitivity. There is a danger, as the author can testify from his own experience, that the amount of shape distortion in a test may be too small to clearly identify a locking problem. For this reason, if for no other, we should try to determine the parameters which intensify the shape sensitivity of locking. This can best be done, and perhaps can only be done, by analysis.* Analysis will, in addition, provide insights to possible cures for locking.

6.6.1 Locking of the Four-Node Parallelogram

The parallelogram is a particularly important shape in finite element analysis because, as noted in Figure 5.5, it is the shape approached when an arbitrary quadrilateral is progressively refined by uniform subdivision of its edges. Consider the parallelogram element shown in Figure 6.18. Its Cartesian position coordinates, measured from the center of the element, are related to its parametric coordinates by

$$x = \Lambda\xi + \eta \tan\delta \quad , \quad y = \eta \tag{6:45}$$

---

*Barlow (Reference 6.10) has used analysis to produce a comprehensive classification of errors due to shape distortions for eight-node membranes and twenty-node bricks.

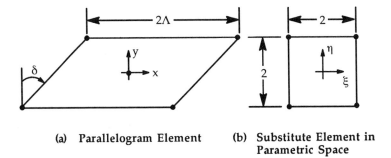

(a) **Parallelogram Element**   (b) **Substitute Element in Parametric Space**

Figure 6.18 Four-Node Parallelogram.

or

$$\xi = \frac{1}{\Lambda}\left(x - y \tan \delta\right) \quad, \quad \eta = y \tag{6:46}$$

The inverse of the Jacobian matrix, which is needed to compute strains, is

$$[J]^{-1} = \begin{bmatrix} \xi_{,x} & \eta_{,x} \\ \xi_{,y} & \eta_{,y} \end{bmatrix} = \begin{bmatrix} \frac{1}{\Lambda} & 0 \\ \frac{-\tan\delta}{\Lambda} & 1 \end{bmatrix} \tag{6:47}$$

As in previous examples, we consider the displacement field for in-plane bending, $u = xy$, $v = -x^2 / 2$. Substitution of the expressions for x and y from Equation 6:45 gives, for the displacement field and its alias within the element,

$$u = \Lambda\xi\eta + \eta^2 \tan\delta \Rightarrow \Lambda\xi\eta + \tan\delta = u^a$$

$$v = -\frac{1}{2}\left(\Lambda^2\xi^2 + 2\Lambda\xi\eta\tan\delta + \eta^2\tan^2\delta\right) \Rightarrow -\frac{1}{2}\left(\Lambda^2 + \tan^2\delta + 2\Lambda\xi\eta\tan\delta\right) = v^a$$

$$\tag{6:48}$$

The strains in the element, obtained by application of the chain rule (see Equation 6:9), are

$$\varepsilon_x = \eta \quad, \quad \varepsilon_y = \tan\delta\left(-\Lambda\xi + \eta\tan\delta\right) \quad, \quad \gamma_{xy} = \Lambda\xi \tag{6:49}$$

Since the correct strains are $\varepsilon_x = y = \eta$, $\varepsilon_y = 0$, $\gamma_{xy} = 0$, it is seen that both ε_y and γ_{xy} are in error and that locking will result if Λ is large. The errors could be eliminated by evaluating ε_y and γ_{xy} at $\xi = \eta = 0$, but since isotropy would require that ε_x also be measured at $\xi = \eta = 0$, the element would have no stiffness at all for in-plane bending.

Another solution to the locking problem is to use skewed strain components. If skewed axes $\bar{x}$, $\bar{y}$ are taken parallel to the edges of the element, it can be shown that the skewed components of strain are:

$$
\left\{ \begin{array}{c} \varepsilon_{\bar{x}} \\ \varepsilon_{\bar{y}} \\ \gamma_{\overline{xy}} \end{array} \right\} = \left[\begin{array}{c|c|c} 1 & 0 & 0 \\ \hline \sin^2 \delta & \cos^2 \delta & \sin \delta \cos \delta \\ \hline 2 \sin \delta & 0 & \cos \delta \end{array} \right] \left\{ \begin{array}{c} \varepsilon_x \\ \varepsilon_y \\ \gamma_{xy} \end{array} \right\}
\tag{6:50}
$$

(see Section 6.A at the end of this chapter).

Substitution of the Cartesian strain components from Equation 6:49 then gives, for the errors in the skewed strain components,

$$
E\left(\varepsilon_{\bar{y}} \right) = \varepsilon_{\bar{y}} - \varepsilon_{\bar{y}}^c = \eta \sin^2 \delta
$$

$$
E\left(\varepsilon_{\overline{xy}} \right) = \gamma_{\overline{xy}} - \gamma_{\overline{xy}}^c = \Lambda \xi \cos \delta
\tag{6:51}
$$

Here we note that zero error is achieved by measuring $\varepsilon_{\bar{y}}$ at $\eta = 0$ and $\gamma_{\overline{xy}}$ at $\xi = 0$. By implication we should also measure $\varepsilon_{\bar{x}}$ at $\xi = 0$. These restrictions are acceptable because measurement of $\varepsilon_{\bar{x}}$ at two points along the line $\xi = 0$ will provide the desired bending stiffness.

The use of skewed coordinates for parallelogram elements was introduced in 1971 for assumed stress hybrid elements,[21] but apparently not until 1985 for assumed displacement elements.[22] The latter design extended the concept to general quadrilateral shapes by computing strains in coordinate systems locally parallel to lines of constant ξ and η (the so-called *natural covariant* strain components). The MSC/NASTRAN QUAD4 element does not use skewed coordinates. As shown in Figure 1.4, it locks severely for parallelogram shapes. The term *shear locking* should not be applied to this example because ε_y is also spurious and has the same order of magnitude as the shear strain. The author has introduced the term *parallelogram locking*[23] for this case.

6.6.2 Locking of the Four-Node Trapezoid

Consider the four-node element of trapezoidal shape shown in Figure 6.19.

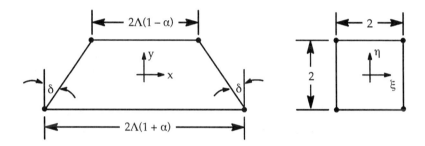

Figure 6.19 Four-Node Isosceles Trapezoid, $\alpha\Lambda = \tan \delta$.

Its Cartesian position coordinates, measured from the center of the element, are related to its parametric coordinates by

$$x = \Lambda\xi(1 - \alpha\eta) \qquad y = \eta \qquad (6:52)$$

or

$$\xi = \frac{x}{\Lambda(1-\alpha y)} \qquad \eta = y \qquad (6:53)$$

The inverse Jacobian matrix is

$$[J]^{-1} = \begin{bmatrix} \xi_{,x} & \eta_{,x} \\ \xi_{,y} & \eta_{,y} \end{bmatrix} = \left[\begin{array}{c|c} \dfrac{1}{\Lambda(1-\alpha\eta)} & 0 \\ \hline \dfrac{\alpha\xi}{1-\alpha\eta} & 1 \end{array} \right] \qquad (6:54)$$

Note that, in contrast to the case of a parallelogram shape, the terms in $[J]^{-1}$ are not constants. Let us assume, as before, the quadratic displacement field $u = xy$, $v = -x^2 / 2$. The components of this field and its aliases are, when expressed in terms of parametric coordinates,

$$u = \Lambda\left(\xi\eta - \alpha\xi\eta^2\right) \Rightarrow \Lambda\left(\xi\eta - \alpha\xi\right) = u^a$$

$$v = -\tfrac{1}{2}\Lambda^2\left(\xi^2 - 2\alpha\xi^2\eta + \alpha^2\xi^2\eta^2\right) \Rightarrow -\tfrac{1}{2}\Lambda^2\left(1 - 2\alpha\eta + \alpha^2\right) = v^a$$

$$(6\!:\!55)$$

The corresponding strains in the element, obtained with the aid of Equation 6:9, are

$$\varepsilon_x = \frac{\eta - \alpha}{1 - \alpha\eta} \quad , \quad \varepsilon_y = \alpha\Lambda^2 \quad , \quad \gamma_{xy} = \Lambda\xi\left(1 + \frac{\alpha(\eta - \alpha)}{1 - \alpha\eta}\right) \qquad (6\!:\!56)$$

Since the correct strains are $\varepsilon_x = \eta$, $\varepsilon_y = 0$, $\gamma_{xy} = 0$, we note that all strain components are in error for $\alpha \neq 0$. The error in ε_x is, however, seen to be small for α small. The error in the shear term can, as in the case of rectangular elements, be eliminated by evaluating γ_{xy} at $\xi = \eta = 0$. The error in ε_y is larger than the error in γ_{xy} for $\alpha\Lambda = \tan\delta > 1$. Furthermore, since the error in ε_y is constant over the element's surface, it cannot be eliminated by a judicious selection of strain evaluation points. Indeed, since it cannot be distinguished from constant strain due to $v = y$, there does not appear to be any practical way in which it can be eliminated. We will discover in Section 8.4.2 that the error in ε_y can be eliminated, but only at the cost of violating the constant strain patch test.

This type of locking, which we might call *trapezoidal locking*,[23] is exhibited in Figure 1.4 by the MSC/NASTRAN QUAD4 element. Its locking parameter is $\alpha\Lambda^2 = \Lambda\tan\delta$. The parameter's value is 5.0 in the example shown in Figure 1.4.

6.6.3 Locking of the Eight-Node Trapezoid

Consider an eight-node element with the same trapezoidal shape as that of the four-node element shown in Figure 6.19 but with four centered edge nodes. Let the displacement state again be $u = xy, v = -x^2 / 2$. This displacement field does not lock for rectangular shapes because $\xi\eta$ and ξ^2 are in the element's displacement basis. For the trapezoidal case the displacement field is expressed in terms of parametric coordinates by Equation 6:55 where it is seen that the only term not in the displacement basis is the $\xi^2\eta^2$ term in v. Since the alias of $\xi^2\eta^2$ is $\xi^2 + \eta^2 - 1$ for the eight-node element, the resulting errors in the displacement field are

$$E(u) = 0 \quad , \quad E(v) = v^a - v = -\frac{1}{2} \Lambda^2 \alpha^2 \left(\xi^2 + \eta^2 - 1 - \xi^2\eta^2 \right) \quad (6:57)$$

The corresponding errors in strain are

$$E\left(\varepsilon_x\right) = 0 \quad , \quad E\left(\varepsilon_y\right) = -\Lambda^2\alpha^2 \left(\eta\left(1 - \xi^2\right) + \frac{\alpha\xi^2\left(1-\eta^2\right)}{1-\alpha\eta} \right) \quad ,$$

$$(6:58)$$

$$E\left(\gamma_{xy}\right) = \frac{-\Lambda\alpha^2\xi\left(1-\eta^2\right)}{1-\alpha\eta}$$

The dominant error for large aspect ratio, Λ, is $E\left(\varepsilon_y\right)$. Its locking parameter, for small α, is $\Lambda^2\alpha^2 = \tan^2\delta$. We note that the errors cannot be eliminated by evaluating strains at 2 x 2 Gauss quadrature points, where $\xi^2 = \eta^2 = \frac{1}{3}$. The errors can, however, be eliminated by adding a ninth central node which provides $\xi^2\eta^2$ as a basis function. This fact has already been noted in Section 3.5.

Figure 6.20 illustrates the shape sensitivity of the HEX20 and HEX20R elements which, it will be recalled, are twenty-node solid elements with 3 x 3 x 3 and 2 x 2 x 2 Gauss integration respectively. The cantilever beam example is exactly the same as that shown in Figure 1.4 and the error in tip deflection, where the load is applied, is exactly equal to the error in strain energy. The solid elements behave like membrane elements in this example. It is seen that the

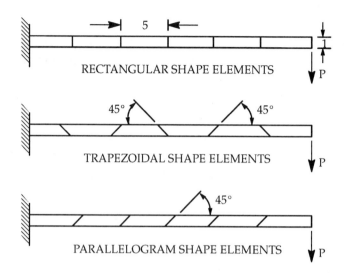

ELEMENT	RECTANGULAR SHAPE	PARALLELOGRAM SHAPE	TRAPEZOIDAL SHAPE
	←——————— TIP DEFLECTION ———————→		
HEX20	.970	.967	.886
HEX20R	.984	.994	.964
Exact	1.000	1.000	1.000

Figure 6.20   Solutions of End-Loaded Cantilever Beam Problem.

errors for rectangular and parallelogram shapes are quite small, as might be expected when locking does not occur.[*] The main locking parameter for the trapezoidal case is $\tan^2 \delta = 1$, while the auxiliary locking parameter $\alpha = (\tan \delta) / \Lambda = 0.2$. Thus the locking effect should be weaker than in the case of the four-node trapezoid where the locking parameter $\Lambda \tan \delta = 5.0$.

---

[*]The complex three-dimensional stress field at the clamped end causes the observed errors.

This conclusion is confirmed by the results which give tip deflections of .071 for the QUAD4 element and .886 for the HEX20 element.

The locking effect for the HEX20R element appears to be surprisingly small, only two percent larger than for the case of rectangular elements. There is, in fact, a mitigating effect which explains the small size of the error. We observe from Equation 6:58 that the dominant term in $E\left(\varepsilon_y\right)$ for small α is $-\Lambda^2\alpha^2\eta\left(1-\xi^2\right)$, which is equal to $-2\Lambda^2\alpha^2\eta / 3$ at $2 \times 2 \times 2$ Gauss integration points. This functional form of ε_y can be mimicked by vertical motion at edge nodes as shown in Figure 6.21. Note that the motion of edge nodes is uncontrolled in the example of Figure 6.20 and may indeed go unnoticed in the solution. In addition, the factor $\Lambda^2\alpha^2$ is the same for (most) pairs of adjacent elements in the example, so that vertical motion of a particular edge node will have the same beneficial effect on adjacent elements.

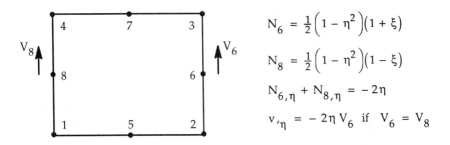

$$N_6 = \tfrac{1}{2}\left(1 - \eta^2\right)\left(1 + \xi\right)$$

$$N_8 = \tfrac{1}{2}\left(1 - \eta^2\right)\left(1 - \xi\right)$$

$$N_{6,\eta} + N_{8,\eta} = -2\eta$$

$$v_{,\eta} = -2\eta\, V_6 \quad \text{if} \quad V_6 = V_8$$

Figure 6.21   Strain Field Due to Motion of Edge Nodes.

The mitigating effect of edge node motion is present, to a lesser degree, with $3 \times 3 \times 3$ integration. In general, locking effects in fields of finite elements may be less than predicted by the rigid imposition of displacements on the nodes of an individual element. The actual displacements will, after all, be those that minimize the strain energy.

6.6.4   Locking of a Six-Node Triangle with an Offset Edge Node

We have encountered, in Figure 3.20, an example where the location of an edge node away from its normal centered position appeared to cause shear locking. We will attempt here to explain the observed result.   To simplify the calculation we will consider the right triangle shown in Figure 6.22.

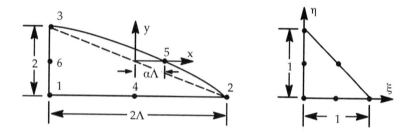

Figure 6.22   A Six-Node Triangle with Offset Edge Node.

The usual bending displacement field $u = xy$, $v = -x^2/2$ is assumed.   The first task is to determine the parametric form of the position coordinates, x and y.   A convenient way to do this is to use the factored form of the element's shape functions.   For example,

$$y = \lfloor X \rfloor \{b_j\} \qquad (6{:}59)$$

where the coefficients of the basis functions in $\lfloor X \rfloor$ are given by

$$\{b_j\} = \left[A_{ji}\right]\{y_i\} \qquad (6{:}60)$$

$\lfloor X \rfloor$ and $\left[A_{ji}\right]$ are recorded in Table 4.6(c) for the six-node triangle.

The nodal values of x_i and y_i are listed below.

NODE	1	2	3	4	5	6
x_i	$-\Lambda$	Λ	$-\Lambda$	0	$\alpha\Lambda$	$-\Lambda$
y_i	-1	-1	1	-1	0	0

The results are

$$x = \Lambda\left(-1 + 2\xi + 4\alpha\xi\eta\right) \qquad y = 2\eta - 1 \qquad (6:61)$$

The inverse Jacobian matrix is

$$[J]^{-1} = \begin{bmatrix} \xi_{,x} & \eta_{,x} \\ \xi_{,y} & \eta_{,y} \end{bmatrix} = \frac{1}{|J|}\left[\begin{array}{c|c} 2 & 0 \\ \hline -4\alpha\Lambda\xi & 2\Lambda(1 + 2\alpha\eta) \end{array}\right] \qquad (6:62)$$

where the Jacobian determinant $|J| = 4\Lambda(1 + 2\alpha\eta)$.

Parametric expression of the displacement field, $u = xy$ and $v = -x^2/2$ includes the terms $\xi^2\eta$, $\xi\eta^2$, and $\xi^2\eta^2$, which are not in the element's basis vector. The aliases for these terms are, from Table 6.1,

$$\xi^2\eta \Rightarrow \xi\eta/2 \quad , \quad \xi\eta^2 \Rightarrow \xi\eta/2 \quad , \quad \xi^2\eta^2 \Rightarrow \xi\eta/4 \qquad (6:63)$$

With these changes, the strains are computed in the usual fashion. Since the resulting formulas are rather long, we omit all terms proportional to α^2 and higher powers of α. With this simplification, the errors are

$$E\left(\varepsilon_x\right) = 2\alpha\eta(1 - 2\eta) \quad , \quad E\left(\varepsilon_y\right) = 2\alpha\Lambda^2\xi(2\xi - 1) \quad , \quad E\left(\gamma_{xy}\right) = 2\alpha\Lambda(\xi - \eta)$$

$$(6:64)$$

For large Λ the dominant error term is $E\left(\varepsilon_y\right)$. The value of its locking parameter, $\alpha\Lambda^2$, is, for the example of Figure 3.20, equal to 5.0. Thus we expect that the locking effect will be severe, on the order of that observed for the trapezoidal locking of the QUAD4 element in Figure 1.4. The tip deflection observed in Figure 3.20 for the TRIA6 is .391 compared to .071 for the QUAD4 in Figure 1.4. While the locking of the TRIA6 is obviously severe, the difference in these results indicates the mitigating influence of some higher-order deformation mode.

On a practical level, we observe once again the deleterious effect of edge node offset. There does not appear to be any easy way to avoid its consequences for the six-node triangle. The root cause of the problem is parametric interpolation. If it were replaced by metric interpolation the element would not lock because it could correctly interpolate any quadratic displacement field regardless of element shape. The down side would be nonconforming displacements and a resulting failure to pass the constant strain patch test with offset edge nodes. This matter will be continued in Section 8.3.

We have examined the locking problem in this chapter by the straightforward approach of computing the errors in strains for various element shapes. The analytical results have indicated easy fixes to locking problems in some cases and apparently severe limitations in other cases. As is true generally of interpolation failure, solution of the locking problem seems to become more difficult as element shape becomes more general. In the case of the four-node quadrilateral, for example, selective integration was the indicated cure for rectangular shape; skewed strain components were additionally needed for parallelogram shapes (where the transformation between metric and parametric coordinates is still linear); and no solution to locking was apparent for trapezoidal shapes. The same trend was evident for the eight-node quadrilateral, where it was shown that 2 x 2 integration could solve locking problems for rectangular shapes but not for trapezoidal shapes.

Given the severity of the locking problem, these results and similar conclusions for other cases define the main agenda pursued by element designers over the past twenty-five years. We will discover that some of the easy fixes are not as easy as they look. At a minimum they involve variational crimes (inexact integration and nonconforming displacement states) which curl the hair of mathematicians. At the practical level they can result in patch test failure and spurious mechanisms. Finding fixes for such side effects, if possible, and deciding which are the least of alternative evils, if not possible, is the stuff of finite element design.

6.A SKEWED STRAIN COMPONENTS

A transformation between rectangular and skewed Cartesian coordinates is shown in Figure 6.23(a). The position coordinates are computed according to the parallelogram rule, just like force components. Thus,

$$x = \bar{x} + \bar{y} \sin \delta \quad , \quad y = \bar{y} \cos \delta \tag{6:65}$$

or

$$\bar{x} = x - y \tan \delta \quad , \quad \bar{y} = y \sec \delta \tag{6:66}$$

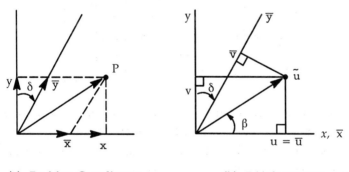

| (a) Position Coordinates | (b) Displacements |

Figure 6.23 Transformation Between Rectangular and Skewed Coordinates.

The displacement components follow a projection rule, i.e., the component of displacement in any direction equals the projection of the total displacement vector, $\tilde{u}$, on that direction. The parallelogram rule and the projection rule are the same for rectangular coordinates but not for skewed coordinates. From Figure 6.23(b),

$$u = \bar{u} = |\tilde{u}| \cos \beta \qquad\qquad v = |\tilde{u}| \sin \beta$$

$$\bar{v} = |\tilde{u}| \cos\left(\frac{\Pi}{2} - \beta - \delta\right) = |\tilde{u}|\left(\sin \beta \cos \delta + \cos \beta \sin \delta\right) = u \sin \delta + v \cos \delta$$

$$(6\!:\!67)$$

Thus we can also say that $\bar{u}$ and $\bar{v}$ are found by summing the projections of the rectangular displacement components in the $\bar{x}$ and $\bar{y}$ directions respectively.

The skewed components of strain are defined by analogy with the rectangular components. Thus

$$\varepsilon_{\bar{x}} = \bar{u}_{,\bar{x}} \quad , \quad \varepsilon_{\bar{y}} = \bar{v}_{,\bar{y}} \quad , \quad \gamma_{\overline{xy}} = \bar{u}_{,\bar{y}} + \bar{v}_{,\bar{x}} \qquad (6\!:\!68)$$

The results of these operations, expressed in matrix form, are

$$\left\{\begin{matrix} \varepsilon_{\bar{x}} \\[1.5ex] \varepsilon_{\bar{y}} \\[1.5ex] \gamma_{\overline{xy}} \end{matrix}\right\} = \left[\begin{array}{c:c:c} 1 & 0 & 0 \\ \hdashline \sin^2 \delta & \cos^2 \delta & \sin \delta \cos \delta \\ \hdashline 2 \sin \delta & 0 & \cos \delta \end{array}\right] \left\{\begin{matrix} \varepsilon_x \\[1.5ex] \varepsilon_y \\[1.5ex] \gamma_{xy} \end{matrix}\right\} \qquad (6\!:\!69)$$

Skewed components of stress are defined as the skewed components of force acting on the unit rhombus shown in Figure 6.24. The inner product of the skewed stress components with the skewed strain components produces the strain energy density multiplied by $\cos \delta$, the area of the unit rhombus.

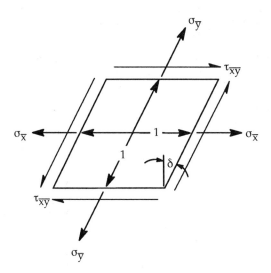

Figure 6.24 Skewed Components of Stress.

REFERENCES

6.1   B. Fraeijs de Veubeke, "Displacement and Equilibrium Models in the Finite Element Method," Chap. 9, *Stress Analysis*, O. C. Zienkiewicz and G. Holister (Eds.), Wiley, 1965.

6.2   B. M. Irons, "Engineering Application of Numerical Integration in Stiffness Methods," *J. AIAA*, 14, pp 2035-7, 1966.

6.3   W. P. Doherty, E. L. Wilson, and R. L. Taylor, "Stress Analysis of Axisymmetric Solids Using Higher Order Quadrilateral Finite Elements," U. of Calif. Berkeley, Struct. Eng. Lab. Report SESM 69-3, 1969.

6.4   J. Robinson and S. Blackham, "An Evaluation of Lower Order Membranes as Contained in the MSC/NASTRAN, ASAS, and PAFEC FEM Systems," Robinson and Associates, Dorset, England, 1979.

6.5   R. H. MacNeal and R. L. Harder, "A Proposed Standard Set of Problems to Test Finite Element Accuracy," *Finite Elem. Analysis & Design*, 1, pp 3-20, 1985.

6.6   T. Belytschko, H. Stolarski, W. K. Liu, N. Carpenter, and J. S.-J. Ong, "Stress Projection for Membrane and Shear Locking in Shell Finite Elements," *Comput. Methods Appl. Mech. Engrg*, 51, pp 221-58, 1985.

6.7   I. Babuška, and M. Suri, "On Locking and Robustness in the Finite Element Method," Report BN-1112, Inst. for Physical Sci. and Tech., U. of Maryland, College Park Campus, May 1990.

6.8   R. H. MacNeal, "The Shape Sensitivity of Isoparametric Elements," *Finite Element Methods in the Design Process*, Proc. 6th World Congress on Finite Element Methods, Banff, Alberta, Canada, 1990.

6.9   M. Utku, E. Citipitioğlu, and G. Özkan, "Isoparametric Elements with Unequally Spaced Edge Nodes," *Comput. Struct.*, 41, pp 455-60, 1991.

6.10   J. Barlow, "More on Optimal Stress Points—Reduced Integration, Element Distortions, and Error Estimation," *Intl. J. Numer. Methods Eng.*, 28, pp 1487-504, 1989.

6.11   E. L. Wilson, R. L. Taylor, W. P. Doherty, and J. Ghaboussi, "Incompatible Displacement Models," *Numer. Computer Meth. in Struc. Mech.*, S. T. Fenves, et al. (Eds.), Academic Press, pp 43-57, 1973.

6.12   T. J. R. Hughes, "Equivalence of Finite Elements for Nearly Incompressible Elasticity," *J. Appl. Mech.*, 44, pp 181-3, 1977.

6.13   L. R. Herrmann, "Elasticity Equations for Nearly Incompressible Materials by a Variational Theorem," *J. AIAA*, 3, pp 1896-900, 1965.

6.14   J. C. Nagtegaal, D. M. Parks, and J. R. Rice, "On Numerically Accurate Finite Element Solutions in the Fully Plastic Range," *Comput. Methods Appl. Mech. Engrg.*, 4, pp 153-78, 1974.

6.15   T. J. R. Hughes, *The Finite Element Method*, Prentice Hall, Englewood Cliffs, NJ, p. 209, 1987.

6.16   O. C. Zienkiewicz, J. Too, and R. L. Taylor, "Reduced Integration Technique in General Analysis of Plates and Shells," *Intl. J. Numer. Methods Eng.*, 3, pp 275-90, 1971.

6.17   J. R. Brauer, private communication, 1989.

6.18   R. H. MacNeal, "Limitations on the Performance of Electromagnetic Brick Elements," MSC/EMAS Memo RHM-4, The MacNeal-Schwendler Corp., Los Angeles, 1990.

6.19   J. Robinson, "Distortion Measures for Quadrilaterals with Curved Boundaries," *Finite Elem. Analysis & Design*, 4, pp 115-31, 1988.

6.20   B. M. Irons, "Numerical Integration Applied to Finite Element Methods," Conf. on Use of Digital Computers in Structural Engineering, Univ. of Newcastle, 1966.

6.21   J. Robinson and G. W. Haggenmacher, "Some New Developments in Matrix Force Analysis," *Recent Advances in Matrix Methods of Structural Analysis & Design*, U. of Alabama Press, pp 183-228, 1971.

6.22   K. J. Bathe and E. N. Dvorkin, "A Formulation of General Shell Elements—The Use of Mixed Interpolation of Tensorial Components," Proc. NUMETA 85, Balkema Rotterdam, pp 551-65, 1985.

6.23   R. H. MacNeal, "Toward A Defect-Free Four-Node Membrane Element," *Finite Elem. Analysis & Design*, 5, pp 31-7, 1989.

7

Reduced Integration and
Spurious Modes

The analysis of interpolation failure and locking in Chapter 6 showed that the error in strain sometimes disappears at Gauss integration points of reduced order. In this chapter we will expand upon the benefits of reduced integration and upon the singularities caused by a possible reduction in matrix rank. The latter go by the names of *spurious mechanisms* or *spurious modes*. We prefer the term spurious modes because it is discipline neutral.

First, we need a working definition of reduced integration. Various objective criteria could be used to decide when integration is full and when it is reduced, such as the exactness of certain integrals, the satisfaction of certain tests, or the onset of reduction in rank of the element's stiffness matrix. We will instead take the simple rule that, if p is the degree of complete polynomials in an element's displacement basis, Gauss integration with p + 1 or more points in each direction is full and Gauss integration with p points or less in each direction is reduced. Usually the designer's practical choice is between p and p + 1 integration points. The definition can, if desired, be extended to elements with different polynomial degrees in each direction. Note, however, that the definition does not extend to triangular elements.

7.1   THE BENEFITS OF REDUCED INTEGRATION

The first obvious benefit of reduced integration is a reduction in computer time. Since the number of multiplications employed in the calculation of an element's stiffness matrix is proportional to the number of integration points, the ratio of computer times for this operation by full and reduced Gauss integration is $\left(\frac{p+1}{p}\right)^d$, where d = 1, 2, 3 is the number of dimensions. Thus for p = 1, d = 2 the ratio is 4, and for p = 2, d = 3 the ratio is 3.375.   These improvements in efficiency may or may not be significant.

In nonlinear analysis the time used in repeated evaluation of stress components at Gauss points *is* significant. In this context, a successful reduction from four Gauss points to one would be a major accomplishment.

In Chapter 6 we identified reduced integration as a potential remedy for many locking problems. To be specific, the summary data in Table 6.4 shows that reduced integration eliminates nearly all cases of shear and dilatation locking for rectangular plane and solid isoparametric elements of polynomial orders one and two. The only exception found was a shear locking mode of the twenty-node brick in which $w = x^2 y^2$.

The examination of shape sensitivity in Section 6.6 showed that reduced integration provides less than complete relief from the locking modes introduced by shape distortions. Likewise, the examination of edge node displacement in Section 6.1 found that reduced integration reduces the dependence of the error on node displacement, α, from $0(\alpha)$ to $0(\alpha^2)$. In his systematic study of the shape sensitivity of the eight-node quadrilateral, Barlow[1] found that, in general, reduced integration lowers the critical (largest) error term for linear strain states from $0(\alpha)$ to $0\left(\alpha^2\right)$. Thus, with these qualifications, reduced integration is an important potential remedy for errors due to interpolation failure.

More generally speaking, it has long been known that strains computed at the reduced-order Gauss points are more accurate than strains computed at other points. Barlow[2,3] was the first to notice this (1968) and has recommended that such points be used for stress output regardless of whether they are used for calculation of the stiffness matrix. The existence of similar optimal points for triangular elements has been sought without much success, except to note that the errors in some but not all components of strain are minimized at the reduced-order Gauss points of edges.[4] The existence of optimal locations for stress recovery, or *Barlow points*, is made plausible by the mathematical demonstration[5,6] that the error in average strain converges more rapidly than the errors at typical points. This implies that the error does not have the same sign everywhere and hence that it must be zero at some locations. The term *superconvergence* has been used to describe the reduction in error at optimal locations. Typically, if the error in stress converges as h^k at typical points, it converges as h^{k+1} at the optimal points.

It is not difficult to discover why Gauss points have a special relationship to error reduction. An important property of Gauss points, which may even be taken as a defining property, is that the Legendre polynomial of order p vanishes at a set of p Gauss points, i.e.,

$$P_p\left(\xi_p\right) = 0 \qquad (7:1)$$

Let a particular strain state, $\varepsilon(\xi)$, be expanded in a series of Legendre polynomials

$$\varepsilon(\xi) = \sum_{s=0}^{p} \varepsilon_s P_s(\xi) \qquad (7:2)$$

where p is the highest term present in $\varepsilon(\xi)$. If ε is derived from $p + 1$ nodal displacements, the strains will be of order $p - 1$ and the coefficient ε_p will be zero regardless of the shape of the true displacement field. The term $\varepsilon_p P_p(\xi)$ will, nevertheless, have the correct value (zero) at Gauss points, ξ_p, by virtue of Equation 7:1. If $p+1$ Gauss points are used, $\varepsilon_p P_p\left(\xi_{p+1}\right) = 0$ will not, in general, be the correct value at Gauss points. This admittedly narrow point is the basis for the claimed superiority of reduced integration.

The argument just presented does not take into account the lower-degree aliases that represent ξ^{p+1} and other high-degree terms within the element. We know, from the material in Chapter 6, that the aliases do not degrade the accuracy of strain representation at reduced-order Gauss points for $p = 1$ and $p = 2$. It is fair to ask whether the same is true for $p \geq 3$. Consider, for example, a one-dimensional element with $p + 1$ equally spaced nodes. It is an easy matter to work out the alias, u^a, for $u = \xi^{p+1}$ and to compare the resulting strains with the true strains. This is done in Table 7.1 for $p = 1, 2, 3$.

Table 7.1

Barlow Points and Gauss Points for One-Dimensional Elements

p	NODE LOCATIONS	u	u^a	$u_{,\xi}$	$u^a_{,\xi}$	BARLOW POINTS (WHERE $u_{,\xi} = u^a_{,\xi}$)	GAUSS POINTS
1	± 1	ξ^2	1	2ξ	0	0	0
2	$0, \pm 1$	ξ^3	ξ	$3\xi^2$	1	$\pm 1/\sqrt{3}$	$\pm 1/\sqrt{3}$
3	$\pm\frac{1}{3}, \pm 1$	ξ^4	$\dfrac{10\xi^2-1}{9}$	$4\xi^3$	$\dfrac{20\xi}{9}$	$0, \pm\dfrac{\sqrt{5}}{3}$	$0, \pm\sqrt{\dfrac{3}{5}}$

The Barlow points (points where $u_{,\xi} = u^a_{,\xi}$) are identical to the Gauss points for $p = 1, 2$ but not for $p = 3$ where the Barlow points are at $0, \pm\sqrt{5}/3 = \pm.74536$ and the Gauss points are at $0, \pm\sqrt{.6} = .77460$. The difference is small, not enough to tempt us to alter the Gauss points for numerical integration but perhaps enough to induce us to change the locations for stress recovery.

Similar conclusions are reached if hierarchical degrees of freedom are employed along the lines suggested in Section 4.5. As will be recalled, the hierarchical shape functions were taken to be

$$\phi_n(\xi) = P_n(\xi) - P_{n-2}(\xi) \qquad 2 \le n \le p \qquad (7:3)$$

The derivative $\phi_n'(\xi) = P_n'(\xi) - P_{n-2}'(\xi) = P_{n-1}(\xi)$, the latter being a known property of Legendre polynomials. Thus, if the actual displacement is taken to be $u = \phi_{p+1}(\xi)$, the actual strain, $\varepsilon = u_{,\xi} = P_p(\xi)$, will vanish at Gauss points, ξ_p. The difficulty is that a least squares fit of $\phi_{p+1}(\xi)$ to the shape functions produces a nonzero result because, unlike Legendre polynomials, the functions $\phi_n(\xi)$ are not orthogonal to each other. As a consequence the derivative of the alias of $\phi_{p+1}(\xi)$, $u^a_{,\xi} = \sum_{n=0}^{p} u_n \phi_n'(\xi)$, will not vanish at Gauss points for $p > 2$.

So much for the advantages of reduced integration. They are considerable, particularly as a simple, inexpensive solution to many locking problems. The disadvantages of reduced integration begin with the reduced accuracy of integration. Table 4.12 shows, for example, that full integration produces exact integrals under less restrictive shape conditions than reduced integration. Still, it was shown in Section 5.4 that plane quadrilaterals can pass constant strain patch tests with reduced integration under the most general conditions on shape. Brick elements, on the other hand, require full integration to pass similar tests. Barlow[1] concludes, in his study of eight-node quadrilaterals and twenty-node bricks, that the advantage of accurate strain evaluation outweighs the disadvantage of less accurate integration. In all fairness, the issue would appear to remain open.

We will, in the next section, begin to examine a more serious disadvantage, spurious modes.

7.2 THE SPURIOUS MODES OF AN ELEMENT

We consider first the spurious modes of a single element. Later we will examine the spurious modes of a field of elements. Spurious modes are eigenvectors of the stiffness matrix, i.e.,

$$[K]\{\phi_s\} = 0 \qquad (7:4)$$

A spurious mode, $\{\phi_s\}$, differs from a rigid body mode, $\{\phi_o\}$, in that the modal strain state

$$\{\varepsilon_s\} = [B]\{\phi_s\} \qquad (7:5)$$

is *not* zero everywhere.[*] It is, however, zero at integration points with the result that Equation 7:4 holds.

Spurious modes will clearly occur if the number of strain evaluations at integration points is less than the number of independent strain states provided by the nodal displacements. If this is the case, certain linear combinations of the strain states will produce zero strains at all integration points. Thus the difference between the number of independent strain states and the number of strain evaluations provides a lower limit to the number of spurious modes. (There may be more because some strain evaluations may be redundant.)

It is not difficult to compute this lower limit for a particular element. Consider the four-node plane quadrilateral. Remembering to exclude rigid body modes, we know that it has $4 \times 2 - 3 = 5$ independent strain states. With reduced integration there is one integration point which provides evaluation of three strain components. Thus the element has, as a minimum, two spurious modes. Full integration, on the other hand, provides $4 \times 3 = 12$ strain evaluations so that the minimum number of spurious modes is zero.

---

[*]We will encounter cases where $\{\varepsilon_s\}$ is zero everywhere, but the mode shape $\{\phi_s\}$ is not a standard rigid body mode. We may or may not choose to label such modes spurious, depending on how much we deplore their existence.

It is instructive to extend this sort of bookkeeping to the general class of quadrilateral plane elements and hexahedral solid elements which use regular arrays of Gauss integration points. The number of strain evaluations is

for reduced integration, 2-D: $\quad 3p^2$

for full integration, 2-D: $\quad 3(p + 1)^2$

for reduced integration, 3-D: $\quad 6p^3$

for full integration, 3-D: $\quad 6(p + 1)^3$

The number of strain states depends on the nodal arrangements. We consider two arrangements: serendipity elements which have nodes at corners and along edges only, and Lagrange elements which have a full lattice of $(p + 1)^n$ nodes. The number of strain states is

for serendipity elements, 2-D: $\quad 8p - 3$

for Lagrange elements, 2-D: $\quad 2(p + 1)^2 - 3$

for serendipity elements, 3-D: $\quad 36p - 18$

for Lagrange elements, 3-D: $\quad 3(p + 1)^3 - 6$

It can be seen right away that with full integration the number of strain evaluations exceeds the number of strain states, even for the more populous Lagrange elements. (The excess equals $(p + 1)^2 + 3$ for 2-D Lagrange elements and $3(p + 1)^3 + 6$ for 3-D Lagrange elements). As a result, full integration produces a net redundancy in strain evaluations. Spurious modes do not, in fact, occur when full integration is used. The reason is that $p + 1$ Gauss points can correctly integrate a polynomial of degree $2p + 1$, which exceeds, for elements with one, two, or three dimensions and constant Jacobians, the highest degree term in the strain energy density. Thus, for such elements, a strain term derived from displacements cannot be zero at all Gauss points if it is not zero everywhere.

Reduced integration, on the other hand, is accompanied by spurious modes. The data in Table 7.2, obtained by differencing the formulas presented above, show that spurious modes exist in all cases with $p = 1$ and $p = 2$, and that spurious modes exist for Lagrange elements with $p = 3$. As indicated by

Table 7.2

The Minimum Number of Spurious Modes
Which Accompany Reduced Integration

DIMENSION:	2	2	3	3
ELEMENT TYPE:	SEREN-DIPITY	LAGRANGE	SEREN-DIPITY	LAGRANGE
POLYNOMIAL DEGREE	MINIMUM NUMBER OF SPURIOUS MODES			
1	2	2	12	12
2	1	3	6	27
3	$\langle 6 \rangle$	2	$\langle 72 \rangle$	24
4	$\langle 19 \rangle$	$\langle 1 \rangle$	$\langle 258 \rangle$	$\langle 15 \rangle$

brackets, $\langle \ \rangle$, higher polynomial degrees produce net redundancies in strain evaluation, even when reduced integration is used. Whether or not spurious modes exist at higher polynomial orders remains to be established.

Knowing that spurious modes exist in certain circumstances, we feel obliged to find out what they look like. In the case of the four-node ($p = 1$) quadrilateral, for example, we know that there are at least two spurious modes when a single integration point is used. To deduce their shapes we note first that the displacements of the spurious modes must conform to the element's displacement basis $(1, \xi, \eta, \xi\eta)$ and, second, that the three constant strain states do not vanish anywhere. Thus the two spurious modes have displacement states $u = \xi\eta$ or $v = \xi\eta$. The strains due to either of these displacement states vanish at the integration point, $\xi = \eta = 0$ since, for example,

$$u_{,x} = u_{,\xi}\, \xi_{,x} + u_{,\eta}\, \eta_{,x} = \eta\xi_{,x} + \xi\eta_{,x} \qquad (7:6)$$

We conclude that the five independent strain states of the four-node quadrilateral with reduced integration consist of three constant strain states and two spurious modes whose displacement states are $u = \xi\eta$ and $v = \xi\eta$.

This analysis of the four-node quadrilateral extends immediately to an eight-node brick with one integration point. The shapes of the spurious modes are given by

$$\left(u, v, w\right) = \left(\xi\eta, \xi\zeta, \eta\zeta, \xi\eta\zeta\right) \tag{7:7}$$

which accounts for all twelve of the spurious modes identified in Table 7.2. The six remaining strain states are constant.

Similar reasoning can be employed to find spurious modes for more complex elements. Consider the displacement state

$$u = P_p(\xi)\,P_p(\eta) \tag{7:8}$$

where $P_p(\xi)$ is a Legendre polynomial of degree p. The derivative of u with respect to x or y will vanish at a set of p x p Gauss points; for example,

$$u_{,x} = P_p'(\xi)\,P_p(\eta)\,\xi_{,x} + P_p(\xi)\,P_p'(\eta)\,\eta_{,x} \tag{7:9}$$

vanishes because $P_p\left(\xi_p\right) = 0$ and $P_p\left(\eta_p\right) = 0$. The highest degree monomial in the expansion of Equation 7:8 is $\xi^p\eta^p$. This term exists in the bases of Lagrange elements, which include the terms $\left(1, \xi, \cdots, \xi^p\right)\left(1, \eta, \cdots, \eta^p\right)$, but not in the base of serendipity elements. Consequently we conclude that Lagrange elements have spurious modes of the type indicated by Equation 7:8 regardless of their polynomial degree. In Lagrange quadrilateral elements they account for two spurious modes. In Lagrange brick elements any mode of the form

$$u = P_p(\xi)\,P_p(\eta)\,f(\zeta) \tag{7:10}$$

where $f(\zeta)$ is any function of ζ, is clearly spurious. For p = 2 there are twenty-one such modes given by

$$(u, v, w) = P_2(\xi) P_2(\eta)\left(1, \zeta, \zeta^2\right) + P_2(\xi) P_2(\zeta)(1, \eta) + P_2(\eta) P_2(\zeta)(1, \xi)$$

$$(7{:}11)$$

Even this large number of modes does not account for all of the spurious modes identified for the 27-node Lagrange brick in Table 7.2. Nor do they account for any of the spurious modes of the twenty-node serendipity brick or for the single spurious mode of the eight-node quadrilateral. The shapes for these latter modes are not easily determined if the element shapes are distorted. For the special case of plane quadrilateral elements with linear geometry (i.e., parallelograms) the shapes of this second type of spurious mode are given by

$$u = P'_p(\xi) P_p(\eta) \xi_{,x} - P_p(\xi) P'_p(\eta) \eta_{,x}$$
$$v = P'_p(\xi) P_p(\eta) \xi_{,y} - P_p(\xi) P'_p(\eta) \eta_{,y}$$

$$(7{:}12)$$

where $\xi_{,x}, \xi_{,y}, \eta_{,x},$ and $\eta_{,y}$ are constant by assumption. It can easily be verified that $\varepsilon_x = u_{,x}, \varepsilon_y = v_{,y},$ and $\gamma = u_{,y} + v_{,x}$ all vanish at the set of $p \times p$ Gauss points. For the special case of an eight-node rectangular element, $\xi_{,y} = \eta_{,x} = 0$ and Equation 7:12 reduces to

$$u = \xi\left(\eta^2 - \tfrac{1}{3}\right)\xi_{,x}$$
$$v = -\eta\left(\xi^2 - \tfrac{1}{3}\right)\eta_{,y}$$

$$(7{:}13)$$

The shape of this spurious mode is sketched in Figure 7.1. This shape also accounts for all six of the spurious modes of a twenty-node rectangular brick with reduced integration. There are two such modes in each direction, one as indicated in Equation 7:13 and the other similar but with u and v multiplied by ζ.

The second type of spurious mode, identified by Equation 7:12, does not exist for serendipity elements with $p \geq 3$ for the simple reason that monomials of the form $\xi^p \eta^{p-1}$ do not exist in the basis of such an element. They do, however, exist in the basis of an isoparametric Lagrange element.

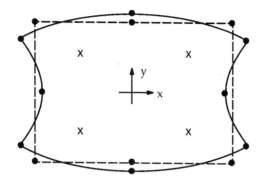

Figure 7.1 The Spurious Mode of an Eight-Node Rectangular Element with Reduced Integration.

In summary, we have identified spurious modes of two types which account for all of the minimum numbers of spurious modes identified in Table 7.2 for $p = 1$ and $p = 2$. We are confident that no other spurious modes exist for $p = 1$ and $p = 2$ since we can easily verify that the remaining lower-order strain states do not vanish at integration points. We know, in addition, that spurious modes of both types exist for all orders of Lagrange elements, even though Table 7.2 indicates a net redundancy in strain evaluations for $p \geq 4$. With slightly less confidence we assert that spurious modes do not exist for reduced-integration serendipity elements with $p \geq 3$.

So far we have considered only elements which use Gauss integration. Spurious modes do, of course, occur for any pattern of integration points when the number of strain evaluations at those points is less than the number of strain states derived from displacements. Triangular elements are an important example but there is no reason to employ reduced integration with them because it conveys no advantage in accuracy.

In the case of solid pentahedral elements, the most obvious integration pattern is a combination of an appropriate triangular array in two dimensions and a Gauss distribution in the third direction (n layers of triangular arrays). Spurious modes can occur in these elements either by design or through inadvertence. Consider for example the lowest order element of this type, PENTA6. Each triangular face requires only a single integration point to evaluate its constant strain states so we are led to consider a 1×2 pattern (two points in the ζ direction). There are $6 \times 3 - 6 = 12$ strain states and $2 \times 6 = 12$ strain evaluations. Everything looks fine, but the element has, in fact, a spurious mode, corresponding to in-plane rotation of one triangular face relative to the other (see Figure 7.2). The spurious mode is balanced, in the strain state inventory, by redundant evaluation of ε_z (if z is parallel to ζ). The minimum practical pattern of integration points for the PENTA6 which avoids spurious modes is a 3×2 pattern.

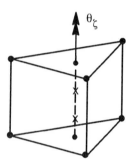

Figure 7.2   Spurious Mode of a PENTA6 Element with Two Integration
             Points.

The same 3×2 pattern is attractive for the PENTA15 element. In this case there are three spurious modes similar in shape (on one face) to the type 2 mode of the QUAD8 shown in Figure 7.1, but shear locking due to displacements of the form $u = x^2 z, w = -x^3 / 3$ is avoided.

7.3 SPURIOUS MODES IN FIELDS OF ELEMENTS

In going from a single element to a field of similar elements we know that several degrees of freedom per element are eliminated by the connections between elements. It is, therefore, natural to ask whether a particular spurious mode of a single element can exist in a field of elements and, if it does, how many copies there will be. Spurious modes which exist in fields of elements are said to be *communicating* modes. If only one copy of a particular spurious mode exists within the field, we will say that is a *global* spurious mode. If more than one copy exists we will say that they are *local* spurious modes. Thus there are at least three possible classifications of spurious modes: noncommunicating modes (which cannot exist in fields of elements), global modes, and local modes.

Global spurious modes are the simplest. An example is the type 1 spurious mode in the four-node membrane element for which, as we recall, $u = \xi\eta$ (or $v = \xi\eta$). The mode shape for a field of elements is plotted in Figure 7.3 where it is seen that the pattern is reversed in adjacent elements. This mode is called the *hourglass* [7] mode* from its shape. Sometimes all spurious modes are called hourglass modes, regardless of their shape. We note that this mode can exist in fields of nonrectangular or even highly distorted elements because the mode shape for any element, $u = \pm\xi\eta$, has unit magnitude at its nodes regardless of its shape.

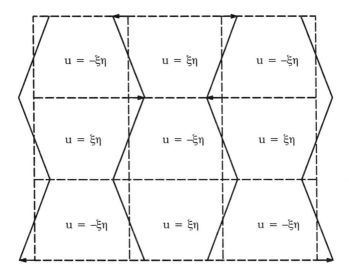

Figure 7.3   **Hourglass Mode of the Four-Node Quadrilateral.**

---

*It has been called *keystoning*, also from its shape, a term attributed to Sam Key.

It is easy to see that the hourglass mode is a global mode. If the mode shape for a particular element is specified, the mode shapes for elements adjacent to it are uniquely determined by the requirement of displacement continuity, and so are the mode shapes of the elements next to them, etc.

The type 1 spurious modes of the nine-node membrane element are also global modes. In this case $\left(u \text{ (or } v) = (\xi^2 - 1/3)(\eta^2 - 1/3) \right)$. The mode shape is sketched in Figure 7.4 where it is seen that displacements in all elements are in phase and the mode shape vaguely resembles a field of hourglasses. Again, element distortion does not affect the existence of this mode.

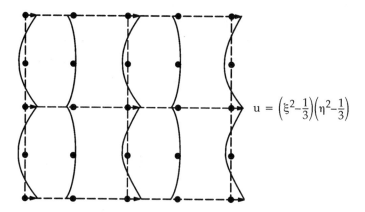

$$u = \left(\xi^2 - \frac{1}{3}\right)\left(\eta^2 - \frac{1}{3}\right)$$

Figure 7.4   Type 1 Spurious Mode in a Field of Nine-Node Quadrilaterals.

In three dimensions many of the type 1 spurious modes are local modes. The spurious mode shapes of the eight-node brick, given by Equation 7:7, can, for example, be combined to form

$$v = \xi\eta(1 - \zeta) \tag{7:14}$$

which vanishes at $\zeta = 1$ and has the value $v = 2\xi\eta$ at $\zeta = -1$.

When such an element is joined from below by an element with $v = \xi\eta(1 + \zeta)$, a layer two elements deep is formed in which v has nonzero value, while outside the layer v is zero. The mode shape is sketched in Figure 7.5. Clearly we can form similar local modes in other layers. The total number of independent spurious modes is $3\left(N_\xi + N_\eta + N_\zeta + 1\right)$ where N_ξ, N_η, and N_ζ are the numbers of elements in the ξ, η, and ζ directions. The factor of three accounts for the three independent displacement components.

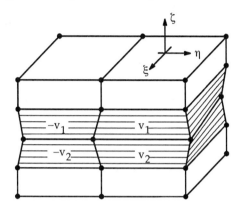

Figure 7.5   **Local Spurious Mode in a Field of Reduced-Order HEXA8 Elements** $v_1 = \xi\eta(1 - \zeta)$ $v_2 = \xi\eta(1 + \zeta)$.

So far we have considered only type 1 spurious modes. We have found them to be global in two-dimensional fields and local in three-dimensional fields. Type 2 spurious modes, on the other hand, are noncommunicating. Consider, for example, the spurious mode of the eight-node rectangle sketched in Figure 7.1. The top edge stretches and bows outward. To conform, an adjacent element must stretch and bow inward, a clear impossibility unless a second spurious mode with that shape exists. (It does not.)

An investigation of the requirements for the conformity of mode shapes in fields of identical elements can shed light on the general question of mode communication. Consider, for example, a two-dimensional lattice of identical rectangular elements and assume further that the displacement fields within each element are identical except for a possible change in sign. Such a mode is classified as global according to our definition of the term. The additional requirement of displacement continuity imposes conditions on the displacement components along element boundaries that are quite different from the familiar conditions of reflective symmetry.

Consider the element shown in Figure 7.6. To satisfy displacement continuity, the displacement vectors at corresponding points on opposite boundaries must either be the same in magnitude and direction or equal in magnitude and opposite in direction. Thus all components of displacement must be the same (even) or all must be opposite (odd) at corresponding points on opposing boundaries. These requirements coincide with the requirements of reflective symmetry only if there is a single nonzero component of displacement.

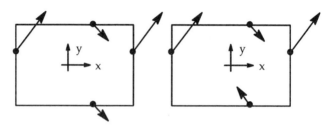

(a) Even in x, Even in y (EE)   (b) Even in x, Odd in y (EO)

Figure 7.6   Two of Four Possible Symmetry Types for Global Communicating Modes in a Field of Rectangles. The Other Two Types are (OE) and (OO).

The general form for the type 2 spurious mode shape of a rectangle is, from Equation 7:12,

$$u = P'_p(\xi) P_p(\eta) \xi'_x$$
$$v = -P_p(\xi) P'_p(\eta) \eta'_y$$

$$(7:15)$$

If p is an even number, then u is an odd function of ξ and an even function of η (symmetry type OE) while v is an even function of ξ and an odd function of η (symmetry type EO). Thus u and v do not have the same symmetry types, with the result that Equation 7:15 cannot be the mode shape of a communicating global mode in a field of rectangular elements. We conclude that all type 2 spurious modes are noncommunicating.

There is at least one exception to this general conclusion. That exception is the case of a beam made up of reduced-order, twenty-node brick elements. The nonzero displacement components are normal to the axis of the beam and the number of spurious modes equals the number of elements plus or minus one, depending on end conditions.

We have seen that, with regard to their behavior in fields of elements, spurious modes come in three varieties: noncommunicating, global, and local. Of these we expect that noncommunicating spurious modes have the smallest effect on element performance and that local modes have the greatest effect. The effects of spurious modes depend in part on the boundary conditions applied to the field of elements, a subject taken up in the next section. After that we will consider ways to avoid the harmful effects of spurious modes while retaining the advantages of reduced integration.

7.4 NEAR MECHANISMS

As we have just seen, spurious modes in fields of elements are of two types: global modes which involve every element in the field, and local modes which involve only some of the elements. Even in the latter case the modes propagate to the edges of the field so that constraints applied to the boundary of the field (i.e., Dirichlet boundary conditions) can remove the singularities. Under these conditions it is appropriate to ask whether the boundary constraints will stiffen the spurious modes enough to make their softening effect unimportant. Certainly that is what the designer of such elements would hope, but the evidence is discouraging.

In 1979, Bićanić and Hinton [8] published data which show that the near mechanisms present in restrained fields of reduced integration four- and nine-node membrane elements can greatly degrade performance. They computed the effects of reduced integration on the vibration modes, thereby avoiding the question of how loads excite the near mechanisms. For example, Figure 7.7 shows the six lowest eigenvectors of a cantilever beam analyzed with four-, eight-, and nine-node elements and with reduced integration. Since the beam is not particularly slender (length/width = 6.0), the support at the base should be enough to supply significant stiffness for even the higher modes. The presence of spurious modes is, however, clearly evident beginning with the lowest

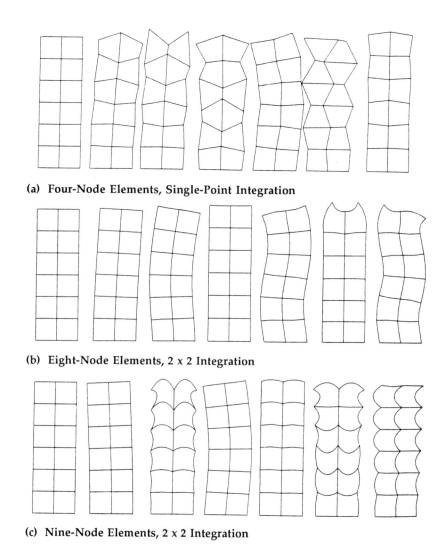

(a) Four-Node Elements, Single-Point Integration

(b) Eight-Node Elements, 2 x 2 Integration

(c) Nine-Node Elements, 2 x 2 Integration

Figure 7.7   Lowest Six Eigenvectors for Simple Cantilever–Reduced Integration, All Support Nodes Fixed.

mode of the beam with four-node elements and with the second mode of the beam with nine-node elements. In contrast, the effects of reduced integration on the beam with eight-node elements are small—principally small changes in the eigenvalues of the three highest modes.

Bićanić and Hinton also computed the effect of reduced integration on a 6 x 6 field of square nine-node membrane elements that was fully constrained on the boundaries. The results indicated a pair of spurious modes with identical eigenvalues equal to only 2.5 times the lowest eigenvalue.

The conclusion is inescapable, from just these few results, that reduced integration of four- and nine-node membrane elements is unacceptably risky for practical application. For eight-node elements, on the other hand, the effect appears to be quite small. The difference which undoubtedly accounts for these separate results is that the type 1 spurious modes of QUAD4 and QUAD9 produce global spurious modes in fields of elements while the type 2 spurious modes of QUAD8 are noncommunicating. Thus, while the incompatibility of the type 2 modes stiffens the modes in all elements, the boundary constraints only stiffen spurious modes in elements adjacent to the boundary. More generally, we conclude that reduced integration cannot be used reliably with Lagrange elements of any order in either two or three dimensions.

The issue of whether or not reduced integration can safely be used with serendipity elements requires further study. The most important elements (apart from the lowest order element which is also a Lagrange element) in the serendipity family are the QUAD8 and the HEXA20. We observed in Chapter 6 that reduced integration cured most of the locking disorders of these elements. We will also see, in Chapters 9 and 10, that these beneficial effects extend to the bending of flat plates and curved shells. The motivation for using reduced integration with these elements is well founded.

An example frequently cited to discourage the use of reduced integration with eight-node membrane or twenty-node brick elements is the case of a single stiff element resting on a very soft foundation (see Figure 7.8). This obviously contrived example is not very convincing. A better example, which was encountered by Irons and Hellen, [9] is shown in Figure 7.9. It represents a cylindrical shell modeled with four twenty-node brick elements per quadrant and internally pressurized. The shell is quite thin. R/t values of 200 and 500 were used. The radial displacement, which should be constant, instead exhibited severe oscillation. The cause of the oscillation proved to be the

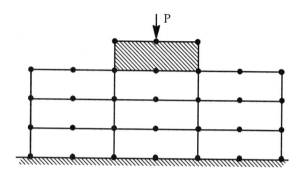

Figure 7.8   A Stiff Element Resting on a Soft Foundation.

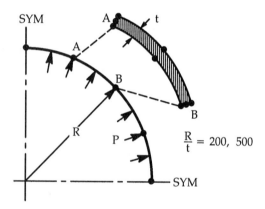

Figure 7.9   Pressurized Shell Modeled with Twenty-Node Brick Elements.

type 2 spurious mode exhibited in Figure 7.1. The mode is excited by constant pressure acting on the slight difference between the element's parabolic shape and a circular arc. Equation 7.13 shows that, since R/t is large, the circumferential displacements in the spurious mode are much smaller than the radial displacements.[*] As a result the strains required to maintain displacement continuity by circumferential motion are small and the spurious mode becomes a near mechanism.

---

[*]If u is (more or less) parallel to the circumferential direction, then $\xi_{,x} \ll \eta_{,y}$.

This example is a severe one for solid elements and not only because a near mechanism is involved. Since the effect of round off is multiplied by a factor[*] of $(R/t)^4$, the larger of the two radius-to-thickness ratios (500) is near the limit of numerical precision. Further, the ratio of hoop stiffness to bending stiffness is so large that any slight variation of internal pressure causes a large variation of radial displacement. There are good reasons to have shell elements in finite element programs, and this example exhibits at least two of them (avoidance of near mechanisms and round-off accumulation). By the same token this is not a good application of solid elements. We should take note of the limitation it imposes on solid elements, particularly on those which use reduced integration, but that should not deter us from using the same elements in better applications.

7.5 ZERO ENERGY MODES IN MAGNETOSTATICS

It is almost always useful to study a topic in more than one discipline. It is like learning French. The differences between French and English teach us something about the structure of language in general. We propose here to extend the discussion of spurious modes to magnetostatics where we will discover both similarities and differences with their manifestations in elasticity.

As discussed in Section 2.1.4, useful analogies exist between the equations of elasticity and the equations of magnetostatics. According to Table 2.1, if the magnetic vector potential $\bar{A}$ is taken to be analogous to displacement, then the magnetic induction $\bar{B}$ is analogous to strain and the inverse of the permeability matrix, $[\mu]^{-1}$, is analogous to the elastic modulus matrix, $[D]$. The magnetic induction is the curl of the vector potential, i.e., $\bar{B} = \nabla \times \bar{A}$, which also makes it, within a factor of two, analogous to rotation. A curious difference between the two disciplines was noted in Section 2.1.4, namely that the derivatives of the potential function which contain the energy in magnetostatics have no energy in elasticity, and vice versa. More to the point of the topics covered in the present chapter, the magnetic induction has only three components compared

---

[*]This factor is proportional to the ratio of extensional stiffness in the *radial* direction to bending stiffness.

to six components for the elastic strain. We might expect, therefore, that a magnetic field has nine zero energy states (three constant states for $\bar{A}$ plus six constant strain fields) which are analogous to the six rigid body modes of an elastic solid. The actual situation is more complicated, as we shall see.

Let us consider, as a concrete example, the simulation of a magnetic field by eight-node rectangular brick elements. Each such element has $8 \times 3 = 24$ components of $\bar{A}$ as external degrees of freedom. In matrix notation

$$\{A\}_i = \begin{Bmatrix} A_x \\ A_y \\ A_z \end{Bmatrix}_i \qquad i = 1, 2, \ldots, 8 \qquad (7:16)$$

We will tentatively accept the following classifications to describe internal degrees of freedom.

> translation modes: $\{A\}$ = constant
>
> curl-free modes: $\nabla \times \bar{A} = 0$
>
> curl modes: all others

The displacement basis vector for the element is

$$\lfloor X \rfloor = \lfloor 1, \ x, \ y, \ z, \ xy, \ xz, \ yz, \ xyz \rfloor \qquad (7:17)$$

The linear terms give nine constant derivatives of the components of $\bar{A}$ which produce, according to previous discussion, three curl modes (the constant components of magnetic induction) and six curl-free modes (the "strains"). The four incomplete higher order terms in the basis yield only a single curl-free mode:

$$A_x = yz, \ A_y = xz, \ A_z = xy \qquad (7:18)$$

The fact that this mode is curl-free may be verified by computing the components of curl

$$\nabla \times \bar{A} = \bar{i}\left(A_{z,y} - A_{y,z} \right) + \bar{j}\left(A_{x,z} - A_{z,x} \right) + \bar{k}\left(A_{y,x} - A_{x,y} \right) \qquad (7:19)$$

A useful rule for finding curl-free modes results from the observation that $\nabla \times \nabla \phi = 0$ where ϕ is a scalar. Thus we need only postulate a scalar, such as $x^a y^b z^c$, and take $\bar{A}$ as its gradient. The result must be checked however to see whether some of the components of $\bar{A}$ are excluded by the incompleteness of the basis vector. That is the case here for all higher order candidates except $\phi = xyz$ which produces the vector described in Equation 7:18.

Referring to the classification assumed above, we conclude that the eight-node brick has three translation modes, seven curl-free modes, and fourteen curl modes. In fields of elements the translation modes behave just like rigid body translation modes in elasticity, i.e., they are global communicating modes. Some of the curl-free modes, on the other hand, are local modes. Consider, for example, the constant "strain" mode, $A_x = x$. This mode can be combined with the translation mode $A_x = 1$ to produce $A_x = 0$ along a line of nodes, $x = x_1$. Then, as described in Section 7.3, a second linear combination can be added in an adjacent set of elements to produce a local mode which vanishes outside the disk $x_1 < x < x_2$ as depicted in Figure 7.10.

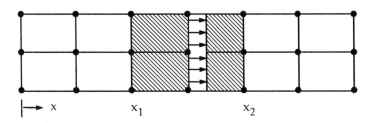

Figure 7.10   Local Mode in a Field of Eight-Node Electromagnetic Elements.

The presence of local energy-free modes is very undesirable for practical computation.[*] Fortunately there exists a firm basis in electromagnetic theory for their elimination. That basis stems from the observation that the divergence of $\bar{A}$ appears nowhere in Maxwell's equations and hence may be freely

---

[*]Consider the small motions of a fluid treated as a degenerate solid with zero shear modulus. In this case only one of the nine displacement derivatives has energy. Attempts to use this approach are confounded by swarms of local "spurious" modes.

assigned. A very common assumption, known as the Coulomb gauge, is to set the divergence, $\nabla \bullet \bar{A} = A_{x,x} + A_{y,y} + A_{z,z}$, to zero everywhere. Other assumptions, such as the Lorenz gauge, are sometimes used to simplify the equations of motion in particular contexts, but the Coulomb gauge is well suited to the needs of finite element analysis. [10]

To implement the Coulomb gauge we evaluate $\nabla \bullet \bar{A}$ at Gauss points and assign it a *reluctivity*, α / μ_o, where μ_o is a characteristic permeability and α is an arbitrary dimensionless parameter. In effect, the element's "stiffness" matrix is augmented by a penalty stiffness assigned to the divergence of $\bar{A}$. The penalty parameter, α, is assigned a value that will make the stiffness of the divergence much larger than that of the curl, $B = \nabla \times \bar{A}$.

Assuming that penalty stiffness is applied to the divergence, the element's curl-free modes may be separated into divergence modes and divergence-free modes. These modes correspond, respectively, to the dilatation states and deviatoric strain states of elasticity. We will call the divergence-free modes, which are also curl-free, *shear* modes, [11] for want of any name supplied to them by electromagnetic theory. A shear mode is energy free (in magnetostatics) and has the property that it can be derived as the gradient of a scalar, $\bar{A} = \nabla\phi$, where ϕ satisfies Laplace's equation, i.e., $\nabla \bullet \bar{A} = \phi_{,xx} + \phi_{,yy} + \phi_{,zz} = 0$. The five lowest shear modes are recorded in Table 7.3. The mode described by Equation 7:18 is also a shear mode.

When it comes to classifying the status of shear modes in fields of elements, we observe that none of them qualifies as a communicating mode according to the lattice symmetry requirements described in Section 7.3. Thus in mode S1 of Table 7.3, A_x is even in x and odd in y while A_y is odd in x and even in y. Nevertheless all six shear modes of the HEXA8 element are communicating modes. This may be shown by assuming that their spatial dependencies, such as those given for the lowest five modes in Table 7.3, extend over the entire field of elements. Each such mode is clearly energy free and just as clearly satisfies displacement continuity. It therefore qualifies as a global energy-free mode. As measured from the center of any element its shape includes linear terms from Table 7.3 and constant terms due to translation.

Table 7.3

The Five Lowest Shear Modes in Magnetostatics

MODE	ϕ	A_x	A_y	A_z
S1	xy	y	x	0
S2	yz	0	z	y
S3	xz	z	0	x
S4	$\frac{1}{2}\left(x^2 - z^2\right)$	x	0	$-z$
S5	$\frac{1}{2}\left(y^2 - z^2\right)$	0	y	$-z$

The shear modes in magnetostatics behave in the same way as rigid body rotations in static elasticity. They and the translation modes can be suppressed by supplying a minimum number of point constraints to a field of elements (nine constraints for a field of HEXA8 elements). There are, however, differences. The most important difference is that the number of shear modes depends on the polynomial order of the elements used. Thus, for a field of solid elements which are complete to quadratic degree, there are a minimum of twelve shear modes. Incomplete higher order terms in the displacement basis occasionally add shear modes. This has been observed for serendipity brick elements with $p = 1$ and $p = 3$, but not for higher values of p. It has also been observed that the shear modes due to the incomplete terms disappear (pick up energy) when the elements are distorted.

The numbers of modes of each of the classes which have been defined are recorded in Table 7.4 for the HEXA8 and HEXA20 elements. It is noted that while, by definition, divergence modes have no curl and shear modes have neither divergence nor curl, divergence modes may have shear and some curl modes may have both shear and divergence.

Table 7.4

**Number of Modes of Different Types
in Two Electromagnetic Elements**

TYPE	HEXA8	HEXA20
Translation	3	3
Shear	6	12
Divergence	1	10
Curl	14	35
TOTAL	24	60

Quite clearly, a divergence penalty reduces the confusion of "spurious" local energy-free modes to the comparative order of a few "shear" modes which behave like rigid body modes in mechanics. At the same time, the divergence penalty can cause locking disorders similar to the dilatation locking of nearly incompressible materials. The reason is that, in the desire to shift the divergence modes above the curl modes, we may stiffen the divergence modes to such a degree (α is typically of the order of 10^5)[(12)] that a small coupling between curl and divergence may significantly stiffen the curl modes. This is bound to happen if full integration is used to compute the divergence penalty stiffness because many of the curl modes derived from incomplete terms in the displacement basis have unavoidable parasitic divergences.

Consider, for example, that $A_x = xyz$, $A_y = 0$, $A_z = 0$. The divergence of $\bar{A}$ is equal to yz. To negate the divergence would require $A_y = -y^2z / 2$ or $A_z = -yz^2 / 2$, but neither of these functions is in the basis of the HEXA8 element. This situation is exactly analogous to that recorded in Table 6.4 for dilatation locking of the HEXA8 element. The remedy is also the same — reduced integration of the divergence/dilatation term, in this case at a single point.

While reduced integration causes spurious modes when applied to the dominant energy terms (magnetic induction in magnetostatics and deviatoric strains in elasticity), it is much less likely to cause such difficulty when applied to the divergence/dilatation term. In fact the optimal number of integration points for the divergence/dilatation term, as predicted by the constraint ratio argument mentioned in Section 6.3, may well be less than the p^3 integration points provided by standard reduced-order integration.[13]

7.6 SELECTIVE UNDERINTEGRATION

As we have seen in previous sections of this chapter, reduced integration causes spurious modes which in many cases will destroy the accuracy of a finite element analysis. We should also remember that reduced integration was introduced as a remedy for locking and that locking is the result of the incorrect interpolation of some, but not all, components of strain. Thus the evaluation of some, but not all, of the strain components at a reduced set of points should eliminate locking and may retain a sufficient number of strain evaluations to prevent spurious modes. This procedure, which is called *selective underintegration*, does in fact produce the desired result. It also has unpleasant side effects. These side effects and clues to *their* elimination will emerge from the examination of particular cases.

7.6.1 Selective Underintegration of the Four-Node Quadrilateral

Selective underintegration of the four-node quadrilateral was the first ever application (1969)[14] of reduced integration, uniform or selective. The element is simple and serviceable and may still be found in commercial finite element codes such as MSC/NASTRAN. The intent was to eliminate shear locking which, independent of element shape, results from spurious shear strains of the form $\gamma_{xy} = a\xi + b\eta$ (see Section 6.2.1, 6.6.1, and 6.6.2). Clearly, evaluation of γ_{xy} at $\xi = \eta = 0$ eliminates shear locking but, as we have seen, uniform evaluation of all strain components at that point produces spurious modes. The obvious remedy is to evaluate ε_x and ε_y at 2 x 2 Gauss points, as shown in Figure 7.11. The energy in ε_x and ε_y will, as a result, provide stiffness for the spurious hourglass modes, $u = \xi\eta$ and $v = \xi\eta$.

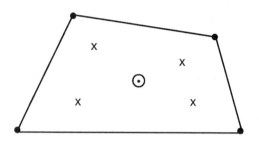

x Points for evaluation of ε_x and ε_y ; $\xi = \eta = \pm\, 1\,/\sqrt{3}$

⊙ Point for evaluation of γ_{xy} ; $\xi = \eta = 0$

Figure 7.11 Selective Underintegration of a Quadrilateral Element.

If shear strain is not coupled elastically to the direct strains, as in the case of isotropic materials, we can compute the contributions of γ_{xy} and ε_x, ε_y separately and add them. Thus the ij partition of the stiffness matrix is

$$\left[K_{ij}\right] = \sum_{g=1}^{4} J_g \left[B_{gi}^{12}\right]^T \left[D_g^{12}\right]\left[B_{gj}^{12}\right] + 4J_o\left\{B_{oi}^3\right\}D_o^3\left\lfloor B_{oj}^3\right\rfloor \qquad (7{:}20)$$

where the matrices are the values at particular points of portions of the strain-displacement and modulus matrices, i.e., particular values of the partitions of

$$\left[B_j\right] = \begin{bmatrix} B_j^{12} \\ \cdots \\ B_j^3 \end{bmatrix} \quad ; \quad \left[D\right] = \begin{bmatrix} D^{12} & \vdots & 0 \\ \cdots & \cdots & \cdots \\ 0 & \vdots & D^3 \end{bmatrix} \qquad (7{:}21)$$

A similar tactic alleviates dilatation locking for plane strain. In this case the deviatoric strains are integrated at 2×2 Gauss points and the dilatation is integrated at the center. Deviatoric strains and stresses are defined as follows.

$$\{\bar{\varepsilon}\}^T = \left\lfloor \varepsilon_x - \tfrac{1}{3}e, \quad \varepsilon_y - \tfrac{1}{3}e, \quad \varepsilon_z - \tfrac{1}{3}e, \quad \gamma_{xy}, \quad \gamma_{xz}, \quad \gamma_{yz} \right\rfloor$$

$$\{\bar{\sigma}\}^T = \left\lfloor \sigma_x + p, \quad \sigma_y + p, \quad \sigma_z + p, \quad \tau_{xy}, \quad \tau_{xz}, \quad \tau_{yz} \right\rfloor \qquad (7{:}22)$$

where the dilatation, e, and mean pressure, p, are

$$e = \varepsilon_x + \varepsilon_y + \varepsilon_z \quad , \quad p = \frac{-1}{3}\left(\sigma_x + \sigma_y + \sigma_z\right) \qquad (7{:}23)$$

The constitutive relationships are

$$\{\bar{\sigma}\} = \left[\bar{D}\right]\{\bar{\varepsilon}\} \quad , \quad p = -Ke \qquad (7{:}24)$$

where, for isotropic material, $\left[\bar{D}\right]$ is a diagonal matrix and $K = E / 3(1 - 2\upsilon)$ is the bulk modulus. The terms on the diagonal of $\left[\bar{D}\right]$ are equal to 2G for $\varepsilon_x, \varepsilon_y, \varepsilon_z$ and equal to G for $\gamma_{xy}, \gamma_{xz}, \gamma_{yz}$.

Both forms of locking can be removed by measuring both in-plane shear and dilatation at the element's center. It should be noted that forms of locking due to shape sensitivity, such as parallelogram locking and trapezoidal locking, are not removed by selective underintegration.

A complication arises if the shear strain is elastically coupled to the direct strains so that the null partitions of $[D]$ in Equation 7:21 are no longer null. An effective procedure for this case is to substitute the shear strain at the center for the shear strain at the 2 x 2 Gauss points and then to integrate strain energy in the standard way. This unorthodox integration tactic should immediately arouse our suspicions regarding satisfaction of constant strain patch tests. To check it, we appeal to the argument in Section 5.3 which holds that constant strain patch tests will be satisfied if the nonconforming part of the strain-displacement matrix integrates to zero over the volume of the element. Thus if

$$\left[B_i\right] = \left[B_i^c\right] + \left[B_i^n\right] \qquad (7{:}25)$$

where $\left[B_i^c\right]$ is the conforming part and $\left[B_i^n\right]$ is the nonconforming part, and if

$$\int_{V_e}\left[B_i^n\right]dV = \int_{V_s}J\left[B_i^n\right]d\xi\,d\eta = 0 \qquad (7{:}26)$$

then the nodal force vector

$$\{F_i\} = \int_{V_e}\left[B_i\right]^T\{\sigma\}\,dV = \int_{V_e}\left[B_i^c\right]^T\{\sigma\}\,dV \qquad (7{:}27)$$

for constant values of $\{\sigma\}$. In the present case

$$\left[B_i^n\right] = \left[\begin{array}{c} 0 \\ \hline B_{oi}^3 - B_i^3 \end{array}\right] \qquad (7{:}28)$$

where $\left[B_{oi}^3\right]$ is the value of $\left[B_i^3\right]$ at the center of the element. From the discussion in Section 4.3.3, we know that a typical term of the integrand in the second form of Equation 7:26 is $X_{,\xi}\,y_{,\eta}$ where X is any one of the element's elementary basis functions. Thus, since $\lfloor X \rfloor = \lfloor 1, \xi, \eta, \xi\eta \rfloor$ for the four-node quadrilateral, $X_{,\xi} = a + b\eta$ and $y_{,\eta} = c + d\xi$. The linear and bilinear terms integrate to zero in parametric space leaving only the constant term which, from the form of Equation 7:28, must also be zero. Consequently, since the integral of the nonconforming part of the strain-displacement matrix is zero, we conclude that the four-node quadrilateral with selective strain *substitution* passes constant strain patch tests.

A variation of this procedure is to replace $\lfloor B_{gi}^3 \rfloor$ with its Jacobian-weighted average.

$$\lfloor \overline{B}_{gi}^3 \rfloor = \frac{\displaystyle\sum_{g=1}^{4} J_g\lfloor B_{gi}^3 \rfloor}{\displaystyle\sum_{g=1}^{4} J_g} \qquad (7{:}29)$$

Since the Jacobian is known to be a linear function of ξ and η, the denominator is just four times the value of J at $\xi = \eta = 0$. The numerator likewise sums to four times its value at the center, giving $\left\lfloor \overline{B}^3_{gi} \right\rfloor = \left\lfloor B^3_{oi} \right\rfloor$. Quite independently, the form of Equation 7:29 guarantees that

$$\sum_{g=1}^{4} J_g \left\lfloor \overline{B}^3_{gi} - B^3_{gi} \right\rfloor = 0 \qquad (7:30)$$

which satisfies Equation 7:26.

The selective *underintegration* procedure, treated first, also passes the patch test, since it has been established in Section 5.4 that single-point integration of the four-node quadrilateral is sufficient. The separate integration of uncoupled terms in Equation 7:20 does not alter that conclusion.

Another important issue relates to the requirement that the stiffness matrix be invariant to the way in which the element's coordinate system is selected. Let us assume that the nodal displacements and forces, and hence stiffness, are recorded in some external, *global* coordinate system and that element properties, such as stresses and strains, are recorded in an *element* coordinate system that is fixed in the element. We wish to establish that stiffness in the global system is invariant to selection of the element coordinate system, or at least to the alternatives allowed for its selection by a particular algorithm. For example, an element's x-axis is frequently taken to be parallel to the line joining the first two nodes identified in the element's input data, so that the x-axis could be parallel to any of the element's four sides.

We have noted, in Section 3.1, that using the same shape functions for all components of displacement guarantees the functional relationship between field quantities (u, v, w) and nodal quantities $\left(u_i, v_i, w_i\right)$ to be invariant to coordinate system rotation. That is all that is required to guarantee invariance of the stiffness matrix to rotation of the element system, provided that anisotropies are not introduced into the calculation. This issue was faced in Section 3.2 where the question of isotropy was found to influence the selection of basis functions and node locations. It also occurs in connection with selective underintegration (or substitution) if the quantity selectively underintegrated (or substituted) is not invariant to coordinate system rotation. In the two examples we have examined, one selectively underintegrated quantity (the dilatation) is invariant to coordinate system rotation and the other (in-plane shear) is not. As a consequence, the method of orienting the element coordinate system is an important concern for reduced integration of in-plane shear but not for reduced integration of dilatation.

For the case of in-plane shear, the important result is that the orientation of the coordinate system not change, or at most change by multiples of 90°, when nodes are renumbered or called in a different order by an element. This protection will ensure, for example, that resequencing the nodes to minimize computer time will not change the solution. Any number of schemes will satisfy this requirement; for example, we could take the longest edge to be the x-axis. The selection is, however, narrowed by the further requirement that the solution not change abruptly when a small change is made to the finite element model, as for example when the relative sizes of two nearly equal edges are interchanged.

A method[15] which satisfies both requirements for the four-node quadrilateral is shown in Figure 7.12. It uses the bisectors of the angles between the element's diagonals as the element's x- and y-axes. The bisector most nearly parallel to side ①—② is used as the x-axis. Note that substitution of side ②—③ for side ①—② merely rotates the element's coordinate system through 90°, which leaves the global stiffness matrix unchanged.

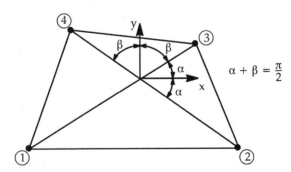

Figure 7.12 Method for Selecting the Coordinate Axes of a Quadrilateral Element.

7.6.2 Selective Underintegration of the Eight-Node Brick

As discussed in Section 7.2, the eight-node brick with single point integration has twelve spurious modes whose shapes are described by Equation 7:7. It was further shown in Section 7.3 that these modes produce numerous local mechanisms in fields of elements. We should recall, however, that reduced integration cures a number of locking problems associated with the element's bilinear and trilinear basis functions (see Section 6.4.2) and would be useful if the spurious modes could be suppressed.

As in the case of the four-node quadrilateral, the basic tactic is to evaluate the direct strains $\left(\varepsilon_x, \varepsilon_y, \varepsilon_z\right)$ at second order Gauss points which form a 2 x 2 x 2 array and to evaluate shear strains $\left(\gamma_{xy}, \gamma_{xz}, \gamma_{yz}\right)$ at a reduced set of points. Single point integration of the latter components is not sufficient because three of the spurious modes involve twisting about one of the element's axes. This is most easily demonstrated for a rectangular brick using the displacement field

$$u = 0 \quad , \quad v = -xz \quad , \quad w = xy \qquad (7{:}31)$$

which yields strains

$$\varepsilon_x = \varepsilon_y = \varepsilon_z = \gamma_{yz} = 0 \quad , \quad \gamma_{xy} = -z \quad , \quad \gamma_{xz} = y \qquad (7{:}32)$$

Clearly this spurious mode cannot be suppressed by measuring shear strains at $\xi = \eta = \zeta = 0$. If, however, we measure γ_{xy} at $(\xi, \eta, \zeta) = \left(0, 0, \pm 1 / \sqrt{3}\right)$ and γ_{xz} at $(\xi, \eta, \zeta) = \left(0, \pm 1/ \sqrt{3}, 0\right)$, the spurious mode will be suppressed and locking will be avoided. We can, in addition, obtain the correct stiffnesses for constant shear and twist by treating these points like Gauss integration points with weighting factors of four rather than the weighting factor of eight that would be assigned to a single integration point at the origin. Figure 7.13 illustrates the locations of integration points.

If shear strains are elastically coupled to direct strains, the shear strains and direct strains must somehow use the same integration points. There are important enough applications of such coupling, for example in the analysis of laminated composites, that general purpose codes are obliged to provide it.

The obvious ploy is to transfer the value of shear strain at one of the special axial points to the four adjacent regular Gauss points—for example to substitute the value of γ_{xy} at $\left(\xi, \eta, \zeta\right) = \left(0, 0, 1/\sqrt{3}\right)$ for Gauss point values at $\left(\xi, \eta, \zeta\right) = \left(\pm 1/\sqrt{3}, \pm 1/\sqrt{3}, 1/\sqrt{3}\ \right)$. Unfortunately this tactic does not satisfy constant strain patch tests for general element shapes.   An equivalent procedure which does satisfy constant strain patch tests is to replace the value of shear strain at each Gauss point by its Jacobian-weighted average over four points. Translated into an operation on the strain displacement matrix, this means that if $\left\lfloor B_{gi}^{xy} \right\rfloor$ is the row of the strain-displacement matrix for γ_{xy} at Gauss point g, then in Figure 7.13, the Jacobian-weighted average value

$$\left\lfloor \overline{B}_{gi}^{xy} \right\rfloor = \frac{\displaystyle\sum_{g=5}^{8} J_g \left\lfloor B_{gi}^{xy} \right\rfloor}{\displaystyle\sum_{g=5}^{8} J_g} \tag{7:33}$$

replaces $\left\lfloor B_{gi}^{xy} \right\rfloor$ at Gauss points 5, 6, 7, and 8.

This operation clearly satisfies the patch test requirement

$$\int_{V_s} J \left\lfloor B_i^n \right\rfloor d\xi\, d\eta \ = \ \sum_g J_g \left\lfloor B_{gi}^n \right\rfloor \ = \ 0 \tag{7:34}$$

where, in this case, the nonconforming part of the strain-displacement matrix is $\left\lfloor B_{gi}^n \right\rfloor = \left\lfloor \overline{B}_{gi}^{xy} - B_{gi}^{xy} \right\rfloor$.

Unlike the selective *substitution* procedure just described, the selective *underintegration* procedure described first does not satisfy the patch test for the simple reason that reduced integration does not satisfy constant strain patch tests for solid elements with general shape. The expansion from one point to two points for each component of shear strain provides no relief from this conclusion.

As in the case of the four-node quadrilateral, the orientation of the element coordinate system must be invariant to the order of node identification. There appears to be no easy way to satisfy this requirement and the further requirement of no abrupt changes of stiffness in response to small changes in the dimensions of the finite element model. It is, on the other hand, relatively easy to find a suitable set of skewed Cartesian axes, namely those which are tangent to the $\xi, \eta,$ and ζ directions at $\xi = \eta = \zeta = 0$. A suitable set of rectangular axes might then be one with minimum angular differences from these skewed axes.

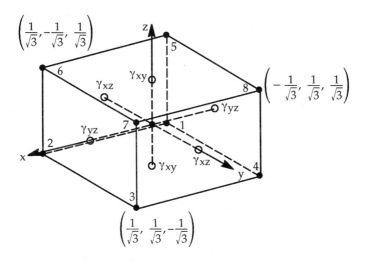

●   Integration points for $\varepsilon_x, \varepsilon_y, \varepsilon_z$

○   Integration points for shear strains

Figure 7.13   **Location of Integration Points for the Eight-Node Brick with Selective Underintegration.**

7.6.3   Selective Integration of Eight- and Nine-Node Quadrilaterals

As was shown in Sections 7.2 and 7.3, the eight-node quadrilateral with reduced integration has a single spurious mode which is noncommunicating in fields of elements. It was further demonstrated in Section 7.4 that, while examples can be found where the spurious mode causes poor results, the performance is generally quite good. Thus there is little reason to consider suppression of the eight-node quadrilateral's spurious mode. Nevertheless, many eight-node elements use selective underintegration in their designs. For example, the design of the MSC/NASTRAN QUAD8, which dates from 1976,[17] did not have access to the analysis and experimental evidence presented in earlier sections of this chapter. Our policy, conceived in fear and sustained by ignorance, was: no spurious modes.

The nine-node quadrilateral has three spurious modes, including two which are global modes in element fields and which unquestionably require suppression. The two type 1 global modes have the form

$$(u, v) = \left(\xi^2 - \frac{1}{3} \right)\left(\eta^2 - \frac{1}{3} \right) \qquad (7{:}35)$$

while the type 2 noncommunicating mode, shared with the eight-node element, has the following form for rectangular shapes.

$$u = \xi\left(\eta^2 - \frac{1}{3} \right)\xi'_x$$

$$v = \eta\left(\xi^2 - \frac{1}{3} \right)\eta'_y \qquad (7{:}36)$$

Clearly, all three modes can be suppressed by evaluating ε_x and ε_y at points other than the 2 x 2 Gauss points, $(\xi, \eta) = \pm 1/\sqrt{3}, \pm 1/\sqrt{3}$, in particular at the 3 x 3 Gauss points. The shear strain γ_{xy} must continue to be measured at the 2 x 2 Gauss points to avoid shear locking (see Table 6.4).

Membrane locking, which will be studied in Chapter 10, imposes an additional requirement. This type of locking is associated with curved shell elements or with curved membrane and solid elements, such as QUAD8 and HEXA20, which are used to model curved beams and shells. It does not occur in the simpler QUAD4 and HEXA8 elements. The nature of the membrane locking phenomenon is that large membrane strains appear in situations where only bending strains should be present. The cause is failure to interpolate cubic in-plane displacements such as $u = x^3$. The indicated remedy, from the point

of view of reduced integration, is to measure ε_x at only two values of ξ. Thus
2×2 integration of ε_x and ε_y appears to be needed to suppress membrane
locking in eight-node and nine-node quadrilateral elements, in conflict with the
proposed strategy for spurious mode suppression.

Note, however, that if we measure ε_x at a 3×2 mesh in ξ, η space and ε_y at a
2×3 mesh, the spurious modes will be suppressed because, in any of the
modes described by Equations 7:35 and 7:36, either the factor $\xi^2 - \frac{1}{3}$ or the
factor $\eta^2 - \frac{1}{3}$ remains intact after differentiation and cannot be zero at three
points simultaneously.

Figure 7.14 illustrates a distribution of strain evaluation points which satisfies
all locking and mode suppression requirements. It evaluates ε_x at six points,
ε_y at six points, and γ_{xy} at four points for a total of sixteen strain evaluations,
more than enough to provide independent stiffnesses for the thirteen strain
states of QUAD8 or the fifteen strain states of QUAD9.

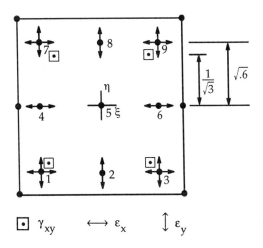

Figure 7.14 Strain Evaluation Points for Selective Integration of Eight- and Nine-Node Elements.

For integration, the direct strains, ε_x and ε_y, must be provided at all nine of the
3×3 Gauss points. This requires interpolation from evaluation points. The
value of ε_x at point 2 can, for example, be obtained by averaging the values at

points 1 and 3.[*] If shear strains are uncoupled from direct strains, they may be integrated at the 2 x 2 Gauss points where they are evaluated. Otherwise the shear strains must be extrapolated to the 3 x 3 Gauss points. The shape function concept is convenient for this purpose. Thus, if γ_i is the value of γ_{xy} at 2 x 2 Gauss point i, we might take the value of γ_{xy} at any other point to be

$$\gamma_{xy} = \frac{1}{4} \sum_{i=1}^{4} \left(1 + 3\xi\xi_i\right)\left(1 + 3\eta\eta_i\right)\gamma_i \qquad (7:37)$$

where $\xi_i, \eta_i = \pm 1/\sqrt{3}, \pm 1/\sqrt{3}$.

The strain evaluation procedure just described is conceptually simple and computationally efficient. It does not, unfortunately, pass constant strain patch tests for general element shapes. The reason for the failure is the usual reason in such cases—failure of the equilibrium condition due to the presence of nonconforming terms in the strain-displacement matrix, terms which are introduced in this case by the interpolation of strains. Remedies are available for equilibrium failure of eight- and nine-node elements with selective integration, but they are not transparently simple and they will lead us to connections with other important concepts in finite element design. This is the subject matter of the next section.

It is a straightforward matter to extend the selective strain evaluation technique just described to the twenty-node brick element. The shear strains are evaluated at 2 x 2 x 2 Gauss points while the direct strains are evaluated at 3 x 3 x 2 Gauss points (or at permutations thereof). The MSC/NASTRAN HEXA20 element[17] uses this procedure. Compared to straight reduced integration, the selective version eliminates six noncommunicating spurious modes at some loss of efficiency and some overall reduction in accuracy.

---

[*]As shown in Figure 7.14, ε_x and ε_y are evaluated (selectively) at 3 x 3 Gauss points. Alternatively, ε_x and ε_y can be evaluated at 2 x 2 Gauss points and points between the 2 x 2 Gauss points and then extrapolated to the 3 x 3 Gauss points. The latter procedure is a little more accurate.

Neither version passes constant strain patch tests for general element shapes,[*] but the selective version can be fixed so that it will pass, using methods described in the next section.

7.7 ASSUMED STRAIN HYBRID FORMULATIONS

The preceding treatment of selective integration began by separating the strain vector into uncoupled parts which were integrated with different sets of Gauss points. It then proceeded to consider the extrapolation of strains from one set of Gauss points to another, or the replacement of a set of strains by a weighted average value, followed by the integration of strain energy at a common set of points. Thus the focus shifted from selective underintegration to selective *substitution*, which was shown to be more versatile with respect to applications (anisotropic materials) and error control. At the same time the principal reason for patch test failure shifted from integration failure to equilibrium failure, and in the last example (eight- and nine-node quadrilaterals) no easy fix for equilibrium failure was found.

We consider in this section a general class of element formulations in which strains computed from displacements are modified prior to the integration of strain energy at a common set of points. Such formulations are sometimes called B-bar methods[18] because if a strain vector computed from displacement is

$$\left\{ \varepsilon^u \right\} = [B] \left\{ u_i \right\} \tag{7:38}$$

then the modified strain vector is simply

$$\left\{ \hat{\varepsilon} \right\} = [\bar{B}] \left\{ u_i \right\} \tag{7:39}$$

---

[*]As will be recalled, solid elements require *full* integration to pass constant strain patch tests.

Such formulations become *projection* methods if the modified strain is obtained from the unmodified strain by

$$\{\hat{\varepsilon}\} = [S]\{\varepsilon^u\} \qquad\qquad (7{:}40)$$

in such a way that

$$[S]\{\hat{\varepsilon}\} = \{\hat{\varepsilon}\} \qquad\qquad (7{:}41)$$

This property is shared with geometric projections, $(\bar{x}) = P(x)$, since $P(\bar{x}) = P(P(x)) = \bar{x}$ is just the projection of a projection on itself. It should be noted that not all B-bar formulations are projection methods.

Perhaps the most insightful way to characterize formulations based on strain modification is to call them *assumed strain methods*. In all such cases, a strain field is assumed, quite independently of the displacement field, which has its own spatial distribution.

$$\hat{\varepsilon} = \lfloor \hat{X} \rfloor \{C\} \qquad\qquad (7{:}42)$$

where $\lfloor \hat{X} \rfloor$ is a row vector of basis functions and $\{C\}$ is a column vector of coefficients. To complete the calculation, the coefficients $\{C\}$ must somehow be related to nodal displacements,

$$\{C\} = [A]\{u_i\} \qquad\qquad (7{:}43)$$

If the relationship is derived through the intermediacy of an assumed displacement field

$$\{u\} = [N]\{u_i\} \qquad\qquad (7{:}44)$$

the formulation is designated as an assumed strain *hybrid* method. The strain substitution techniques described in the preceding section can be classified as examples of the assumed strain hybrid approach. They included the technique of *collocation*, in which assumed strains are equated at specific points to strains derived from the displacement field, and also the technique of averaging. In examples where both techniques were applied, the averaging technique proved superior in that we could easily demonstrate the conditions under which it would pass patch tests. One of the two forms of the Hellinger-Reissner variational principle (see Section 2.5) also represents an assumed strain hybrid approach. Assumed strain methods which do not also employ an assumed displacement field are taken up in later chapters.

In 1978 Malkus and Hughes[19] clarified the relationship between strain substitution methods and mixed energy principles, thereby providing a theoretical foundation for reduced and selective integration. During the 1980s, many new elements, and particularly shell elements, have appeared which employ strain modification methods under one or another of the designations described above.

7.7.1 Least Squares Smoothing

With least squares smoothing, the integral of the squared difference between assumed strains and strains due to displacements is minimized with respect to the coefficients of the assumed strain field. Usually the assumed strain field has a smoother set of basis functions than the strain field due to displacements. Razzaque[20] (1973) was probably the first to use least squares smoothing in a finite element. The Jacobian weighted average, employed in Sections 7.6.1 and 7.6.2 to modify shear strains for QUAD4 and HEXA8 elements, is a primitive form of least squares smoothing with a single, constant basis function.

The first step, and the only one requiring ingenuity, is to select the basis functions for the assumed strain field. Consider the example of the eight- and nine-node elements treated in Section 7.6.3. We know that shear locking is avoided if γ_{xy} is evaluated at 2×2 Gauss points. Hence an appropriate set of basis functions for γ_{xy} is $\left\lfloor \hat{X}^3 \right\rfloor = \left\lfloor 1, \xi, \eta, \xi\eta \right\rfloor$. Likewise, we know that membrane locking is avoided if ε_x is evaluated at a 3×2 set of points. The corresponding set of basis functions for ε_x is $\left\lfloor \hat{X}^1 \right\rfloor = \left\lfloor 1, \xi, \eta, \xi\eta, \eta^2, \xi\eta^2 \right\rfloor$ and for ε_y it is $\left\lfloor \hat{X}^2 \right\rfloor = \left\lfloor 1, \xi, \eta, \xi\eta, \xi^2, \xi^2\eta \right\rfloor$.

The least squares smoothing is applied to each strain component independently so there is no need to group the basis vectors into a matrix. We can even apply local smoothing to parts of each strain component's field. For example, it was noted in connection with Figure 7.14 that ε_x at point 2 could be evaluated as the average of ε_x at points 1 and 3. Thus an appropriate set of assumed strain basis functions for ε_x at points 1, 2, and 3 is $\lfloor \hat{X}^1 \rfloor = \lfloor 1, \xi \rfloor$.

There is no reason to restrict $\lfloor \hat{X} \rfloor$ to functions of the parametric coordinates. We will, for example, use $\lfloor \hat{X} \rfloor = \lfloor 1, x, y \rfloor$ to define a strain field in Section 8.2. The only firm requirement is that a constant term be one of the basis functions; otherwise we could not represent a constant strain field. In addition, the number of assumed strain basis functions should be less than the number of integration points because the least squares fit becomes an identity if the numbers are equal[*] and it becomes singular if the number of basis functions exceeds the number of integration points.

We will assume the following form for the functional which is minimized by least squares smoothing.

$$\Pi_3 = \frac{1}{2} \left\{ \hat{\varepsilon} - \varepsilon^u \right\}^T \left[\hat{W} \right] \left\{ \hat{\varepsilon} - \varepsilon^u \right\} \tag{7:45}$$

In this expression, $\left\{ \hat{\varepsilon} - \varepsilon^u \right\}$ is a vector representing the difference at integration points between a component of the assumed strain field and the corresponding component of the strain field due to displacements. $\left[\hat{W} \right]$ is a symmetric matrix of weighting factors, to be determined. The functional is minimized with respect to the coefficients, $\{C\}$, of the assumed strain field, defined by Equation 7:42. Thus

$$\left\{ \frac{\partial \Pi_3}{\partial C} \right\} = \left[\hat{X} \right]^T \left[\hat{W} \right] \left\{ \left[\hat{X} \right] \{C\} - \varepsilon^u \right\} = 0 \tag{7:46}$$

---

[*]This can be regarded as an example of Fraeijs de Veubeke's limitation principle[21] which states that the results of hybrid and assumed displacement formulations will be identical unless portions of the stress (or strain) field are suppressed.

so that, solving for $\{C\}$,

$$\{C\} = \left[\hat{X}^T\hat{W}\hat{X}\right]^{-1}\left[\hat{X}\right]^T\left[\hat{W}\right]\{\varepsilon^u\} \qquad (7{:}47)$$

The resulting relationship between the assumed strain field and the field due to displacements is

$$\{\hat{\varepsilon}\} = \left[\hat{X}\right]\{C\} = [S]\{\varepsilon^u\} \qquad (7{:}48)$$

where the *smoothing matrix*

$$[S] = \left[\hat{X}\right]\left[\hat{X}^T\hat{W}\hat{X}\right]^{-1}\left[\hat{X}\right]^T\left[\hat{W}\right] \qquad (7{:}49)$$

The substitute $\left[\overline{B}\right]$ matrix is

$$\left[\overline{B}\right] = [S][B] \qquad (7{:}50)$$

It should be noted that, in the present context,

$$\{\hat{\varepsilon}\} = \left[\overline{B}\right]\{u_i\} \qquad (7{:}51)$$

refers to values of the same component of strain at all integration points rather than to all components of strain at one integration point. Construction of the $\left[\overline{B}\right]$ matrix in the latter sense requires sorting and piecing together the $\left[\overline{B}\right]$ matrices for the separate components of strain.

We have, as yet, no assurance that the least squares smoothing process will permit the element to pass constant strain patch tests. To pass such a test, the least squares smoothing operation must exactly reproduce a constant strain field and its accompanying nodal forces.

It turns out that a proper selection of the weighting matrix $\left[\hat{W}\right]$ can guarantee patch test satisfaction. In fact we can exceed patch test requirements and produce exactly the same results as the unmodified element for any strain field that is a linear combination of the basis functions of the assumed strain field. To accomplish this result, we require that[*]

$$\{\hat{\varepsilon}\} = \{\varepsilon^u\} \text{ and } \{\hat{F}_i\} = \{F_i^u\} \text{ for } \{\varepsilon^u\} = [\hat{X}] \qquad (7{:}52)$$

In these relationships $\{\hat{F}_i\}$ is the vector of nodal forces due a single component of strain and $[\hat{X}]$ is the collection of integration point values of the basis functions. It is seen that $\{\hat{\varepsilon}\} = \{\varepsilon^u\}$ produces, by virtue of Equation 7:48, the requirement that

$$[S][\hat{X}] = [\hat{X}] \qquad (7{:}53)$$

or, in other words, the requirement that the smoothing process be a projection method. Substitution of $[S]$ from Equation 7:49 shows that the requirement is satisfied and also that it places no restrictions on $[\hat{W}]$ other than that $[\hat{X}^T \, \hat{W} \, \hat{X}]$ be non-singular.

The requirement that nodal forces remain unmodified for a strain state equal to one of the basis vectors is developed as follows. We know that

$$\{F_i^u\} = \sum_g w_g J_g \{B_{gi}\} \sigma_g^u \qquad (7{:}54)$$

where σ_g^u is a single component of stress at integration point g and $\{B_{gi}\}$ is the ith column of the transposed strain-displacement matrix for the particular stress component. Replacing the summation on g by a matrix operation we obtain

$$\{F_i^u\} = [B_{gi}]^T [W] \{\sigma_g^u\} \qquad (7{:}55)$$

where $[W]$ is a diagonal matrix of terms $w_g J_g$. The stress due to a particular component of strain will have the spatial distribution of that component of

---

[*]Credit goes to Robert L. Harder for the (previously unpublished) proofs which follow.

strain. Hence we can replace $\left\{ \sigma_g^u \right\}$ by $\left[\hat{X} \right]$, the matrix of basis functions for the particular strain component. Thus we can write

$$\left[F_i^u \right] = \left[B_{gi} \right]^T [W] \left[\hat{X} \right] \tag{7:56}$$

In like manner

$$\left[\hat{F}_i \right] = \left[\overline{B}_{gi} \right]^T [W] \left[\hat{X} \right]$$

$$= \left[B_{gi} \right]^T [S]^T [W] \left[\hat{X} \right] \tag{7:57}$$

The equality of modified and unmodified nodal forces requires

$$\left[B_{gi} \right]^T \left([S]^T [W] \left[\hat{X} \right] - [W] \left[\hat{X} \right] \right) = 0 \tag{7:58}$$

which is satisfied if

$$[S]^T [W] \left[\hat{X} \right] = [W] \left[\hat{X} \right] \tag{7:59}$$

Substitution for $[S]$ from Equation 7:49 produces, since $\left[\hat{W} \right]$ is symmetric,

$$\left[\hat{W} \right] \left[\hat{X} \right] \left[\hat{X}^T \hat{W} \hat{X} \right]^{-1} \left[\hat{X} \right]^T [W] \left[\hat{X} \right] = [W] \left[\hat{X} \right] \tag{7:60}$$

This requirement is satisfied if $\left[\hat{W} \right] = [W]$, i.e., if the weighting matrix $\left[\hat{W} \right]$ is taken to be a diagonal matrix of weighted Jacobians, $\left[\smallsetminus wJ \smallsetminus \right]$.

What we have achieved here is a method which retains the flexibility of selective integration (or substitution) and which also satisfies patch tests to the same degree as the standard, unmodified isoparametric formulation. Note, in particular, that full integration should be used. Locking is also mitigated to the same extent as with selective integration. We note, from Table 6.4, that the shear locking terms for QUAD8, and also QUAD9, have $\xi^2 - 1/3$ or $\eta^2 - 1/3$ as a factor. The least squares fit to these terms is zero if the assumed strain basis is $\left[\hat{X} \right] = \lfloor 1, \xi, \eta, \xi\eta \rfloor$. Dilatation locking is also avoided if dilatation is computed separately with the same assumed basis.

The down side of the assumed strain hybrid formulation relative to reduced integration is increased computational cost, largely because full rather than reduced integration is used. This is not a disadvantage for Lagrange elements, such as QUAD9, because reduced integration produces severe mechanisms in these elements. The best, most economical solution for the eight-node membrane element, and perhaps also for the twenty-node brick element, may well be straight reduced integration, which produces only noncommunicating spurious modes in these elements.

Selective substitution works well for the simple four-node membrane and eight-node brick elements.[*] Shear locking is eliminated, spurious modes are avoided, and the patch test is passed. Dilatation locking and shape sensitivity, on the other hand, remain unresolved issues which will be addressed in the next chapter.

7.7.2 The Hellinger-Reissner Variational Principle

The Hellinger-Reissner variational principle provides a means to derive the equations of an element when two independent fields are assumed. As shown by Equations 2:81 and 2:82, the functional to be minimized has two closely related forms, depending on whether the assumed fields are displacement and strain, or displacement and stress. We will show here, for the case of an assumed strain field, that the Hellinger-Reissner variational principle is equivalent to least squares smoothing with a particular choice of weighting functions.

The starting point is Equation 2:81, restated below with current notation.

$$\Pi_R = \int_{V_e} \left(\{\hat{\varepsilon}\}^T [D]\{\varepsilon^u\} - \frac{1}{2}\{\hat{\varepsilon}\}^T [D]\{\hat{\varepsilon}\} \right) dV - W \qquad (7:61)$$

---

[*]Provided that Jacobian-weighted averaging is used, as needed, to ensure patch test satisfaction.

where

$$W = \int_{V_e} \{p\}^T \{u\} dV + \int_{S_e} \{t\}^T \{u\} dS \qquad (7:62)$$

is the potential energy due to applied loads.

The variation of Π_R, as expressed in terms of the variations of $\hat{\varepsilon}$, ε^u, and W, is set to zero.

$$\delta \Pi_R = \int_{V_e} \left(\{\delta\hat{\varepsilon}\}^T [D]\{\varepsilon^u\} + \{\hat{\varepsilon}\}^T [D]\{\delta\varepsilon^u\} - \{\delta\hat{\varepsilon}\}^T [D]\{\hat{\varepsilon}\} \right) dV - \delta W = 0$$

$$(7:63)$$

The symmetry of $[D]$ has been used in this result. We note, from prior definitions, that

$$\{\delta\hat{\varepsilon}\} = \left[\hat{X}\right]\{\delta C\} \quad , \quad \{\delta\varepsilon^u\} = [B]\{\delta u_i\} \qquad (7:64)$$

and that, since W is a function of u only,

$$\delta W = \lfloor Q \rfloor \{\delta u_i\} \qquad (7:65)$$

The matrix $\left[\hat{X}\right]$ in Equation 7:64 is the collection of basis functions for all components of assumed strain. Substitution of Equations 7:64 and 7:65 into Equation 7:63 gives

$$\delta \Pi_R = \{\delta C\}^T \int_{V_e} \left[\hat{X}\right]^T [D]\{\varepsilon^u - \hat{\varepsilon}\} dV + \left(\int_{V_e} \{\hat{\varepsilon}\}^T [D][B] dV - \lfloor Q \rfloor \right) \{\delta u_i\} = 0$$

$$(7:66)$$

Since $\{C\}$ and $\{u_i\}$ are vectors of independent variables, the coefficients of $\{\delta C\}$ and $\{\delta u_i\}$ in Equation 7:66 must vanish independently. Consequently the Euler equations for the Hellinger-Reissner variational principle can be written as

$$\int\limits_{V_e} \left[\hat{x}\right]^T [D]\left\{\varepsilon^u - \hat{\varepsilon}\right\} dV = 0 \qquad (7{:}67a)$$

$$\int\limits_{V_e} [B]^T [D]\left\{\hat{\varepsilon}\right\} dV - \left\{Q\right\} = 0 \qquad (7{:}67b)$$

We wish to show that these two equations can also be derived from the least squares fit of an assumed strain field to strains derived from displacements. For this purpose, let the least squares fit be governed by minimization of the functional

$$\Pi_3 = \frac{1}{2} \int\limits_{V_e} \left\{\hat{\varepsilon} - \varepsilon^u\right\}^T [D]\left\{\hat{\varepsilon} - \varepsilon^u\right\} dV \qquad (7{:}68)$$

with respect to the coefficients of the assumed strain field. We note that Equation 7:68 differs from the functional used in Section 7.7.1. The method described in Section 7.7.1 is equivalent to Equation 7:68 if $[D] = [I]$ and if all integration points are included in the summation.

Setting the variation of Π_3 to zero gives, with $[D]$ symmetric,

$$\delta\Pi_3 = \int\limits_{V_e} \left\{\delta\hat{\varepsilon}\right\}^T [D]\left\{\hat{\varepsilon} - \varepsilon^u\right\} dV = \left\{\delta C\right\}^T \int\limits_{V_e} \left[\hat{x}\right]^T [D]\left\{\hat{\varepsilon} - \varepsilon^u\right\} dV = 0$$

$$(7{:}69)$$

which yields Equation 7:67a exactly.

The least squares fit is completed by minimizing

$$\Pi_1 = \frac{1}{2} \int\limits_{V_e} \left\{\hat{\varepsilon}\right\}^T [D]\left\{\hat{\varepsilon}\right\} dV - W \qquad (7{:}70)$$

with respect to $\{u_i\}$. The assumed strains are related to nodal displacements by

$$\{\hat{\varepsilon}\} = [\hat{x}]\{C\} = [\hat{x}][A]\{u_i\} \tag{7:71}$$

where $[A]$ is evaluated as follows by the least squares fit.

$$[A] = \left[\int_{V_e} [\hat{x}]^T [D][\hat{x}] dV \right]^{-1} \int_{V_e} [\hat{x}]^T [D][B] dV \tag{7:72}$$

Minimization of Π_1 with respect to $\{u_i\}$ then gives

$$\delta\Pi_1 = \{C\}^T \left[\int_{V_e} [\hat{x}]^T [D][\hat{x}] dV \right] [A]\{\delta u_i\} - \lfloor Q \rfloor \{\delta u_i\}$$

$$\tag{7:73}$$

$$= \left(\{C\}^T \int_{V_e} [\hat{x}]^T [D][B] dV - \lfloor Q \rfloor \right) \{\delta u_i\} = 0$$

or, since $\{\delta u_i\}$ is a vector of independent variations,

$$\int_{V_e} \{\hat{\varepsilon}\}^T [D][B] dV - \lfloor Q \rfloor = 0 \tag{7:74}$$

which is just Equation 7:67b transposed.

We have shown that the Hellinger-Reissner variational principle is equivalent to a particular least squares fit of an assumed strain field to strains derived from an assumed displacement field. We have also shown that least squares fits, including this one, can pass patch tests and thus yield convergent results.

7.8 SPURIOUS MODE STABILIZATION

As we have seen, selective integration and its surrogates employ either full integration or a combination of full and reduced integration points. In so doing they give away some or all of the economic advantage of reduced integration. That advantage is particularly important in the case of material nonlinearity where the evaluation of stresses at integration points can consume half or more of the computer time. For such cases it would be highly beneficial to retain reduced integration while finding some simple way to restrain or *stabilize* the resulting spurious modes.

The basic idea behind spurious mode stabilization is to separate the element's modes into low order modes which are evaluated at integration points and high order modes whose stiffnesses are approximated analytically.[7] If the intent is only to prevent large responses of spurious modes, we can assign penalty stiffnesses to the higher order modes without much care. Whether this is our intent or whether it is to add the full information content of the higher order modes to the solution, care must be taken to preserve the accuracy of lower order modes by ensuring patch test satisfaction.

Spurious mode stabilization occupies a middle position between reduced integration and full integration with strain modification (by a mixed energy principle or otherwise). It retains some of the economic advantage of reduced integration and avoids spurious modes. Its chief disadvantage is that it places a burden on the user to select a level of penalty stiffness. A level that is too low may cause the spurious modes to be evident in the solution, while a level that is too high may allow locking to return. An acceptable level may not always be possible.

Belytschko and his colleagues[22-26] have developed a family of stabilized, reduced integration elements which they call γ-elements. Although these elements are intended primarily for plate and curved shell applications, their principles can be explained in the context of four- and nine-node membrane elements. Belytschko, et al., are careful to preserve their elements' rigid body and constant strain properties. They do this by "orthogonalizing" the higher order modes with respect to the lower order ones. We can conveniently explain their procedure in terms of the basis function theory developed in Section 3.1.

Let us begin by considering a four-node isoparametric element with a single integration point. The basis vector for this element is, of course, $\lfloor X \rfloor = \lfloor 1, \xi, \eta, \xi\eta \rfloor$. As we have seen in Section 7.2, the $\xi\eta$ term accounts for the element's spurious modes because strains computed from it will be zero at the integration point, $\xi = \eta = 0$. Thus $\xi\eta$ is the mode shape of both higher order modes (one for u, the other for v).

If we select $\lfloor 1, \xi, \eta \rfloor$ to be the basis vector for the lower order modes, we run into trouble because the representation of linear displacement states in elements with general quadrilateral shapes requires all four terms. We can, however, select the element's basis vector to be $\lfloor \overline{X} \rfloor = \lfloor 1, x, y, \xi\eta \rfloor$, which is exactly equivalent to the original basis (x and y are linear combinations of $1, \xi, \eta, \xi\eta$) and then isolate the last term. The displacement field is expressed as

$$u = \lfloor \overline{X}_i \rfloor \left[\overline{X}_{ij} \right]^{-1} \{ u_i \} \tag{7:75}$$

where, as we recall, $\overline{X}_{ij}$ is the value of $\overline{X}_j$ at node i. If $\{ u_i \}$ is selected to be any one of the basis functions, $\{ \overline{X}_{ij} \}$, u will exactly equal $\overline{X}_j$. Consequently, if $\{ u_i \} = \{ 1 \}, \{ x_i \}$, or $\{ y_i \}$, u will equal 1, x, or y. Thus, since the higher order mode will not be excited by lower order nodal displacements, rigid body and patch test properties will be preserved.

Put formally, we separate u into lower and higher order parts

$$u = u_\ell + u_h \tag{7:76}$$

where

$$u_\ell = \lfloor 1, x, y \rfloor \left[A_{\ell i} \right] \{ u_i \}$$
$$u_h = \xi\eta \lfloor \gamma \rfloor \{ u_i \} \tag{7:77}$$

and

$$\begin{bmatrix} A_{\ell i} \\ ---- \\ \gamma \end{bmatrix} = \begin{bmatrix} \overline{X}_{ij} \end{bmatrix}^{-1} \qquad (7\text{:}78)$$

i.e., $\begin{bmatrix} A_{\ell i} \end{bmatrix}$ and $\begin{bmatrix} \gamma \end{bmatrix}$ are just partitions of $\begin{bmatrix} 1, x_i, y_i, \xi_i \eta_i \end{bmatrix}^{-1}$. Belytschko, et al., provide[26] explicit formulas for these partitions. There is, in fact, no need to compute $\begin{bmatrix} A_{\ell i} \end{bmatrix}$. The value of u_ℓ at the integration point is just the same, for any $\{u_i\}$, as the value obtained by the standard isoparametric formulation. The important result of the revised formulation is the assurance that u_h is not excited when $\{u_i\} = \{u_{\ell i}\}$ or, to put it a little differently, assurance that $\lfloor \gamma \rfloor$ is orthogonal to $\{u_{\ell i}\}$.

The derivatives of $u_\ell \left(\text{and } v_\ell \right)$ provide constant strains which are evaluated at the single integration point or, as noted, we can simply use the standard formulation to compute them. Penalty stiffness is assigned to the higher order modes as follows. Let u_h and v_h be combined into a vector

$$\begin{Bmatrix} u_h \\ v_h \end{Bmatrix} = \xi\eta \begin{bmatrix} \gamma & \vdots & 0 \\ ---&-&--- \\ 0 & \vdots & \gamma \end{bmatrix} \begin{Bmatrix} u_i \\ v_i \end{Bmatrix} \qquad (7\text{:}79)$$

Then the vector of higher order strains

$$\{\varepsilon_h\} = [L]\begin{Bmatrix} u_h \\ v_h \end{Bmatrix} = [A]\begin{bmatrix} \gamma & \vdots & 0 \\ ---&-&--- \\ 0 & \vdots & \gamma \end{bmatrix} \begin{Bmatrix} u_i \\ v_i \end{Bmatrix} \qquad (7\text{:}80)$$

where $[A]$ is a function of position.

The corresponding matrix of higher order stiffness terms is

$$\left[K^h \right] = \int_{V_e} [B]^T \left[D^h \right] [B]\, dV$$

$$= \left[\begin{array}{c|c} \gamma^T & 0 \\ \hline 0 & \gamma^T \end{array} \right] \int_{V_e} [A]^T \left[D^h \right] [A]\, dV \left[\begin{array}{c|c} \gamma & 0 \\ \hline 0 & \gamma \end{array} \right]$$

$$= \left[\begin{array}{c|c} C_1 \gamma^T \gamma & C_2 \gamma^T \gamma \\ \hline C_2 \gamma^T \gamma & C_3 \gamma^T \gamma \end{array} \right] \tag{7:81}$$

where C_1, C_2, C_3 are scalar constants. We can choose to ignore the values which the derivation gives to C_1, C_2, C_3 and assign them arbitrary values to suit our requirements. In fact, Belytschko observes that "the plethora of papers which have appeared on the four-node quadrilateral ultimately have all revolved on the selection of these three constants."[26] This, of course, assumes that patch test satisfaction is a fixed requirement. The assignment which Belytschko gives to these constants ensures freedom from locking for parallelogram shapes (see Section 6.6.1). It has been shown[27] that it is not possible to go farther without violating the patch test.

It is not difficult to extend the stabilization concept just described to the eight-node solid element with single point integration.[22] In that case the appropriate basis vector is $\lfloor 1, x, y, z, \xi\eta, \xi\zeta, \eta\zeta, \xi\eta\zeta \rfloor$ where the last four terms constitute the higher order displacement field. The γ row vector becomes a 4×8 matrix. The economic advantage of a single integration point for the lower order terms may compensate for the complexity of the higher order terms. Of course, if the intent is only to stabilize the higher order terms, it will be sufficient to replace the equivalent of the volume integral in Equation 7:81 with a diagonal matrix of small terms.

The nine-node quadrilateral with reduced integration has three spurious modes. Belytschko chooses[25] to stabilize only the two type 1 communicating modes. Their mode shapes are, as will be recalled, $u \text{ (or } v) = \left(1 - \frac{1}{3}\xi^2\right)\left(1 - \frac{1}{3}\eta^2\right)$. The appropriate equivalent basis vector is

$$\lfloor \overline{X} \rfloor = \left\lfloor 1, \; x, \; y, \; \xi\eta, \; \xi^2, \; \eta^2, \; \xi^2\eta, \; \xi\eta^2, \; \left(1 - \frac{1}{3}\xi^2\right)\left(1 - \frac{1}{3}\eta^2\right)\right\rfloor$$

where the last term constitutes the higher order displacement field. Thus

$$u_h = \left(1 - \frac{1}{3}\xi^2\right)\left(1 - \frac{1}{3}\eta^2\right)\lfloor \gamma \rfloor\{u_i\} \qquad (7:82)$$

where, as in the case of the four-node quadrilateral, $\lfloor \gamma \rfloor$ is the bottom row of $\left[\overline{X}_{ij}\right]^{-1}$. The stabilized nine-node quadrilateral retains all of the properties of its standard reduced integration counterpart plus stabilization of the type 1 spurious modes. The type 2 spurious mode is not stabilized but, because it is noncommunicating, it is relatively benign.

REFERENCES

7.1   J. Barlow, "More on Optimal Stress Points—Reduced Integration, Element Distortions, and Error Estimation," *Intl. J. Numer. Methods Eng.*, 28, pp 1487-504, 1989.

7.2   J. Barlow, "A Stiffness Matrix for a Curved Membrane Shell," Conf. Recent Advances in Stress Analysis, Royal Aeron. Soc., 1968.

7.3   J. Barlow, "Optimal Stress Locations in Finite Element Models," *Intl. J. Numer. Methods Eng.*, 10, pp 243-51, 1976.

7.4   R. D. Lazarov and A. B. Andreev, "Superconvergence of the Gradient for Quadratic Triangular Finite Elements," *Numer. Methods for Partial Differ. Equations*, 4, pp 15-32, 1988.

7.5   G. Strang and G. J. Fix, *An Analysis of the Finite Element Method*, Prentice-Hall, Englewood Cliffs, NJ, p. 168, 1973.

7.6   M. Zlamal, "Superconvergence and Reduced Integration in the Finite Element Method," *Math. Computations*, 32, pp 663-85, 1978.

7.7   D. Kosloff and G. A. Frazier, "Treatment of Hourglass Patterns in Low Order Finite Element Codes," *Numer. Analyt. Methods in Geomechanics*, 2, pp 57-72, 1978.

7.8   N. Bićanić and E. Hinton, "Spurious Modes in Two-Dimensional Isoparametric Elements," *Intl. J. Numer. Methods Eng.*, 14, pp 1545-57, 1979.

7.9   B. M. Irons and T. K. Hellen, "On Reduced Integration in Solid Isoparametric Elements when Used in Shells with Membrane Modes," *Intl. J. Numer. Methods Eng.*, 10, pp 1179-82, 1978.

7.10   B. E. MacNeal, J. R. Brauer, and R. N. Coppolino, "A General Finite Element Vector Potential Formulation of Electromagnetics Using a Time-Integrated Electric Scalar Potential," *IEEE Trans. on Magnetics*, 26, p. 1768, 1990.

7.11   B. E. MacNeal, R. H. MacNeal, and R. N. Coppolino, "Spurious Modes of Electromagnetic Vector Potential Finite Elements," *IEEE Trans. On Magnetics*, 25, p. 4141, 1989.

7.12   B. E. MacNeal, L. A. Larkin, J. R. Brauer, and A. O. Cifuentes, "Elimination of Finite Element Spurious Modes Using a Modal Transformation Technique," *IEEE Trans. on Magnetics*, 26, p. 1765, 1990.

7.13   R. H. MacNeal, "Selective Penalization of Divergence," MacNeal-Schwendler Corp. MSC/EMAS Memorandum RHM-6, 1991.

7.14   W. P. Doherty, E. L. Wilson, and R. L. Taylor, "Stress Analysis of Axisymmetric Solids Using Higher Order Quadrilateral Finite Elements," U. of Calif. Berkeley, Struct. Eng. Lab. Report SESM 69-3, 1969.

7.15   R. H. MacNeal, "A Simple Quadrilateral Shell Element," *Comput. Struct.*, 8, pp 175-83, 1978.

7.16   R. H. MacNeal, "Specifications for the QUAD8 Quadrilateral Curved Shell Element," MacNeal-Schwendler Corp. MSC/NASTRAN Memorandum RHM-46B, 1976.

7.17   R. H. MacNeal, "Specifications for the HEXA Element," MacNeal-Schwendler Corp. MSC/NASTRAN Memorandum RHM-38C, 1976.

7.18   T. J. R. Hughes, "Generalization of Selective Integration Procedures to Anisotropic and Nonlinear Media," *Intl. J. Numer. Methods Eng.*, 15, pp 1413-8, 1980.

7.19   D. S. Malkus and T. J. R. Hughes, "Mixed Finite Element Methods—Reduced and Selective Integration Techniques: A Unification of Concepts," *Comput. Methods Appl. Mech. Engrg.*, 15, pp 68-81, 1978.

7.20   A. Razzaque, "Program for Triangular Bending Elements with Derivative Smoothing," *Intl. J. Numer. Methods Eng.*, 6, pp 333-43, 1973.

7.21   B. Fraeijs de Veubeke, "Displacement and Equilibrium Models in the Finite Element Method," *Stress Analysis* (Ed. O. C. Zienkiewicz and G. S. Holister), John Wiley, London, 1965.

7.22   D. P. Flanagan and T. Belytschko, "A Uniform Strain Hexahedron and Quadrilateral with Orthogonal Hourglass Control," *Intl. J. Numer. Methods Eng.*, 17, pp 679-706, 1981.

7.23   T. Belytschko, C. S. Tsay, and W. K. Liu, "A Stabilization Matrix for the Bilinear Mindlin Plate Element," *Comput. Methods Appl. Mech. Engrg.*, 29, pp 313-27, 1981.

7.24   T. Belytschko, J. S.-J. Ong, and W. K. Liu, "A Consistent Control of Spurious Singular Modes in the Nine-Node Lagrange Element for the Laplace and Mindlin Plate Equations," *Comput. Methods Appl. Mech. Engrg.*, 44, pp 269-95, 1984.

7.25   T. Belytschko, W. K. Liu, J. S.-J. Ong, and D. Lam, "Implementation and Application of a Nine-Node Lagrange Shell Element with Spurious Mode Control," *Comput. Struct.*, 20, pp 121-8, 1985.

7.26   T. Belytschko, W. K. Liu, and B. E. Engelman, "The Gamma-Elements and Related Developments," *Finite Element Methods for Plate and Shell Structures* (Ed. T. J. R. Hughes and E. Hinton), Pineridge Press, Swansea, pp 316-47, 1986.

7.27   R. H. MacNeal, "A Theorem Regarding the Locking of Tapered Four-Noded Membrane Elements," *Intl. J. Numer. Methods Eng.*, 24, pp 1793-9, 1987.

8
More Remedies for Locking

We have seen that reduced integration and the related tactics introduced in Chapter 7 can very often eliminate or greatly reduce locking symptoms. Cases exist, however, where reduced integration is ineffective. Examples include a shear locking mode of the twenty-node brick (Section 6.4.3) and trapezoidal locking of the four-node quadrilateral (Section 6.6.2). We have also seen that, while underintegration of volumetric expansion can prevent dilatation locking of the four-node rectangle, it does not entirely eliminate the error in deviatoric strains (Section 6.2.2).

Even in cases where reduced integration is an effective remedy for locking, the attendant spurious modes create difficulties, particularly for Lagrange elements. This fact provided motivation for the introduction of *selective* underintegration and the related assumed strain hybrid formulations. The computational advantage of reduced integration is sacrificed thereby since all strains must be extrapolated to full integration points.

In this chapter we consider methods which are not derivatives of reduced integration. They are of two types. In one type, additional degrees of freedom provide the missing spatial distributions which will relax locking. Such additional degrees of freedom include Wilson's incompatible modes[1] (Section 8.1) and drilling freedoms[2] (Section 8.2). The contrast of these remedies with reduced integration is sharp because relief is achieved by adding rather than removing terms in the strain field.

In the other type of remedy, changes are made to the way in which displacements are interpolated to form strains. These methods include the substitution of metric interpolation for standard parametric interpolation (Section 8.3) and, more radically, the calculation of strains from nodal displacements without the intermediation of a continuous displacement field (Section 8.4).

8.1   BUBBLE FUNCTIONS

We take the terms *bubble function* and *incompatible mode* to be equivalent. Each term is roughly descriptive of a nodeless displacement state which is added to supplement an element's displacement basis. The basic idea may be expressed in the form

$$\{u\} = \sum_i N_i\{u_i\} + \sum_o N_o\{\bar{u}_o\} \qquad (8{:}1)$$

where (i) refers to nodal values and (o) refers to nodeless degrees of freedom. A typical nodeless shape function for a four-node quadrilateral has the form $N_o = 1 - \xi^2$. Displacements due to this function are incompatible, i.e., nonconforming, because they do not vanish on all edges of the element. The curved shape of this function also roughly resembles a bubble.

The center node of a nine-node quadrilateral has the shape function $N_9 = \left(1 - \xi^2\right)\left(1 - \eta^2\right)$ which definitely resembles a bubble but which vanishes on all edges and is therefore compatible. Such a function could be used to supplement the displacement basis of a four-node quadrilateral with interesting if impractical consequences. In this case it is immaterial whether or not we associate $\{\bar{u}_0\}$ with a node because $\{\bar{u}_0\}$ does not communicate directly with adjacent elements. The same can be said for any of the interior nodes of any Lagrange element; in other words, the shape functions of interior nodes are particular cases of (compatible) bubble functions.

Whatever the added functions in Equation 8:1 may be called, their purpose is to augment the element's inventory of strain states. In other words, the strain vector becomes

$$\{\varepsilon\} = \sum_i [B_i]\{u_i\} + \sum_o [B_o]\{\bar{u}_o\} \tag{8:2}$$

We will see later that it is frequently convenient to *assume* the form of $[B_o]$ rather than to derive it from an assumed set of displacement functions. The columns of $[B_o]$ are aptly described as *auxiliary strain states*.

8.1.1 Wilson's Incompatible Four-Node Quadrilateral

The earliest element with incompatible modes was the four-node incompatible membrane quadrilateral published by Wilson, et al. [1] in 1973. The specific form of Equation 8:1 for this element is

$$\{u\} = \sum_{i=1}^{4} N_i^{(4)}\{u_i\} + N_5\{\bar{u}_5\} + N_6\{\bar{u}_6\} \tag{8:3}$$

where $N_i^{(4)}$ is one of the standard four-node isoparametric shape functions, $N_5 = \left(1 - \xi^2\right) / 2$ and $N_6 = \left(1 - \eta^2\right) / 2$.

The functions N_5 and N_6 are sketched in Figure 8.1. They are nonconforming because the incompatible modes of adjacent elements are *not* required to have identical values on common boundaries. An important feature of the formulation is that the work done by boundary tractions on the incompatible modes is ignored. Inclusion of this work would couple the nodeless variables

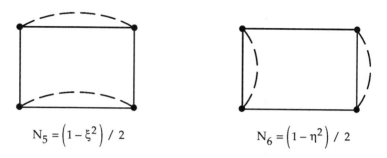

$$N_5 = \left(1 - \xi^2\right) / 2 \qquad\qquad N_6 = \left(1 - \eta^2\right) / 2$$

Figure 8.1   Wilson's Incompatible Modes.

of adjacent elements, thereby violating rule 2 of the basic assumptions listed in Section 2.2:

> "Two finite elements interact with each other only through the common values of a finite set of variables located in their mutual boundary."

The only purpose assigned to functions N_5 and N_6 is to produce the auxiliary strain-displacement matrix $\left[B_o\right]$ in Equation 8:2. A typical term of that matrix is, from Equation 4:51,

$$T\left(B_o\right) = N_{5,x} = -\xi\, y_{,\eta} \, / J \tag{8:4}$$

where J is the Jacobian determinant. The complete set of auxiliary strain states is

$$
\left\{ \begin{matrix} \Delta\varepsilon_x \\ \Delta\varepsilon_y \\ \Delta\gamma_{xy} \end{matrix} \right\}
= \frac{1}{J} \begin{bmatrix} -\xi y_{,\eta} & 0 \\ 0 & \xi x_{,\eta} \\ \xi x_{,\eta} & -\xi y_{,\eta} \end{bmatrix}
\left\{ \begin{matrix} \overline{u}_5 \\ \overline{v}_5 \end{matrix} \right\}
+ \frac{1}{J} \begin{bmatrix} \eta y_{,\xi} & 0 \\ 0 & -\eta x_{,\xi} \\ -\eta x_{,\xi} & \eta y_{,\xi} \end{bmatrix}
\left\{ \begin{matrix} \overline{u}_6 \\ \overline{v}_6 \end{matrix} \right\}
\tag{8:5}
$$

For a rectangle J, $x_{,\xi}$ and $y_{,\eta}$ are constant while $x_{,\eta}$ and $y_{,\xi}$ are zero. We see that, for rectangles, the nodeless variables add independent strain states as follows: $\Delta\varepsilon_x = ax$, $\Delta\varepsilon_y = by$, and $\Delta\gamma_{xy} = cx + dy$ where a, b, c, d are constants proportional to $\overline{u}_5, \overline{v}_6, \overline{v}_5$, and $\overline{v}_6$ respectively. These independent strain states are sufficient to correct shear locking (e.g., to cancel $\gamma_{xy} = x$ in

Equation 6:20) and dilatation locking (e.g., to add $\Delta \varepsilon_y = - \upsilon y /(1 - \upsilon)$ in Equation 6:27). Minimization of strain energy will assure that these corrections occur, or at least that excessively large locking energy does not occur. Tests show [3] that nearly exact results are achieved for in-plane bending of rectangular shapes.

The correction of locking is still virtually perfect for parallelogram shapes but not for general quadrilateral shapes. (Recall from Section 6.6.2 that in-plane bending of a four-node trapezoid induces a *constant* spurious strain in the transverse direction).

The design of the element is completed by forming stiffness matrices and load vectors in the usual manner except that no loads are applied to the nodeless variables. The resulting element equations are

$$
\begin{bmatrix} K_{ii} & K_{oi}^T \\ K_{oi} & K_{oo} \end{bmatrix} \begin{Bmatrix} u_i \\ \bar{u}_o \end{Bmatrix} = \begin{Bmatrix} P_i \\ 0 \end{Bmatrix}
\tag{8:6}
$$

where $\{u_i\}$ includes all nodal variables and $\{\bar{u}_o\}$ includes all nodeless variables.

Static condensation will produce an equilibrium equation in terms of nodal displacements only, i.e.,

$$
\begin{bmatrix} \bar{K}_{ii} \end{bmatrix} \{u_i\} = \{P_i\}
\tag{8:7}
$$

and an equation for displacement at nodeless variables

$$
\{\bar{u}_o\} = \begin{bmatrix} G_{oi} \end{bmatrix} \{u_i\}
\tag{8:8}
$$

where

$$
\begin{bmatrix} \bar{K}_{ii} \end{bmatrix} = \begin{bmatrix} K_{ii} \end{bmatrix} - \begin{bmatrix} K_{oi} \end{bmatrix}^T \begin{bmatrix} K_{oo} \end{bmatrix}^{-1} \begin{bmatrix} K_{oi} \end{bmatrix}
$$

$$
\begin{bmatrix} G_{oi} \end{bmatrix} = - \begin{bmatrix} K_{oo} \end{bmatrix}^{-1} \begin{bmatrix} K_{oi} \end{bmatrix}
\tag{8:9}
$$

An expression for strains in terms of nodal displacements only, obtained by substitution of Equation 8:8 into Equation 8:2, is

$$\{\varepsilon\} = \left[\overline{B}_i\right]\{u_i\} \qquad (8:10)$$

where $\{u_i\}$ includes only nodal displacements and

$$\left[\overline{B}_i\right] = \left[B_i\right] + \left[B_o\right]\left[G_{oi}\right] \qquad (8:11)$$

Thus we see that the method of bubble functions is a $\overline{B}$ method or, in other words, that the net effect of the bubble functions is to alter the matrix relating strains to nodal displacements. In this respect it does not differ from the assumed strain hybrid formulations described in Section 7.7.

In nonlinear analysis it is usually better to keep the nodeless variables as degrees of freedom because static condensation would require specialized iterative procedures.

In its original form the Wilson quadrilateral did not pass constant strain patch tests for general shapes. To see this, we recall from Section 5.3 that a sufficient condition for patch test satisfaction is

$$\int_{V_e} B_i^n dV = \sum_g J_g w_g B_{ig}^n = 0 \qquad (8:12)$$

where B_i^n is any component of the nonconforming part of the strain displacement matrix. The (potentially) nonconforming part is $\left[B_o\right]\left[G_{oi}\right]$ in Equation 8:11. Since $\left[G_{oi}\right]$ is a matrix of constants, we need only apply the test to $\left[B_o\right]$. Substitution of $T(B_o)$ from Equation 8:4 for B_i^n shows that since $w_g = 1$ at the 2 x 2 Gauss points of a four-node element,

$$\int_{V_e} B_i^n dV = \sum_g J_g w_g B_{ig}^n = -\sum_g \xi y_{,\eta} \qquad (8:13)$$

which vanishes if $y_{,\eta}$ is constant. Thus the Wilson element passed constant strain patch tests for parallelogram shapes but not for more general shapes. This fact [4] was brought to the attention of the development team and a correction was issued[5] in 1976. The correction consists simply of replacing $y_{,\eta}$ and the other members of the 2 x 2 Jacobian matrix in Equation 8:5 by their values at the center of the element, $\xi = \eta = 0$. With this change, the element passes constant strain patch tests for all element shapes.

8.1.2 Assumed Auxiliary Strain States

A curious fact which emerges from the preceding discussion is that the added shape functions, N_0, do not appear explicitly in the implementation of the theory. Shape functions usually appear in the conversion of boundary tractions and body forces into nodal forces, but these uses are specifically forbidden for bubble functions. As a result N_0 is used only to compute $\left[B_0 \right]$ and, since N_0 is an assumed function, we might just as well assume $\left[B_0 \right]$ instead. This has the advantage that we can concentrate on the selection of the terms in $\left[B_0 \right]$ to achieve such purposes as the cancellation of the terms which cause locking in the element's strain states. Assuming $\left[B_0 \right]$ has, however, the disadvantage that invariance to orientation of the element coordinate system is no longer guaranteed. (The specification of displacement shape functions has this property so long as the same shape functions are used for all components of displacement.) Lack of such invariance is not serious if orientation of the element coordinate system is invariant, within multiples of 90°, to node sequencing. See Figure 7.12 for one such method.

As an example to illustrate the direct selection of $\left[B_0 \right]$, we recall that the elimination of shear and dilatation locking of the four-node rectangle requires the additional terms , $\Delta\varepsilon_x = ax$, $\Delta\varepsilon_y = by$, and $\Delta\gamma_{xy} = cx + dy$. Since x and y are respectively proportional to ξ and η for rectangular geometry, we surmise that locking will be eliminated if the following bubble functions are added.

$$\left\{ \begin{array}{c} \Delta\varepsilon_x \\ \Delta\varepsilon_y \\ \Delta\gamma_{xy} \end{array} \right\} = \frac{1}{J} \begin{bmatrix} \xi & 0 & 0 & 0 \\ 0 & \eta & 0 & 0 \\ 0 & 0 & \xi & \eta \end{bmatrix} \left\{ \begin{array}{c} \overline{u}_1 \\ \overline{u}_2 \\ \overline{u}_3 \\ \overline{u}_4 \end{array} \right\} \qquad (8:14)$$

These functions are identical to those supplied by the Wilson element (Equation 8:5) for rectangular geometry but not for more general shapes. Which are better? We can easily demonstrate that the additional terms in the Wilson element allow it to solve the shear locking problem for parallelogram shapes. (Examine Equation 6:49 and observe that the coefficients of $\overline{v}_5$ in Equation 8:5 are correct for elimination of shear locking from the example described in Section 6.6.1.) Thus the functions in Equation 8:5 are better than those in Equation 8:14 because the latter provide an exact solution to the shear locking problem for rectangular shapes only.

From a rigorous standpoint, the statements which have been made about the remedial effectiveness of bubble functions are merely conjectural. We have as yet no proof that the added strain states will cancel the locking terms. While we can be sure that the added terms will reduce the strain energy we do not yet know by how much or whether it will be enough to accomplish a cure. Examination of the internal forces will, fortunately, lead us to a true measure of effectiveness. Recall that the internal force vector is related to stresses by

$$\{F\} = \int_V [B]^T \{\sigma\} dV \tag{8:15}$$

It is convenient, in the present case, to partition $\{F\}$ into the forces on nodes and the forces on nodeless variables. Then

$$\{F\} = \begin{Bmatrix} F_i \\ \hline F_o \end{Bmatrix} = \int_V \begin{bmatrix} B_i^T \\ \hline B_o^T \end{bmatrix} \{\sigma\} dV \tag{8:16}$$

Furthermore, we know from Equation 8:6 that $\{F_o\} = 0$ because we have deliberately excluded the application of tractions and body forces to the nodeless variables. As a result

$$\int_V [B_o]^T \{\sigma\} dV = \sum_g w_g J_g [B_{go}]^T \{\sigma_g\} = 0 \tag{8:17}$$

which constitutes a set of constraint conditions on *stresses*. For example, if bubble functions are computed according to Equation 8:14, then

$$\int_V \frac{1}{J} \begin{bmatrix} \xi & 0 & 0 \\ 0 & \eta & 0 \\ 0 & 0 & \xi \\ 0 & 0 & \eta \end{bmatrix} \begin{Bmatrix} \sigma_x \\ \sigma_y \\ \tau_{xy} \end{Bmatrix} dV = 0 \tag{8:18}$$

If we assume rectangular geometry and limit the stresses to constant and linear variations, the operations in Equation 8:18 lead to the conclusion that σ_x can have no variation in the x direction, σ_y can have no variation in the y direction, and τ_{xy} must be constant. These conditions, which imply slightly different conditions on the strains, are clearly sufficient to eliminate both dilatation and shear locking.

Occasions arise where we might wish to impose conditions on the stresses for purposes other than the elimination of locking. For example, we might want the stresses to satisfy equilibrium. While this is the rule for assumed stress elements, it is far from being guaranteed for assumed displacement elements. Bubble functions provide a means to achieve this result. [6]

Consider, for example, the homogeneous stress equilibrium equations in two dimensions

$$\begin{Bmatrix} \sigma_{x,x} + \tau_{xy,y} \\ \sigma_{y,y} + \tau_{xy,x} \end{Bmatrix} = 0 \tag{8:19}$$

Let us assume an element where the stresses are linear functions of x and y. [This requirement can be achieved by least squares smoothing (Section 7.7.1)]. The equilibrium equations are satisfied by Equation 8:17 if

$$\begin{bmatrix} B_{go} \end{bmatrix}^T = \begin{bmatrix} B_1 & 0 & B_2 \\ 0 & B_2 & B_1 \end{bmatrix} \tag{8:20}$$

where

$$\sum_g w_g J_g \begin{Bmatrix} B_1 \\ B_2 \end{Bmatrix} \begin{bmatrix} 1, & x_g - \bar{x}, & y_g - \bar{y} \end{bmatrix} = \begin{bmatrix} 0 & 1 & 0 \\ 0 & 0 & 1 \end{bmatrix} \qquad (8{:}21)$$

and $\bar{x}, \bar{y}$ are the coordinates of the centroid of the element. For example, if $\sigma_x = a + b(x - \bar{x}) + c(y - \bar{y})$, etc., Equation 8:21 implies that

$$\sum_g w_g J_g B_1 \sigma_x = b = \sigma_{x,x}$$

$$(8{:}22)$$

$$\sum_g w_g J_g B_2 \tau_{xy} = \tau_{xy,y}$$

Let

$$\begin{Bmatrix} B_1 \\ B_2 \end{Bmatrix} = [M] \begin{Bmatrix} x_g - \bar{x} \\ y_g - \bar{y} \end{Bmatrix} \qquad (8{:}23)$$

where $[M]$ is a matrix of constants to be determined. Equation 8:23 clearly satisfies the first column of Equation 8:21. The second and third columns yield

$$[M] = \begin{bmatrix} I_{xx} & I_{xy} \\ I_{xy} & I_{yy} \end{bmatrix}^{-1} \qquad (8{:}24)$$

where

$$I_{xx} = \sum_g w_g J_g (x_g - \bar{x})^2$$

$$I_{xy} = \sum_g w_g J_g (x_g - \bar{x})(y_g - \bar{y}) \qquad (8{:}25)$$

$$I_{yy} = \sum_g w_g J_g (y_g - \bar{y})^2$$

are the second moments of the element's area about its centroid. Explicit formulas for B_1 and B_2 are

$$B_1 = \frac{I_{yy}\left(x_g - \bar{x}\right) - I_{xy}\left(y_g - \bar{y}\right)}{I_{xx}I_{yy} - I_{xy}^2}$$

$$B_2 = \frac{I_{xx}\left(y_g - \bar{y}\right) - I_{xy}\left(x_g - \bar{x}\right)}{I_{xx}I_{yy} - I_{xy}^2}$$

(8:26)

Since the integrals of these expressions over the surface of the element vanish, the constant strain patch test is satisfied. In fact, the first column of Equation 8:21 is just a statement of the patch test sufficiency condition.

These results will be applied to the improvement of the design of an element with drilling freedoms in Section 8.2.

8.1.3 Bubble Functions for Other Elements

The Wilson element (Section 8.1.1) extends quite easily to the eight-node hexahedron. In place of Equation 8:3 we have

$$\{u\} = \sum_{i=1}^{8} N_i^{(8)}\{u_i\} + N_9\{\bar{u}_9\} + N_{10}\{\bar{u}_{10}\} + N_{11}\{\bar{u}_{11}\} \quad (8:27)$$

where $N_i^{(8)}$ is a standard eight-node isoparametric shape function, $N_9 = 1 - \xi^2$, $N_{10} = 1 - \eta^2$ and $N_{11} = 1 - \zeta^2$. Note that there are a total of nine bubble functions. These bubble functions are *not* sufficient to remedy all of the locking modes of the HEXA8 isoparametric element. As described in Section 6.4.2, some of the locking modes are two dimensional, i.e., equivalent to the locking modes of QUAD4 in three perpendicular planes, and the bubble functions described above can certainly correct these modes. Table 6.4 shows, however, that the HEXA8 has three additional shear locking modes and three additional dilatation locking modes which involve *cubic* functions of the displacements and which, therefore, require cubic bubble functions.

The HEXA8 element used in MSC/NASTRAN provides relief from locking by a combination of selective underintegration and bubble functions. The selective underintegration (actually selective *substitution*) follows the procedure described in Section 7.6.2 and relieves the shear locking modes. Dilatation locking is corrected by six assumed auxiliary strain states of the following form.

$$
\left\{ \begin{array}{c} \Delta\varepsilon_x \\ \Delta\varepsilon_y \\ \Delta\varepsilon_z \end{array} \right\} = \frac{1}{J} \left[\begin{array}{cccccc} \xi & 0 & 0 & \xi\eta & 0 & \xi\zeta \\ 0 & \eta & 0 & \xi\eta & \eta\zeta & 0 \\ 0 & 0 & \zeta & 0 & \eta\zeta & \xi\zeta \end{array} \right] \left\{ \begin{array}{c} \overline{u}_1 \\ \overline{u}_2 \\ \overline{u}_3 \\ \overline{u}_4 \\ \overline{u}_5 \\ \overline{u}_6 \end{array} \right\}
\qquad (8:28)
$$

Note that the presence of J in the denominator ensures passage of patch tests. Referring to Table 6.4(b) we see that, for the example which illustrates the dilatation locking of HEXA8, an appropriate negative value for $\overline{u}_5$ reduces the dilatation, ε, to zero. The form indicated in Equation 8:28 provides complete relief from locking but only for rectangular element shapes. Extension of this protection to parallelogram shapes can be achieved by applying the correction to skewed Cartesian strain components, $\varepsilon_{\overline{x}}, \varepsilon_{\overline{y}}$, and $\varepsilon_{\overline{z}}$, where the coordinate directions are taken parallel to ξ, η, and ζ at the point where $\xi = \eta = \zeta = 0$.

As we have seen, bubble functions are particularly useful for low order elements. They also have some application to higher order elements. In the case of the eight-node quadrilateral, reduced integration eliminates all common forms of locking in rectangular shapes, including the membrane locking of curved shells which will be studied in Chapter 10. The only drawback to reduced integration for this element is a relatively benign noncommunicating spurious mode. Bubble functions could eliminate locking without introducing this mode, but at the cost of a considerable increase in complexity and computer time. On balance, reduced integration appears to be a better choice for QUAD8.

In the case of the nine-node quadrilateral, reduced integration is accompanied by two *global* spurious modes which are unacceptable for practical calculations (see Section 7.4). Selective underintegration can eliminate the spurious modes, as described in Section 7.6.3, but at the cost of failure to pass the patch test. Replacement of selective underintegration by an equivalent assumed strain hybrid formulation (Section 7.7.1) restores satisfaction of the patch test and produces a satisfactory design which is frequently employed in modern shell elements.

A satisfactory design for the nine-node quadrilateral can also be achieved with bubble functions. A set of auxiliary strain states that will eliminate locking for rectangular elements is

$$
\left\{ \begin{array}{c} \Delta\varepsilon_x \\ \Delta\varepsilon_y \\ \Delta\gamma_{xy} \end{array} \right\} = \frac{1}{J} \left[\begin{array}{cccc} \xi^2 - \frac{1}{3} & 0 & 0 & 0 \\ 0 & \eta^2 - \frac{1}{3} & 0 & 0 \\ 0 & 0 & \xi^2 - \frac{1}{3} & \eta^2 - \frac{1}{3} \end{array} \right] \left\{ \begin{array}{c} \bar{u}_1 \\ \bar{u}_2 \\ \bar{u}_3 \\ \bar{u}_4 \end{array} \right\} \tag{8:29}
$$

Comparison with Table 6.4 shows that $\bar{u}_3$ and $\bar{u}_4$ relieve shear locking and that $\bar{u}_1$ and $\bar{u}_2$ relieve dilatation locking. It will be shown in Chapter 10 that these functions also relieve membrane locking for uniform bending of cylindrical and spherical shells.

The conditions on stress corresponding to Equation 8:29 are

$$
\sum_g w_g J_g \left[B_o \right]^T \{\sigma\} = \sum_g w_g \left[\begin{array}{ccc} \xi^2 - \frac{1}{3} & 0 & 0 \\ 0 & \eta^2 - \frac{1}{3} & 0 \\ 0 & 0 & \xi^2 - \frac{1}{3} \\ 0 & 0 & \eta^2 - \frac{1}{3} \end{array} \right] \left\{ \begin{array}{c} \sigma_x \\ \sigma_y \\ \tau_{xy} \end{array} \right\} = 0
$$

$$\tag{8:30}$$

If σ_x is expanded as the product of Legendre polynomials

$$\sigma_x = \sum_{m,n} a_{mn} P_m(\xi) P_n(\eta) \tag{8:31}$$

then, since $P_2(\xi) = \frac{3}{2}\left(\xi^2 - \frac{1}{3}\right)$ and $\int_{-1}^{1} P_a(\xi) P_b(\xi) d\xi = 0, a \neq b$, Equation 8:30 requires that σ_x contain no term of the form $P_2(\xi) f(\eta)$. Thus the functional forms which cause locking are excluded from the stresses and hence also from the strains.

While bubble functions can relieve locking and are occasionally useful for other purposes, the degrees of freedom which they add increase the complexity of element design and the computational cost. There is also a concern that too many bubble functions will cause spurious modes. In the example of the nine-node quadrilateral just cited, the addition of one more quadratic strain state to Equation 8:29 would cause a spurious mode for the simple reason that the four auxiliary strain states already included in Equation 8:29 plus the four quadratic strain states derived from nodal displacements (see Table 3.6) exhaust the possible number of independent quadratic strain states.

Even when the stability limit is not in danger of being surpassed, more bubble functions do not necessarily imply better accuracy. For example, bubble functions can do nothing to raise the competency of a three-node triangle. The constant strains are complete and the nodal displacements produce no higher order strains to interact with higher order bubble functions. In similar fashion, the four bubble functions of the Wilson element do not, when added to the two linear strain states provided by nodal displacements, make the element complete in the linear strains or even quasi-complete in the sense of the term described in Section 3.3. They just modify the strain states provided by nodal displacements.

8.2   DRILLING FREEDOMS

Drilling freedoms have recently come into vogue as a way to improve the per-formance of low order membrane and solid elements. In the case of membrane elements, drilling freedoms are nodal rotations about an axis normal to the plane of the element. One would not normally think that such rotations would be associated with strains since in-plane rotation, $\theta = \frac{1}{2}\left(v_{,x} - u_{,y}\right)$, is the only combination of displacement first derivatives which does not cause strain. We will, in fact, see that the relationship between drilling freedoms and strains is somewhat contrived. When the concept of drilling freedoms is extended to three dimensions, [7] all three components of nodal rotation are involved.

If we take the standard complement of degrees of freedom at a node to be three components of translation and three components of rotation, we see that the drilling freedoms occupy slots which are frequently not needed for other purposes. From this point of view, drilling freedoms constitute an efficient use of resources. An important exception occurs in the analysis of curved shells where rotation at nodes about an axis normal to an element's surface may be coupled to bending strains by the shell's curvature. In this case the attempt to improve membrane performance by using normal rotations as drilling freedoms can be offset by a loss of bending performance.

Tests show that, with the exception noted above, the accuracy of lowest order elements with drilling freedoms is intermediate between that of lowest order elements without drilling freedoms and that of the next higher order ($p = 2$) elements. The cost of using such elements is also intermediate. Recall from Equation 4:77 that the time to decompose a stiffness matrix is proportional to the cube of the average number of degrees of freedom added to the mesh by a single element (this is true in both two and three dimensions). Figure 8.2 illustrates the number of degrees of freedom added by a four-node element with and without drilling freedoms and by an eight-node element. (Note that the designation, QUADR, is given to a four-node element with drilling freedoms, which is the convention used in MSC/NASTRAN.)

The relative matrix decomposition times for the three elements are as follows:

ELEMENT	QUAD4	QUADR	QUAD8
Matrix Decomposition Time	0.296	1.0	8.0

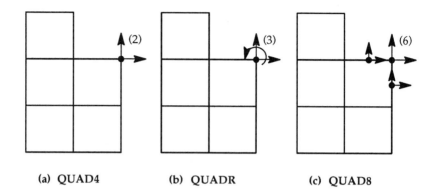

(a) QUAD4	**(b) QUADR**	**(c) QUAD8**

Figure 8.2   Degrees of Freedom Added by One Element in a Mesh of (a) QUAD4, (b) QUADR, and (c) QUAD8 Elements.

It is evident that low order elements with drilling freedoms can be welcome additions to finite element codes if their accuracies approach those of the higher order elements, or at least exceed those of the lower order elements by substantial margins.

Proposals to use drilling freedoms with membrane elements go back to 1965.[8] The performance of the early attempts was, however, so poor that by 1980 Irons and Ahmad[9] tried to head off further disappointment by demonstrating the futility of the concept. Later authors[2,6,10-13] have achieved much better success by modifying the way in which corner rotations are related to deformations.

The issue is illustrated in Figure 8.3. The typical response of an unmodified four-node element to in-plane bending is the keystone deformation pattern shown on the left. The element locks due to the shear strains whose presence is evident at the corners. A possible pattern of deformations due to drilling freedoms is illustrated on the right. When corner rotations are combined with corner translations it is seen that each edge deforms as a general cubic function of position. Note however that the superposition of the deformations due to the drilling freedoms does not relax the shear strains because the drilling freedoms do not change the angle which two edges make at a corner.

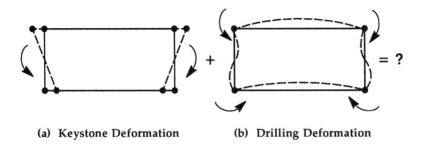

| (a) Keystone Deformation | (b) Drilling Deformation |

Figure 8.3 In-Plane Bending of a Four-Node Element Including the Effect of Drilling Rotations.

The modification which improved the success of drilling freedoms was to ignore the cubic part of edge deformation. Thus, in Figure 8.3(b), the vertical edges remain straight while the horizontal edges bow quadratically. This is precisely what is needed to give correct overall response to in-plane bending.

8.2.1 The Allman/Cook Formulation

The quadratic edge displacements just described can be used to construct special interior strain states, as Allman [2] demonstrated for the case of the membrane triangle. Cook [13] noted further that the quadratic edge displacements are just those provided by elements with midside nodes. Thus, in his modified formulation of a four-node quadrilateral, the corner rotations define normal translations at midside locations which are then used to define strain states via the usual eight-node shape functions. Cook's formulation is appealing because it employs standard isoparametric theory with a little extra work to relate drilling freedoms to edge node displacements.

Figure 8.4 shows the design concept for the four-node element proposed by Cook. Internally the element is an eight-node isoparametric quadrilateral with reduced order integration. Externally, the degrees of freedom at the midpoint of each edge are eliminated in favor of corner point degrees of freedom according to the following formulas.

$$u_{i+0.5} = \frac{1}{2}\left(u_i + u_{i+1}\right) - \frac{y_{i+1} - y_i}{8}\left(\theta_i - \theta_{i+1}\right)$$

$$v_{i+0.5} = \frac{1}{2}\left(v_i + v_{i+1}\right) + \frac{x_{i+1} - x_i}{8}\left(\theta_i - \theta_{i+1}\right)$$

(8:32)

where i is an integer, modulo 4, and all other quantities are defined in Figure 8.4. The part of edge node displacement due to corner rotation is normal to the edge.

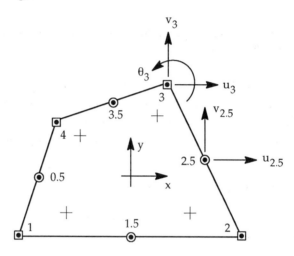

▣	Independent Node
◉	Dependent Node
+	Integration Point (Gauss Point)

Figure 8.4   Cook's Basic Design Concept.

Using Equation 8:32, the relationship between the external degrees of freedom of the underlying eight-node element, $\{u^{(8)}\}$, and the degrees of freedom of the four-node element with rotations, $\{u\}$, can be expressed in condensed matrix form as

$$\{u^{(8)}\} = [T]\{u\}$$

(8:33)

If $\{u^{(8)}\}$ is partitioned into corner degrees of freedom and edge degrees of freedom, and $\{u\}$ is partitioned into translations and rotations, Equation 8:33 expands to

$$
\begin{Bmatrix} u_j \\ v_j \\ --- \\ u_m \\ v_m \end{Bmatrix} = \begin{bmatrix} I & 0 \\ --- & -- \\ L & Q \end{bmatrix} \begin{Bmatrix} u_i \\ v_i \\ -- \\ \theta_i \end{Bmatrix}
\tag{8:34}
$$

where $i = 1 \cdots 4$, $j = 1 \cdots 4$, and $m = 0.5, 1.5, 2.5, 3.5$. The nonzero terms in $[L]$ and $[Q]$, as evaluated from Equation 8:32, are the partitions

$$
\left[L_{i+0.5,\, i} \right] = \begin{bmatrix} \frac{1}{2} & 0 \\ 0 & \frac{1}{2} \end{bmatrix}
$$

$$
\left[L_{i+0.5,\, i+1} \right] = \begin{bmatrix} \frac{1}{2} & 0 \\ 0 & \frac{1}{2} \end{bmatrix}
$$

$$
\left[Q_{i+0.5,\, i} \right] = \frac{1}{8} \begin{bmatrix} y_i - y_{i+1} \\ x_{i+1} - x_i \end{bmatrix}
$$

$$
\left[Q_{i+0.5,\, i+1} \right] = -\frac{1}{8} \begin{bmatrix} y_i - y_{i+1} \\ x_{i+1} - x_i \end{bmatrix}
$$

$$
\tag{8:35}
$$

The strain-displacement relationship is expressed in general as

$$
\begin{Bmatrix} \varepsilon_x \\ \varepsilon_y \\ \gamma_{xy} \end{Bmatrix}_g = \{\varepsilon_g\} = \left[B^{(8)} \right] [T]\{u\}
\tag{8:36}
$$

where $g = 1 \cdots 4$ is the Gauss point index and $[B^{(8)}]$ is the standard isoparametric strain-displacement matrix for an eight-node element. As a result of Equation 8:36, the stiffness matrix can be expressed as

$$[K] = [T]^T \left[K^{(8)} \right] [T] \qquad (8:37)$$

where $\left[K^{(8)} \right]$ is the standard isoparametric stiffness matrix.

Computational efficiency can be improved[6] by writing Equation 8:36 in expanded form and simplifying terms.

$$\{\varepsilon_g\} = \left[B_{gi}^{(8)} \;\middle|\; B_{gm}^{(8)} \right] \begin{bmatrix} I & 0 \\ L & Q \end{bmatrix} \begin{Bmatrix} u_i \\ v_i \\ -- \\ \theta_i \end{Bmatrix}$$

$$\qquad (8:38)$$

$$= \left[B_{gi}^{(8)} + B_{gm}^{(8)} L \;\middle|\; B_{gm}^{(8)} Q \right] \begin{Bmatrix} u_i \\ v_i \\ -- \\ \theta_i \end{Bmatrix}$$

Note that if $\{\theta_i\} = 0$, Equation 8:38 should reduce to the strain-displacement relationship for an isoparametric four-node element. Thus

$$\left[B_{gi}^{(8)} + B_{gm}^{(8)} L \right] = \left[B^{(4)} \right] \qquad (8:39)$$

where $[B^{(4)}]$ is the standard isoparametric strain displacement matrix for a four-node element. As a result

$$\{\varepsilon_g\} = \left[B^{(4)} \;\middle|\; B^{(\theta)} \right] \begin{Bmatrix} u_i \\ v_i \\ -- \\ \theta_i \end{Bmatrix} \qquad (8:40)$$

where $[B^{(\theta)}] = [B_{gm}^{(8)} Q]$. An explicit form for an elementary partition of $[B^{(\theta)}]$ is

$$
\left[B_{gi}^{(\theta)} \right] = \frac{1}{8}
\begin{bmatrix}
\dfrac{\left(y_i - y_{i+1} \right) N_{(i+0.5),x}^{(8)} + \left(y_i - y_{i-1} \right) N_{(i-0.5),x}^{(8)}}{\left(x_{i+1} - x_i \right) N_{(i+0.5),y}^{(8)} + \left(x_{i-1} - x_i \right) N_{(i-0.5),y}^{(8)}} \\[6pt]
\left(y_i - y_{i+1} \right) N_{(i+0.5),y}^{(8)} + \left(y_i - y_{i-1} \right) N_{(i-0.5),y}^{(8)} \\[6pt]
+ \left(x_{i+1} - x_i \right) N_{(i+0.5),x}^{(8)} + \left(x_{i-1} - x_i \right) N_{(i-0.5),x}^{(8)}
\end{bmatrix}
\tag{8:41}
$$

where $N_{(i+0.5),x}^{(8)}$ is the x derivative, measured at the Gauss point, of the displacement shape function for the next mid-side point beyond node i.

In summary, the strain displacement matrix for the four-node element with drilling freedoms is formed by augmenting the isoparametric strain displacement matrix by a partition for rotations whose terms are computed by the formulas shown in Equation 8:41.

Exactly the same approach can be applied to the three-node membrane triangle. The only changes are that $\left[B^{(3)} \right]$ replaces $\left[B^{(4)} \right]$ and $N^{(6)}$ replaces $N^{(8)}$ in Equations 8:40 and 8:41.

The Cook formulation has recently been extended to the eight-node brick. [7] As noted earlier, all three components of rotation at a node are treated as "drilling freedoms." In this case Equation 8:40 becomes

$$
\left\{ \varepsilon_g \right\} = \left[B^{(8)} \mid B_{gm}^{(20)} \, Q_{mi} \right]
\begin{Bmatrix} u_i \\ -- \\ \theta_i \end{Bmatrix}
\tag{8:42}
$$

where $\left[B^{(8)} \right]$ is the standard strain-displacement matrix for an eight-node brick, $\left[B_{gm}^{(20)} \right]$ is the standard strain-displacement matrix for the edge nodes of a twenty-node brick, and both $\left\{ u_i \right\}$ and $\left\{ \theta_i \right\}$ have three components per node. The nonzero elements of $\left[Q_{mi} \right]$ are computed by the method indicated in Figure 8.5. Calculation of the stiffness matrix uses 2 x 2 x 2 Gauss integration.

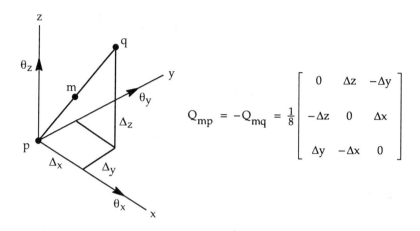

Figure 8.5 Method for Computing Q_{mi}, $i = p, q$.
(Midside node, m, is located between corner nodes p and q.)

The addition of drilling freedoms to an element alters the way in which nodal loads due to edge tractions are formed. Figure 8.6 illustrates the example of uniform pressure. The load on the midside node is transferred to the corners in conformity with the implied equation of constraint

$$u_m = \frac{1}{2}\left(u_1 + u_2\right) + \frac{\ell}{8}\left(\theta_2 - \theta_1\right) \qquad (8{:}43)$$

i.e., the forces of constraint (loads) are proportional to the coefficients in the equation of constraint. Failure to include the resulting moments leads to patch test failures and poor results in practical applications.

When it comes to evaluating the performance of elements with drilling freedoms, the first thing to notice is that they have spurious modes. If, for example, all four corner nodes in Figure 8.4 have the same value of rotation and no translation, the displacements at mid-size nodes will all be zero with the result that the eight-node interior element can have no strain energy.[*] Thus

_____

[*]According to the definition of a spurious mode given in Section 7.2, the mode in question is more like a rigid body mode than a spurious mode. We extend the definition to include this mode.

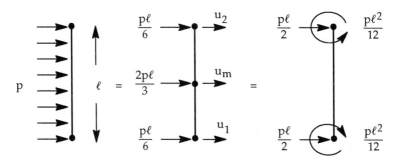

Figure 8.6   Nodal Loads Corresponding to Uniform Pressure.

$\theta_i = \theta_0$, $u_i = v_i = 0$, $i = 1, \cdots n$, where n is the number of corner nodes, describes a spurious mode for membrane elements with drilling freedoms. This mode is *global* in a field of planar elements, i.e., a single spurious mode exists in which all drilling rotations are equal.

A four-node element which uses 2×2 Gauss integration has a second spurious mode for rectangular shapes. This mode corresponds to the noncommunicating spurious mode of the eight-node quadrilateral with reduced integration sketched in Figure 7.1 and described by Equation 7:13. The corner rotations are $\theta_1 = -\theta_2 = \theta_3 = -\theta_4$. The second mode ceases to be spurious (i.e., it picks up energy) when the shape is nonrectangular.

The eight-node rectangular brick with drilling freedoms has six spurious modes which correspond to the global spurious mode of the four-node quadrilateral. In each of these modes all corner translations are zero and the rotations normal to one of the six faces are equal. Three of these modes persist and three vanish for general nonrectangular shapes. They are *local* modes in the sense that each plane in a three-dimensional rectangular array of elements has its own spurious mode. An eight-node rectangular brick with $2 \times 2 \times 2$ Gauss integration also has six noncommunicating modes of the second type, corresponding to the spurious modes of the twenty-node brick with reduced integration.

We have shown in Section 7.4 that global and local spurious modes have serious consequences for practical applications; consequently some means must be provided to suppress the spurious modes of the first type (those with equal rotations). We will address this matter in Section 8.2.3. The noncommunicating modes of the second type can safely be ignored.

It is useful, at this point, to take an inventory of the degrees of freedom of elements with drilling freedoms. The results are shown in Table 8.1. The number of spurious modes for the QUADR and HEXAR elements depends on the element shape and on the order of Gauss integration. The disappearing spurious modes become quadratic or cubic strain states. A more important fact is that drilling freedoms cause the number of *linear* strain states to increase. The number increases from zero to two for TRIAR, from two to four for QUADR, and from nine to fifteen for HEXAR. While in no case do the linear strain states achieve completeness, they do achieve *quasi-completeness*[*] for the QUADR and HEXAR.

The increases in the number of linear strain states provide substantial but incomplete relief from locking. In the case of the triangular membrane element, the two linear strain states reduce but do not completely eliminate shear locking. This is illustrated in Figure 8.7 where the directions of motion for in-plane bending are indicated at the corners and at the midpoints of element sides. The motions are correct at all points except at the center where an unwanted *horizontal* component exists. (Analysis shows that the motion at this point due to drilling freedoms, which is perpendicular to the diagonal, provides only half of the compensation required to negate the horizontal motion due to end point translation.)

In the case of the QUADR element, shear locking is eliminated but dilatation locking persists. The reason is that drilling freedoms only provide motions which are normal to the edges. Consequently, the field component, $v = y^2$, which is required to relieve dilatation locking in a rectangular element (see Section 6.2.2) is not supplied. Additional refinements such as bubble functions can, however, provide such relief.

---

[*]Strain states which are complete when the constraints imposed by equilibrium conditions are taken into account (see Section 3.3 and Table 3.10).

Table 8.1

**Degree-of-Freedom Inventory
for Elements with Drilling Freedoms**

ELEMENT TYPE	TRIAR	QUADR	HEXAR
External Degrees of Freedom	9	12	48
Rigid Body Modes	3	3	6
Constant Strain States	3	3	6
Linear Strain States	2	4	15
Quadratic Strain States	0	0	9
Spurious Modes	1	2*	12**
Internal Degrees of Freedom	9	12	48

*Only one spurious mode if shape is nonrectangular or if 3 x 3 integration is used.

**Only six spurious modes if shape is rectangular and 3 x 3 x 3 integration is used; only three spurious modes for general shapes.

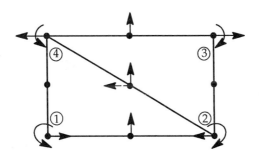

Figure 8.7 In-Plane Bending of a Pair of TRIAR Elements.

8.2.2   Refinements and Performance Data

The Cook variation of the Allman formulation which has just been described relates drilling freedoms to edge nodes and uses the latter's standard isoparametric shape functions to form strain fields. We have seen that some but not all of the locking disorders of the lower order elements are relieved thereby, and that special spurious modes, not present in elements without drilling freedoms, make their appearance. Some of these spurious modes are global in element fields, or even local in three dimensions, and means must be found to suppress them (see Section 8.2.3). We will also find that some of the refinements which have been introduced elsewhere in this book, such as bubble functions and least squares smoothing, can further reduce the locking symptoms.

Consider first the three-node triangle with drilling freedoms, TRIAR. The MSC/NASTRAN variation of this element uses bubble functions to enhance its performance. Several combinations of bubble functions were tried, including the pair which forces satisfaction of equilibrium (see discussion following Equation 8:19), but the most effective combination proved to be one which describes the strains due to *tangential* motion at midside nodes. These motions are, after all, just those displacements of the six-node triangle which drilling freedoms cannot provide. Their shape functions are the standard ones recorded in Table 4.1(b). The constant part of the strains due to these shape functions is removed in order to avoid their coupling with constant strain states.

The in-plane bending performance of the TRIAR element, with and without the bubble functions described above, is compared with standard isoparametric TRIA3 and TRIA6 elements in Figure 8.8. We can see that drilling freedoms substantially improve bending performance over that of the unmodified constant strain triangle but clearly not enough to compete with the linear strain triangle. It is also evident that the addition of bubble functions boosts performance, particularly for slender elements, but not enough to make a critical difference. At bottom, the reason for the lack of spectacular performance for the three-node triangle with drilling freedoms is that the

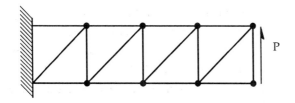

(a) Four Element Pairs; Element Aspect Ratio = 1.0; $\upsilon = 0$

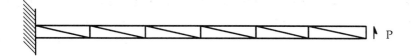

(b) Six Element Pairs; Element Aspect Ratio = 5.0; $\upsilon = 0.3$

ELEMENT TEST:	TIP DISPLACEMENT	
	(a)	(b)
TRIA3	.242	.032
TRIAR (no bubble functions)	.760	.227
TRIAR (with bubble functions)	.796	.554
TRIA6	—	.983
Exact	1.000	1.000

Figure 8.8   Cantilever Beam Tests of Triangular Elements.

drilling freedoms supply only two linear strain states, which is not enough to provide even quasi-completeness.

A difficulty of the QUADR element which requires attention is its sensitivity to dilatation locking. The MSC/NASTRAN QUADR element[6] includes two refinements which together suppress this tendency and which have the bonus feature that they remove the ambiguity of the element's second spurious mode to changes in element shape. The first of these refinements is a least squares smoothing of strains (see Section 7.7.1) which retains only linear terms in x and y, i.e., the assumed strain basis vector is $\lfloor \hat{X} \rfloor = \lfloor 1, x, y \rfloor$ for all components. The second refinement is the addition of auxiliary strain states to enforce stress equilibrium according to the discussion which follows Equation 8:19 in Section 8.1.2. These auxiliary strain states include the linear terms necessary to suppress dilatation locking. If we add them to the seven states derived from nodal freedoms (see Table 8.1) we see that the combined number of strain states is just sufficient to saturate the nine states permitted by least squares smoothing. As a consequence the second spurious mode will remain spurious for all element shapes.

Another consequence of the particular choice of auxiliary strain states is that the element's stiffness matrix is invariant to the orientation of the element's internal coordinate system. This follows because the homogeneous stress equilibrium equations on which the auxiliary strain states are based are the statement of an invariant property (no internal body forces). We can also state with confidence that the resulting element will pass constant strain patch tests if the precautions described in Section 7.7.1 for least squares smoothing and in Section 8.1.2 for forming auxiliary strain states are observed.

The bending performance of the refined and unrefined QUADR elements is compared with other elements in Figure 8.9. These other elements include the QUAD4 and QUAD8 elements which employ selective underintegration for shear. It is observed that the unrefined QUADR gives the same result as the QUAD4 for rectangular element shapes. This is expected because both elements suppress shear locking and neither suppresses the tendency toward dilatation locking. The unrefined QUADR exhibits superior performance only for irregular element shapes.

The performance of the *refined* QUADR is, on the other hand, clearly superior for all element shapes and approaches the performance of the QUAD8 element. We can also point to Figure 1.4 for evidence of the superior performance of the refined QUADR element in an application where the tendency to shear locking

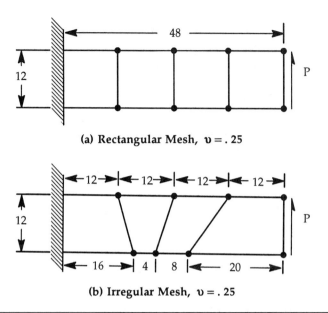

(a) Rectangular Mesh, $\upsilon = .25$

(b) Irregular Mesh, $\upsilon = .25$

ELEMENT	TIP DISPLACEMENT	
TEST:	(a)	(b)
QUAD4*	.9188	.7966
QUADR	.9188	.9457
QUADR (with refinements)	.9776	.9761
QUAD8*	.9846	.9935
Exact	1.000	1.000

*Includes Reduced Integration for Shear

Figure 8.9   Cantilever Beam Tests of Quadrilateral Elements.

is exaggerated by the slenderness of the elements. The element's most impressive achievement is the elimination of locking for trapezoidal shapes which, as we shall see, is not possible for four-node elements without drilling freedoms unless they violate patch tests. It should, incidentally, be noted that spurious modes are not internally suppressed for any of the elements with drilling freedoms in Figures 8.8 and 8.9. Grounding of drilling rotations at the cantilever boundary provides the minimum necessary suppression.

The brick element with drilling freedoms in Reference 8.7 includes a refinement to suppress the tendency toward dilatation locking. That refinement consists of three bubble functions of the form

$$\bar{u}_i = a_i\left(1 - \xi_i^2\right) \quad i = 1, 2, 3 \qquad (8{:}44)$$

where $\xi_1 = \xi$, $\xi_2 = \eta$, $\xi_3 = \zeta$, and $\bar{u}_i$ is displacement in the direction of the ξ_i axis at the origin. The element continues to satisfy constant strain patch tests when these bubble functions are used. As in the case of the four-node quadrilateral, the chief benefit of the bubble functions is to eliminate trapezoidal locking. The overall performance approaches that of a twenty-node brick element at substantially lower computational cost.

8.2.3 Spurious Mode Suppression

The first type of spurious mode identified in Section 8.2.1 has equal values of drilling rotation at all nodes, or at least at all nodes on a face in the case of solid elements. Such modes are global (or possibly local in three dimensions) and should be suppressed. We have seen, in the case of cantilever beam studies, that grounding the drilling freedoms at the built-in end is sufficient to remove this spurious mode without apparent ill effect. This remedy is not always available as, for example, in the case of a simply supported beam. We have seen further, in Section 7.4, that reliance on local conditions to suppress spurious modes is *not* adequate for the hourglass modes of under-integrated four- and nine-node elements. There is, however, an important difference; the equal rotation type of spurious mode for elements with drilling freedoms produces no element deformation.

Whatever the merits of the case, some recent authors [6,7,16] have elected to suppress the first type of spurious mode with a penalty stiffness within each element. The basic idea is to relate the drilling rotations at nodes, θ_i, to the rotation computed from displacement derivatives within the element. At its simplest, in planar elements, this amounts to penalizing the difference between the average drilling rotation at nodes and the in-plane rotation at the center of the element,

$$\Theta_1 = \frac{1}{n} \sum_{i=1}^{n} \theta_i - \theta_o \qquad (8:45)$$

where

$$\theta_o = \frac{1}{2}\left(v,_x - u,_y\right)_o \qquad (8:46)$$

is the in-plane rotation at the center of the element and n is the number of nodes. Any elastic stiffness assigned to Θ_1 will, in the absence of other influences, tend to make the nodal drilling freedoms follow the in-plane rotation. Thus if we assign a "modulus" to Θ_1 equal to $2\gamma G$ where G is the shear modulus and γ is a dimensionless constant, the penalty energy associated with Θ_1 will be $P_1 = V\gamma G\Theta_1^2$ where V is the volume of the element. The penalty stiffness matrix $\left[K_{ij}^p\right] = \left[P_{1,u_i u_j}\right]$ where, u_i, u_j are any two nodal degrees of freedom, is added to the element's elastic stiffness matrix.

The concept which has just been described can be expressed in a variational context [14] which allows for continuous representation of a drilling rotation field and elastic restraint of the difference between in-plane rotation and drilling rotation at any desired number of locations. In the lower order elements considered here, however, there is no need to consider a relationship more complicated than Equation 8:45 nor more than a single location because that is all that is required to suppress the spurious mode.

Suppression of the second (hourglass) type of spurious mode can be achieved by using more integration points or by applying a penalty stiffness along the lines discussed in Section 7.8. Reference 8.6 proposes simply that the function $\Theta_2 = \frac{1}{4}\left(\theta_1 - \theta_2 + \theta_3 - \theta_4\right)$ be penalized elastically to suppress the hourglass mode of QUADR. This will not violate rigid body or constant strain properties because all θ_i are equal in rigid body and constant strain states. Reference 8.7 considers use of Irons' fourteen-point integration rule[15] to suppress the second type of spurious mode for the HEXAR element. In any event, suppression of the noncommunicating hourglass modes of QUADR and HEXAR is not an important issue.

More significance attaches to selection of the magnitude of the parameter, γ, for the first type of spurious mode. Available evidence[6] indicates that the solution of plane stress problems is noticeably degraded for values of γ greater than 0.1 and that for large values of γ the solution approaches that of the element without drilling freedoms. To understand this result, recall from Figure 8.2 that drilling freedoms add about one degree of freedom per element. Thus one (nearly) rigid constraint per element cancels the added freedom.

In the case of curved shells, degradation of the solution starts at very much smaller values of γ. (Reference 8.6 puts the threshold at 10^{-6} for a spherical shell with R / t = 250.) In fact, drilling freedoms are, for reasons discussed earlier, detrimental to the solution of shell problems even without penalty stiffness.

To tell the truth, the issue of penalty stiffness for elements with drilling freedoms has not been resolved at the time of this writing. For example, Ibrahimbegovic,[16] finds no effect of penalty stiffness even for thin shells. This is probably a consequence of the fact that his QUADR element suppresses the constant part of the strain field due to rotations. It also implies that locking must continue to be present in his element for trapezoidal shapes because, as shown in Section 6.6.2, constant transverse strain is a characteristic feature of this form of locking.

8.3   METRIC INTERPOLATION

The exposure of basic concepts at the beginning of this book employed metric interpolation because of its simplicity, but we demonstrated fairly early (Section 3.4) that inter-element displacement continuity fails with metric interpolation in all but the simplest cases. This fact led to the introduction of parametric mapping (Section 3.5) which provides inter-element continuity for all element shapes and which represents linear displacement states exactly, but which cannot properly interpolate higher order displacement states for general shapes. In spite of this shortcoming, parametric mapping dominates the formulation of finite elements. We have employed it consistently with a few exceptions where metric interpolation offers advantages. Those exceptions include the use of mixed metric and parametric interpolation as an aid in the separation of low and high order displacement states (Section 7.8), bubble functions which enforce satisfaction of stress equilibrium (Section 8.1), and a least squares fit which retains constant and linear metric terms in the strain states (Section 8.2).

We consider here additional applications of metric interpolation. As pointed out in Section 3.4, basis functions that are powers of Cartesian coordinates do provide inter-element displacement continuity for triangular and tetrahedral elements with straight edges. This has practical interest for six-node triangles and ten-node tetrahedra because, as demonstrated in Sections 6.1 and 6.6.4, parametric interpolation of quadratic displacement states produces significant errors when edge nodes are offset. Since metric interpolation avoids such errors and also satisfies constant strain patch tests when the edges are straight, it provides a practical alternative to parametric interpolation for TRIA6 and TETRA10 elements. Its drawback is, of course, that patch tests are not satisfied when the edges are curved.

An even better alternative can be provided by offsetting the edge nodes in parametric space[17] as mentioned at the end of Section 6.1. Figure 8.10 illustrates the concept for a TRIA6 element with curved edges.

The projection of an edge node on the subtended chord in metric space subdivides the chord into segments whose proportion is maintained in parametric space. As a result, if the edges are straight in metric space, the relationship between (x, y) and (ξ, η) will be linear with the result that quadratic functions of (x, y) will be properly interpolated. If the edges are curved the interpolation of quadratic functions will no longer be exact but inter-element displacement continuity will continue to be preserved and the element will continue to pass constant strain patch tests.

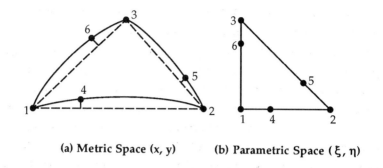

(a) Metric Space (x, y)   (b) Parametric Space (ξ, η)

Figure 8.10   Element with Offset Edge Nodes in Both Metric and Parametric Space.

Metric interpolation has important applications to plate elements because satisfaction of constant curvature patch tests requires correct interpolation of *quadratic* displacement states. As a prelude to these applications, consider the related example of a linear strain patch test for a membrane element. Clearly isoparametric elements cannot satisfy this test for general element shapes but, as indicated in Table 5.1, the QUAD9 element can pass it with bilinear element shape while the QUAD8 element can pass it only for linear geometry (i.e., parallelogram shape). This distinction constitutes the principle advantage of QUAD9 which otherwise suffers computational disabilities, such as global spurious modes, relative to QUAD8. It is, therefore, of some interest to investigate whether the patch test competence of QUAD8 can be raised to that of QUAD9. A simple argument shows that this must be possible because, if the degrees of freedom at the central node of QUAD9 are eliminated by static condensation, the element becomes an eight-node element with all the external properties of the original QUAD9, including its susceptibility to spurious modes. The real issue is whether the competence of the QUAD8 can be raised by a practical approach which avoids the shortcomings of the QUAD9.

Metric interpolation provides a solution in the following manner.[18] Consider a nine-node element in which each component of displacement at the ninth interior node is related to corresponding displacement components at the eight exterior nodes by a rigid equation of constraint. The coefficients in the equation of constraint are selected so that the constrained motion at the ninth node will be correct for any quadratic displacement field imposed on the exterior nodes. Clearly the equation of constraint will not then alter the element's response in a quadratic displacement field. If we use the equation of constraint to eliminate the interior node, we achieve an eight-node element which passes constant and linear strain patch tests under the same conditions as the QUAD9 but which does not inherit QUAD9's susceptibility to spurious modes.

The details are as follows. Let the u component of displacement be represented within the element by nine-node shape functions, i.e., by

$$u = \sum_{j=1}^{9} N_j^{(9)} u_j \tag{8:47}$$

and let the equation of constraint for the interior node be

$$u_9 = \sum_{i=1}^{8} T_i u_i \tag{8:48}$$

The v component uses exactly the same forms. Substitution of Equation 8:48 into Equation 8:47 gives

$$u = \sum_{i=1}^{8} \overline{N}_i u_i \tag{8:49}$$

where

$$\overline{N}_i = N_i^{(9)} + N_9^{(9)} T_i \tag{8:50}$$

This result can also be expressed in terms of standard eight-node shape functions, $N_i^{(8)}$. To accomplish this, we put Equation 8:47 in hierarchical form

$$u = \sum_{i=1}^{8} N_i^{(8)} u_i + N_9^{(9)} \Delta u_9 \tag{8:51}$$

where, assuming that the ninth node is located at $\xi = \eta = 0$,

$$\Delta u_9 = u_9 - \sum_{i=1}^{8} N_i^{(8)}(0,0)u_i \qquad (8:52)$$

and where

$$N_9^{(9)} = \left(1 - \xi^2\right)\left(1 - \eta^2\right) \qquad (8:53)$$

Then, in Equation 8:49,

$$\overline{N}_i = N_i^{(8)} + N_9^{(9)}\left(T_i - N_i^{(8)}(0,0)\right) \qquad (8:54)$$

We see that the shape functions are revised from those of a standard eight-node element to the extent that T_i differs from $N_i^{(8)}(0,0)$.

The constraint coefficients, T_i, are constructed with the aid of a special interpolation formula

$$u = \sum_{i=1}^{8} N_i^* u_i = \left\lfloor N_i^* \right\rfloor \{u_i\} \qquad (8:55)$$

where

$$\left\lfloor N_i^* \right\rfloor = \left\lfloor X_m \right\rfloor \left[A_{mi} \right] = \left\lfloor X_m \right\rfloor \left[X_{im} \right]^{-1} \qquad (8:56)$$

and the elements of $\left\lfloor X_m \right\rfloor$ are a mixed set of metric and parametric basis functions.

$$\left\lfloor X_m \right\rfloor = \left\lfloor 1, x, y, x^2, xy, y^2, \xi^2\eta, \xi\eta^2 \right\rfloor \qquad (8:57)$$

($\left[X_{im} \right]$ is, of course, just the matrix of values of X_m at nodes i.) The constant, linear, and quadratic functions of x and y in $\left\lfloor X_m \right\rfloor$ ensure that the value of u given by Equation 8:55 correctly interpolates any quadratic function of x and y specified at the exterior nodes. This would not be possible if these metric terms were replaced by corresponding powers of ξ and η. The cubic terms, $\xi^2\eta$ and $\xi\eta^2$, which are somewhat arbitrary, are required to ensure the existence of $[A_{mi}] = [X_{im}]^{-1}$.

We require the value of u from Equation 8:55 at only one point, the ninth interior node. If we select the origin of the x, y coordinate system to be at the ninth node, then all terms in $\lfloor X_m \rfloor$ except the first are null at the ninth node, and

$$u_9 = \lfloor 1, 0, 0, 0, 0, 0, 0, 0 \rfloor [A_{mi}]\{u_i\}$$

$$= \lfloor A_{1i} \rfloor \{u_i\} = \sum_{i=1}^{8} A_{1i} u_i$$

(8:58)

Finally, by comparison with Equation 8:48,

$$T_i = A_{1i} \qquad (8:59)$$

It is seen that the evaluation of T_i requires the decomposition of the 8 x 8 matrix $[X_{im}]$. This is the only significant added cost. Computation of the eight-node element's stiffness matrix is altered only by the use of the revised shape functions given by Equation 8:50 or Equation 8:54. We can characterize the method developed here as an N-bar method, in distinction with the B-bar methods developed in Section 7.7.

The revision just described affects performance of the element for non-parallelogram shapes only. The slender cantilever beam with trapezoidal-shaped elements which we have used repeatedly provides a suitable test. The tip deflections are compared in Figure 8.11 for a standard element with full integration (QUAD8F), an element with reduced integration (QUAD8R), and an element with reduced integration and modified shape functions (QUAD8RM). We see that reduced integration provides significant improvement and that the modified shape functions provide a small additional improvement. The improvement which accompanies reduced integration is explained in Section 6.6.3. The additional improvement brought by the modified shape functions is insignificant for this example. We must look to more severe examples of element distortion to find substantial improvement.

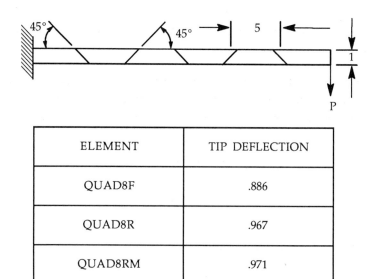

ELEMENT	TIP DEFLECTION
QUAD8F	.886
QUAD8R	.967
QUAD8RM	.971

Figure 8.11 Solutions of Slender Cantilever Beam with Trapezoidal Elements.

Perhaps the greatest benefit of the modification is the knowledge that, in bending applications, the revised element will pass constant curvature patch tests in all cases where the edges are straight and the edge nodes are centered.

Extension of the method just described to a twenty-node brick element is straightforward. The rigid constraints in this case impose correct linear and quadratic displacements at the face nodes and center node of a 27-node brick. The modification will permit twenty-node brick elements to model bending plates with the same patch test competence as similarly modified eight-node plate elements.

8.4   DIRECT ASSUMED STRAIN FORMULATIONS

In this section we continue to develop the concept of an assumed strain field as a remedy for cases in which assumed displacement fields exhibit interpolation failure. The assumed strain hybrid formulations discussed in Section 7.7

improve the displacement field's strain distribution by removing terms which do not have the desired forms; alternatively, the auxiliary strain states introduced in Section 8.1.2 improve the strain distribution by adding terms. Here we consider assumed strain field formulations which do not require a separate displacement field to be defined everywhere within the element. We briefly exposed this concept in Section 2.4.2 as one of three ways to derive the stiffness of the three-node triangle. As described there, the assumed strain field was related to nodal displacement either by line integration over particular paths or by forming indefinite integrals. The examples in this section will either use line integration or a closely related technique where displacement fields assumed along particular line segments are collocated with the assumed strain field.

A significant drawback of these techniques is that it is difficult to determine whether they pass patch tests. Indeed they often fail patch tests where other techniques succeed. For this reason, element designers are inclined to consider *direct* assumed strain formulations, as we shall call them, to be methods of last resort. Nevertheless they have important applications, particularly to plate and shell bending elements as we shall discover in Chapters 9 and 10.

8.4.1   A Six-Node Membrane Shell Element

The six-node membrane triangle has the desirable property that the nodal degrees of freedom are just sufficient to support a complete linear strain field. As a result, the six-node triangle generally exhibits good performance. We have, however, encountered an example (Figure 3.20) where the offset of an edge node causes locking of this element. The cause, as explained in Section 6.6.4, is the aliasing of cubic parametric terms, $\xi^2\eta$, and $\eta\xi^2$, in the displacement field. We will find in Chapter 10 that out-of-plane curvature also causes cubic terms to appear in the displacement field, resulting in the phenomenon called membrane locking.

As discussed in Section 8.3, replacement of parametric interpolation by metric interpolation in a six-node triangle provides relief from locking in some cases. Here we will devise a direct assumed strain formulation which provides similar relief and which is computationally well suited to the analysis of curved shells. In this formulation the set of basis functions for the assumed membrane strain field is taken* to be $\lfloor \hat{X} \rfloor = \lfloor 1, x, y \rfloor$ for all three components $(\varepsilon_x, \varepsilon_y, \gamma_{xy})$. The nine coefficients of the assumed strain field are related to nodal displacements by collocation with *member* strains derived from nodal displacements at the centers of the nine line segments identified in Figure 8.12.

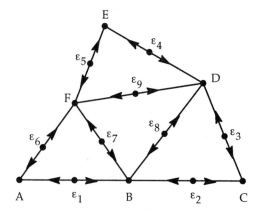

Figure 8.12   The Member Strains of a TRIA6 Element.

Member strain, ε_1 for example, can be expressed in terms of the displacements at nodes A and B with the aid of the line integral

$$\int_A^B \varepsilon \, d\ell = \bar{u}_B - \bar{u}_A \tag{8:60}$$

where $\bar{u}_A$ and $\bar{u}_B$ are displacement components in the direction of the line segment. If ε varies linearly with position, as we have assumed, then

$$\int_A^B \varepsilon \, d\ell = \ell_1 \varepsilon_1 \tag{8:61}$$

---

*The MSC/NASTRAN TRIA6 element[19] uses $\lfloor \hat{X} \rfloor = \lfloor 1, \xi, \eta \rfloor$.

where ε_1 is the value of strain at the center of the segment and ℓ_1 is its length. Thus we have the simple result

$$\varepsilon_1 = \left(\overline{u}_B - \overline{u}_A\right) / \ell_1 \qquad (8:62)$$

We can express each member strain, ε_m, in terms of Cartesian components of strain by

$$\varepsilon_m = \overline{u}_{,\overline{x}} = \varepsilon_x \cos^2 \alpha + \varepsilon_y \sin^2 \alpha + \gamma_{xy} \sin \alpha \cos \alpha \qquad (8:63)$$

where $\overline{u}$ and $\overline{x}$ are directed along the member and α is the angle which the member makes with the x-axis. Replacement of each Cartesian strain component in Equation 8:63 by its assumed spatial distribution then gives us an expression for each member strain in terms of coefficients of the assumed strain field.

The remaining work is straightforward. What we require is a strain-displacement matrix relating strains at integration points to nodal displacements. Put formally, we first assume a strain field

$$\{\varepsilon\} = \left[\hat{X}\right]\{a\} \qquad (8:64)$$

and then we evaluate a vector of member strains $\{\varepsilon_m\}$ in two ways: first in terms of nodal displacements,

$$\{\varepsilon_m\} = [C]\{u_i\} \qquad (8:65)$$

and then in terms of the field coefficients

$$\{\varepsilon_m\} = [D]\{a\} \qquad (8:66)$$

where $[D]$ must be invertible. Finally we require the strains at integration points

$$\{\varepsilon_g\} = \left[\hat{X}_g\right]\{a\} = \left[\hat{X}_g\right][D]^{-1}[C]\{u_i\} \qquad (8:67)$$

which is the desired result.

The procedure just described extends easily to non-flat elements. [19] From Figure 8.12 we see that, if the nodes are displaced in a direction normal to their original plane and if the line segments remain straight, we obtain an object consisting of four flat triangular facets. This object approximates the curved surface passing through the nodes and becomes the geometric model for the assumed strain formulation. The important fact, in any case, is that the member strains represent real deformations of the structure whether curved or flat.

The general procedure described by Equations 8:64 through 8:67 remains intact for curved surfaces. If the angles between the facets and a mean plane are not too large we can continue to assume linear interpolation in the mean (x, y) plane and use the projection of a member onto the mean plane to find the angle α in Equation 8:63. Strict accuracy need only be retained in the relation of member strains to nodal displacements.

We will return to the formulation of six-node shell elements in Section 10.5.3. For the present we restrict our attention to the in-plane bending problem illustrated in Figure 3.20 where it was demonstrated that an isoparametric six-node triangle suffers severe locking when an edge node is offset. Table 8.2 compares these results with those for the MSC/NASTRAN TRIA6 element which uses an assumed strain formulation. The assumed strain formulation clearly eliminates the locking due to edge node displacement.

Table 8.2

**Results for Slender Cantilever Beam in Figure 3.20
Modeled with Two Different Six-Node Element Formulations**

ELEMENT TYPE	TIP DISPLACEMENT	
	ISOPARAMETRIC	ASSUMED STRAIN
Edge Node Centered (A)	.953	.953
Edge Node Displaced (B)	.391	.961
Exact	1.000	1.000

As mentioned earlier, it is not generally an easy matter to determine when an assumed strain element will pass patch tests. In the present case we can establish patch test competence, at least for a flat element, by showing equivalence of the formulation to metric interpolation. For this purpose we need only assure ourselves that the strains at integration points are the same in the two formulations. This must be the case because the calculation of member strains from nodal displacements is exact, given that the strains are linear functions of (x, y), and because no additional approximations are introduced. Thus we conclude, for this example, that the assumed strain formulation will pass flat constant strain patch tests under the same conditions as an element based on metric interpolation, i.e., when the edges of the element are straight.

8.4.2 The Four-Node QUADH Element

As noted repeatedly, four-node elements of trapezoidal shape have a locking problem which stubbornly resists removal by the methods previously considered. The reason, as developed in Section 6.6.2, is that in-plane bending of four-node isoparametric trapezoids is accompanied by a *constant* parasitic strain which is indistinguishable from the legitimate constant strain states. In spite of this difficulty, a direct assumed strain formulation can provide a locking free solution, albeit at the cost of some loss of accuracy for other applications.

Consider the element in Figure 8.13. It has five member strains and has been called[19] the QUADH element because its members somewhat resemble the members of a Hrennikoff[20] framework model. Either diagonal can be used as the fifth member—the results are the same. The element is shown embedded in a beam to which end moments are applied. It is clear, even without defining the assumed strain field, that the element will not lock for uniform in-plane bending. The reason, quite simply, is that for this loading the strain is proportional to y with the result that the average strains in members 2, 4, and 5 are zero.

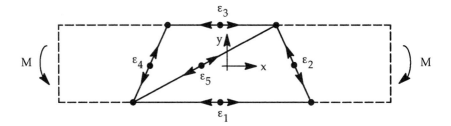

Figure 8.13   The QUADH Element.

We are free to assume any strain dependence we like that has five independent coefficients. For example, we could select

$$\varepsilon_x = a + by$$

$$\varepsilon_y = c + dx \qquad (8:68)$$

$$\gamma_{xy} = e$$

which is comparable to the strain distribution provided by selective underintegration. A better distribution is the one proposed by Turner, et al.,[21] at the dawn of the finite element movement,

$$\left\{ \begin{array}{c} \varepsilon_x \\ \varepsilon_y \\ \gamma_{xy} \end{array} \right\} = \left[D \right]^{-1} \left\{ \begin{array}{c} a + by \\ c + dx \\ e \end{array} \right\} \qquad (8:69)$$

where $\left[D \right]$ is the modulus matrix given, for isotropic materials, by Equation 2:16 or 2:17. Equation 8:69 is equivalent to the assumption that σ_x varies linearly in the y direction, etc., which assumptions are consistent with the loading shown in Figure 8.13. Implementation follows the procedure described in Section 8.4.1.

The performance of the QUADH element is quite good for bending applications, such as that shown in Figure 8.14. The drawback of the QUADH is its inability to satisfy constant strain patch tests. We will provide a proof of

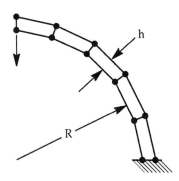

h/R	.03	.006
	TIP DISPLACEMENT	
QUAD4	.615	.163
QUADH	1.008	1.008
Exact	1.000	1.000

Figure 8.14 Curved Cantilever with Five Elements. QUADH Versus Selectively Underintegrated QUAD4 Element.

this assertion in Section 11.4. In the meanwhile we can make it plausible by examining the displacement distribution corresponding to the assumed strain distribution. For example, integration of Equation 8:68 provides

$$u = ax + bxy + \left(e - h\right)y - \tfrac{1}{2}\,dy^2 + j$$

$$v = hx + dxy + cy - \tfrac{1}{2}\,bx^2 + k \tag{8:70}$$

where h, j, k are constants of integration. All eight coefficients can be evaluated in terms of u and v at the four nodes, which demonstrates the equivalence with an assumed displacement formulation. It is, however, a rather peculiar distribution from the viewpoint of assumed displacement theory because u and v do not share the same set of basis functions. Since the presence of the x^2 and y^2 terms makes the displacement field nonconforming for any element shape, patch test satisfaction is not guaranteed. The appearance of similar terms in Wilson's incompatible mode element required special procedures to ensure patch test satisfaction (see Section 8.1.1).

Because it is unable to satisfy patch tests, the QUADH element has not been introduced into MSC/NASTRAN nor, to the author's knowledge, into any other commercial finite element program.

8.4.3   A Nine-Node Element with Assumed Natural Strains

So far we have evaluated member strains at the midpoints of two-node straight line segments where they are correct for any linear strain distribution. The concept readily extends to points on multi-node curved segments where the accuracy of tangential strains can be even higher. Consider, for example, the three-node arc shown in Figure 8.15.

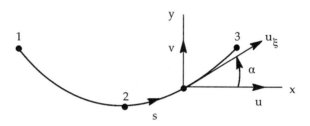

Figure 8.15   Tangential Strain on a Three-Node Arc.

The tangential strain at a point on the arc may be computed from the formula

$$\varepsilon_\xi = u_{,s} \cos \alpha + v_{,s} \sin \alpha = u_{\xi,s} \qquad (8{:}71)$$

where u, v are displacement components in the Cartesian x, y system, s is distance along the arc, and α is the angle which the arc makes with the x-axis. The displacement component u_ξ is in a *fixed* direction, locally tangent to the arc. It is easy to show that Equation 8:71 is equivalent to the more familiar expression

$$\varepsilon_\xi = u_{t,s} - u_n / R \qquad (8{:}72)$$

where u_t and u_n are tangential and normal *curvi-linear* components of displacement and $R = 1 / \alpha_{,s}$ is the radius of curvature.

Since $x_{,s} = \cos \alpha$ and $y_{,s} = \sin \alpha$, we can put Equation 8:71 in the form

$$\varepsilon_\xi = \frac{1}{\left(s_{,\xi}\right)^2} \left(u_{,\xi} x_{,\xi} + v_{,\xi} y_{,\xi} \right) \qquad (8{:}73)$$

where ξ is the familiar parametric coordinate of the arc. If we assume an isoparametric formulation along the arc, then x, y, u, and v employ the same shape functions so that

$$u_{,\xi} = \sum_{i=1}^{3} N_{i,\xi} u_i \quad , \quad \text{etc.} \qquad (8:74)$$

In addition,

$$\left(s_{,\xi}\right)^2 = \left(x_{,\xi}\right)^2 + \left(y_{,\xi}\right)^2 = \left(\sum_{i=1}^{3} N_{i,\xi} x_i\right)^2 + \left(\sum_{i=1}^{3} N_{i,\xi} y_i\right)^2 \qquad (8:75)$$

Evaluation of Equation 8:73 at the reduced order Gauss points, $\xi = \pm 1 / \sqrt{3}$, yields strains which are accurate for displacement fields that are cubic in ξ. This is true because, as demonstrated in Table 7.1, if $u = \xi^3$ then $u_{,\xi}$ is correct at these points and because the same demonstration applies to $x_{,\xi}, y_{,\xi},$ and $v_{,\xi}$.

These results extend easily to higher order curves. They can be applied to a cubic curve in three dimensions by just adding $w_{,\xi} z_{,\xi}$ to Equation 8:73.

An important application of tangential strain evaluation along curved segments occurs in the nine-node assumed natural strain (ANS) element of Park and Stanley.[22] As will be recalled, eight- and nine-node elements have shear locking modes and membrane locking modes which can be eliminated by evaluating all strain components at 2×2 Gauss points. Unfortunately, in the case of the nine-node element, this tactic produces a pair of global spurious modes with disastrous practical implications (see Section 7.4). We showed, in Section 7.6.3, that a selective pattern of strain evaluation points could retain the beneficial properties of reduced integration while removing the spurious modes. That pattern included the evaluation of ε_x and ε_y at six points and the evaluation of γ_{xy} at the four Gauss points. The pattern of strain evaluation points in Park and Stanley's ANS element is similar.

As shown in Figure 8.16, the tangential strains ε_ξ and ε_η are each evaluated at two Gauss points along each of three arc segments passing through the nodes. In previous sections of this chapter we interpolated such member strains by collocating them with an assumed *Cartesian* strain field.   An important innovation in the work of Park and Stanley is the interpolation of member strains by collocation with an assumed *natural covariant* strain field. In two dimensions, natural covariant strains are the set

$$\varepsilon_\xi = u_{\xi,s} \ , \ \varepsilon_\eta = u_{\eta,t} \ , \ \gamma_{\xi\eta} = u_{\xi,t} + u_{\eta,s} \qquad (8\!:\!76)$$

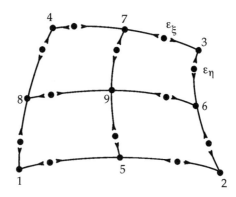

Figure 8.16   Strain Evaluation Points in a Nine-Node ANS Element.

We note that ε_ξ is the tangential strain along a line of constant η and that ε_η is the tangential strain along a line of constant ξ (s and t are arc lengths along these lines). The natural covariant strains are, in fact, just skewed Cartesian components of strain (see Appendix 6A of Chapter 6) oriented with the local directions of ξ and η.

The basis functions used by Park and Stanley for the components of natural covariant strain are as follows:

$$\varepsilon_\xi \ : \left(1, \ \xi\right)\left(1, \ \eta, \ \eta^2\right) = \left(1, \ \xi, \ \eta, \ \xi\eta, \ \eta^2, \xi\eta^2\right)$$

$$\varepsilon_\eta \ : \left(1, \ \eta\right)\left(1, \ \xi, \ \xi^2\right) = \left(1, \ \xi, \ \eta, \ \xi\eta, \ \xi^2, \xi^2\eta\right) \qquad (8\!:\!77)$$

$$\gamma_{\xi\eta} \ : 1, \ \xi, \ \eta, \ \xi\eta$$

The coefficients of the six basis functions for ε_ξ are evaluated by collocation with member strains, ε_{ξ_m}, at the six points previously defined. Note that the use of natural strains rather than Cartesian strains simplifies the calculation in that the projection angles are zero.

Evaluation of the natural shear strain requires a different procedure. We could, perhaps, evaluate Cartesian strains from displacements in the standard isoparametric manner and then extract the skewed shear component at 2×2 Gauss points by means of the transformation given by Equation 6:69. Park and Stanley describe a procedure where $u_{\eta,s}$ is evaluated at the same points as ε_ξ and $u_{\xi,t}$ is evaluated at the same points as ε_η. The formula used to evaluate $u_{\eta,s}$ is

$$u_{\eta,s} = u_{,s}x_{,t} + v_{,s}y_{,t} = \left(u_{,\xi}x_{,\eta} + v_{,\xi}y_{,\eta} \right) / s_{,\xi}\, t_{,\eta} \qquad (8{:}78)$$

where

$$s_{,\xi}\, t_{,\eta} = \left(\left(x_{,\xi} \right)^2 + \left(y_{,\xi} \right)^2 \right)^{\frac{1}{2}} \left(\left(x_{,\eta} \right)^2 + \left(y_{,\eta} \right)^2 \right)^{\frac{1}{2}} \qquad (8{:}79)$$

Shape functions are, as before, used to evaluate the derivatives. Since $\gamma_{\xi\eta}$ has only four basis functions, we can choose to collocate an assumed $u_{\eta,s}$ at only four of the six evaluation points or we can employ least squares smoothing as described in Section 7.7.1. Finally, all of the assumed strain components are evaluated at full 3×3 Gauss integration points and the strain-displacement matrix is computed by the formal procedure described in Section 8.4.1.

An important feature of the ANS element is that it extends directly to curved surfaces. All strain components are defined in the surface so that it is unnecessary to employ projection rules, as would be required if Cartesian components were used.

Nine-node ANS elements perform well, particularly in curved shell applications. A problem with ANS elements, as with any direct assumed strain formulation, is the difficulty in demonstrating patch test satisfaction. Park and

Stanley [22] report that their element does not pass constant strain membrane patch tests with curved edges, but that problems with curved edges tend to converge anyway.

REFERENCES

8.1   E. L. Wilson, R. L. Taylor, W. P. Doherty, and J. Ghaboussi, "Incompatible Displacement Models," *Numerical and Computer Meth. in Struc. Mech.*, S. T. Fenves, et al (Eds.), Academic Press, pp 43-57, 1973.

8.2   D. J. Allman, "A Compatible Triangular Element Including Vertex Rotations for Plane Elasticity Analysis," *Comput. Struct*, 19, pp 1-8, 1984.

8.3   T. J. R. Hughes, *The Finite Element Method*, Prentice-Hall, Englewood Cliffs, NJ, p. 248, 1987.

8.4   P. Lesaint, "On the Convergence of Wilson's Non-Conforming Element for Solving Elastic Problems," *Comput. Methods Appl. Mech. Engrg.*, 7, 1976.

8.5   R. L. Taylor, P. J. Beresford, and E. L. Wilson, "A Nonconforming Element for Stress Analysis," *Intl. J. Numer. Methods Eng.*, 10, pp 1211-9, 1976.

8.6   R. H. MacNeal and R. L. Harder, "A Refined Four-Node Membrane Element with Rotational Degrees of Freedom," *Comput. Struct*, 28, pp 75-84, 1988.

8.7   S. M. Yunus, T. P. Pawlak, and R. D. Cook, "Solid Elements with Rotational Degrees of Freedom: Part 1—Hexahedron Elements," *Intl. J. Numer. Methods Eng.*, 31, pp 573-92, 1991.

8.8   B. N. Abu-Gazaleh, "Analysis of Plate-Type Prismatic Structures," Ph.D. Dissertation, U. of California, Berkeley, 1965.

8.9   B. M. Irons and S. Ahmad, *Techniques of Finite Elements*, Ellis Horwood, Chichester, p. 289, 1980.

8.10   J. Robinson, "Four-Node Quadrilateral Stress Membrane Element with Rotational Stiffness," *Intl. J. Numer. Methods Eng.*, 16, pp 1567-9, 1980.

8.11   G. A. Mohr, "Finite Element Formulation by Nested Interpolations: Application to the Drilling Freedom Problem," *Comput. Struct*, 15, pp 185-90, 1982.

8.12   P. G. Bergan and C. A. Felippa, "A Triangular Membrane Element with Rotational Degrees of Freedom," *Comput. Methods Appl. Mech. Engrg.*, 50, pp 25-69, 1985.

8.13 R. D. Cook, "On the Allman Triangle and a Related Quadrilateral Element," *Comput. Struct*, 22, pp 1065-7, 1986.

8.14 T. J. R. Hughes and F. Brezzi, "On Drilling Degrees of Freedom," *Comput. Methods Appl. Mech. Engrg.*, 72, pp 105-21, 1989.

8.15 B. M. Irons, "Quadrature Rules for Brick-Based Finite Elements," *Intl. J. Numer. Methods Eng.*, 3, pp 239-94, 1971.

8.16 A. Ibrahimbegovic, "A Novel Membrane Finite Element with an Enhanced Displacement Interpolation," *Finite Elem. Analysis & Design*, 7, pp 167-79, 1990.

8.17 M. Utku, E. Citipitioğlu, and G. Özkan, "Isoparametric Elements with Unequally Spaced Edge Nodes," *Comput. Struct*, 41, pp 455-60, 1991.

8.18 R. H. MacNeal and R. L. Harder, "Eight Nodes or Nine?," *Intl. J. Numer. Methods Eng.*, 33, pp 1049-58, 1992.

8.19 R. H. MacNeal, "Derivation of Element Stiffness Matrices by Assumed Strain Distributions," *Nucl. Eng. Design*, 70, pp 3-12, 1982.

8.20 A. Hrennikoff, "Solution of Problems in Elasticity by the Framework Method," *J. Appl. Mech.*, 8, pp 169-75, 1941.

8.21 M. J. Turner, R. W. Clough, H. C. Martin, and L. J. Topp, "Stiffness and Deflection Analysis of Complex Structures," *J. Aeronautical Sci.*, 23, pp 803-23, p. 854, 1956.

8.22 K. C. Park and G. M. Stanley, "A Curved C^0 Shell Element Based on Assumed Natural-Coordinate Strains," *J. Appl. Mech.*, 53, pp 278-90, 1986.

9

Plate Bending Elements

In previous chapters we have, for the most part, treated the design of finite elements from the viewpoint of the two- and three-dimensional theories of elasticity. In this chapter and the next we consider the deformation of plates and shells according to special approximations which permit the two-dimensional treatment of three-dimensional phenomena. The resulting plate and shell elements have an important place in finite element analysis even though it is entirely possible and frequently practical to employ thin solid elements for the same purpose. The advantages of plate and shell elements over thin solid elements are a modest reduction in the number of degrees of freedom and a potential for improved accuracy through the elimination of troublesome terms.

Nowadays finite element designers treat the flat plate element as a special case of the curved shell element. This is a matter of economy: one design to fit all cases.

In the early 1960s, flat plates were about all that an element designer could handle. The design of curved shell elements[1] had to await the development of a practical way to treat curved geometry. [2]

We will separate the treatment of flat plates and curved shells, in part to conform to the historical sequence of development but mainly to simplify the discussion. While the important features of flat plate elements, including their disorders, extend to curved shell elements, the addition of curvature introduces special effects and additional disorders which require separate treatment.

9.1   PLATE THEORY

Plate theory derives from the three-dimensional theory of elasticity by the introduction of simplifying approximations. Consider the region of a flat plate shown in Figure 9.1. The x, y plane is a plane of symmetry for geometric properties and also for material properties. (Later we will relax the latter assumption.) The thickness of the plate, t, is small compared to the scale, L, of dimensions in the plane of symmetry, such as the distance between supports or other structural details, or the wavelengths of load distributions. The translational displacements (u, v, w) are oriented along the (x, y, z) axes. The angles α and β are the "mean" rotational displacements of lines drawn normal to the plane of symmetry. One of the assumptions of first order plate theory is that these lines remain straight during distortion so that the sense of which α and β are "mean" values becomes moot. Note that α and β are related to rotations about the x- and y-axes by

$$\alpha = -\theta_y \quad , \quad \beta = \theta_x \qquad (9:1)$$

The only reason for introducing α and β is that they improve the symmetry of the plate theory equations.

The fundamental assumptions of first order plate theory are, first, that u and v vary linearly through the thickness of the plate and, second, that the normal component of stress, σ_z, is constant through the thickness. Stated in mathematical terms, the fundamental assumptions are

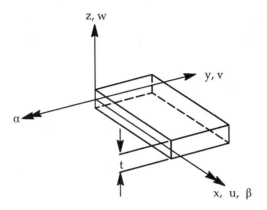

Figure 9.1   Coordinate Definitions for a Plate.

$$u = u^m(x, y) - z\alpha(x, y)$$
$$(9{:}2)$$
$$v = v^m(x, y) - z\beta(x, y)$$

$$\sigma_z = p_z^m(x, y) \qquad (9{:}3)$$

u^m and v^m are called the *membrane* components of displacement. Note that an equivalent form of the first assumption is the statement that normals to the plane of symmetry remain straight during distortion. The second assumption says that the normal stress is equal throughout the thickness to the symmetric part of the normal loading on the top and bottom surfaces, p_z^m, and hence that the normal component of stress does not interact with any of the other components. For most purposes we can simply put $\sigma_z = 0$. This does not, however, imply that the normal component of strain, ε_z, is zero.

So far we have made no assumption about the normal displacement, w. We will, for the moment, limit it to linear dependence on z, i.e.,

$$w = w_0(x, y) + zw_1(x, y) \qquad (9{:}4)$$

and see what develops.

The symmetry of the plate's geometry and material properties allows the separation of displacements and all other variables into uncoupled symmetric and antisymmetric parts. We will designate the symmetric parts of the variables as the *membrane* parts

$$u^m, \ v^m, \ w^m = zw_1, \ \text{and} \ \sigma_z^m = p_z^m \qquad (9:5)$$

Likewise, we will designate the antisymmetric parts of the variables as the *bending* parts

$$u^b = -z\alpha, \ v^b = -z\beta, \ w^b = w_0, \ \text{and} \ \sigma_z^b = 0 \qquad (9:6)$$

The solution for the membrane part consists primarily of the determination of u^m and v^m and the *membrane strains* ($\varepsilon_x^m = u,_x^m, \ \varepsilon_y^m = v,_y^m, \ \gamma_{xy}^m = u,_y^m + v,_x^m$) by using the methods of two-dimensional elasticity, including the use of the two-dimensional finite elements described in previous chapters. The subsequent determination of w^m is trivial[*] and may be ignored if the thickness is small. Or we may simply state the obvious—that w^m is zero in the plane of symmetry.

The solution for the bending part involves new considerations. The normal displacement w^b is coupled to the in-plane displacements, u^b and v^b, through the mechanism of transverse shear. Expressions for the *transverse shear strains* are

$$\gamma_{xz} = w,_x^b + u,_z^b = w,_x^b - \alpha$$
$$\gamma_{yz} = w,_y^b + v,_z^b = w,_y^b - \beta \qquad (9:7)$$

Since w^b, α, and β are independent of z, the transverse shear strains computed from Equation 9:7 are constant through the thickness. This result is contrary to the observation that γ_{xz} and γ_{yz} should be at least quadratic in z in order to have zero values at the top and bottom free surfaces, i.e.,

$$\gamma_{xz} = \gamma_{xz}^0 \left(1 - \left(\frac{2z}{t}\right)^2\right), \ \gamma_{yz} = \gamma_{yz}^0 \left(1 - \left(\frac{2z}{t}\right)^2\right) \qquad (9:8)$$

---

[*] w^m consists of a part due to $\sigma_z^m = p_z^m$ and a part due to the vertical strain, ε_z^m, induced by membrane strains through Poisson's ratio coupling.

Achievement of this effect requires either quadratic variation of the normal displacement or cubic variation of the in-plane displacements.[3]   In Reissner-Mindlin plate theory,[4,5] which uses the first order variations presented here, the effect is accounted for by reducing the transverse shear stiffness by a factor. In the case of material that is homogeneous through the thickness, this *shear effectiveness factor* is equal to 5/6.

The bending part of the displacement field produces *bending strains*

$$\varepsilon_x^b = u_{,x}^b = -z\alpha_{,x} = -z\chi_x$$

$$\varepsilon_y^b = v_{,y}^b = -z\beta_{,y} = -z\chi_y \qquad (9{:}9)$$

$$\gamma_{xy}^b = u_{,y}^b + v_{,x}^b = -z\left(\alpha_{,y} + \beta_{,x}\right) = -z\chi_{xy}$$

It is seen that the definition of *bending curvatures* $\left(\chi_x, \chi_y, \chi_{xy}\right)$ in terms of rotations of the normal $\left(\alpha, \beta\right)$ is formally analogous to the definition of membrane strains $\left(\varepsilon_x^m, \varepsilon_y^m, \gamma_{xy}^m\right)$ in terms of in-plane translations $\left(u^m, v^m\right)$. Note finally that there are five independent degrees of freedom at every point in the plane of symmetry. These include two components of membrane displacement $\left(u^m, v^m\right)$ and three components of bending displacement $\left(w^b, \alpha, \beta\right)$. The superscripts m and b may be dropped because the distinction between membrane and bending components is evident without them. The membrane displacements are uncoupled from the bending displacements if the material properties are symmetrical with respect to the middle plane.

We come next to an important additional assumption which has frequently been made. That assumption, known as the *Kirchhoff hypothesis*, takes the transverse shear strains to be *zero*. Thus, from Equation 9:7, the Kirchhoff hypothesis is equivalent to the assumptions

$$\alpha = w_{,x}^b \quad , \quad \beta = w_{,y}^b \qquad (9{:}10)$$

or, in other words, that the rotations of the normals to the plane of symmetry are just equal to the slopes of the deformed plane of symmetry. Equation 9:10 amounts to a pair of *rigid constraints* which eliminate α and β as independent variables. As a result, and this is the great strength of the Kirchhoff hypothesis, the analysis of plate bending requires only a single dependent variable. As a matter of fact, the Kirchhoff hypothesis reduces the relationships which describe the bending of a homogeneous isotropic plate to a single fourth order partial differential equation

$$\left(\frac{\partial^2}{\partial x^2} + \frac{\partial^2}{\partial y^2} \right)^2 w = \frac{p_z}{D} \tag{9:11}$$

where p_z is the lateral load density and D is the bending modulus $\left(D = Et^3/12\left(1 - v^2\right) \right)$. Equation 9:11 admits of very simple solutions in some cases. For example, the homogeneous solutions for a simply supported rectangular plate of dimensions a, b have the form

$$w = \cos\left(\frac{m\pi x}{a} \right) \cos\left(\frac{n\pi y}{b} \right) \qquad m, n = 1, 3, 5, \cdots \tag{9:12}$$

where x and y are measured from the center of the plate.

While the Kirchhoff hypothesis provides relatively simple analytical solutions for important special cases, difficulties can occur at boundaries, where the undefined transverse shear stresses may be needed, and with relatively thick plates. At any rate, in beginning their study of plate bending, finite element designers trusted that the Kirchhoff hypothesis would prove to be as serviceable as it was for analytic solutions. As we shall see, that trust was misplaced.

We complete the formulation of first order plate theory by consideration of certain stress resultants and their relationships to strains and curvatures. Figure 9.2 shows the bending moments per unit length $\left(m_x, m_y, m_{xy} \right)$ and the transverse shear forces per unit length $\left(Q_x, Q_y \right)$, sometimes called *shear flows*, acting on an infinitesimal rectangular domain. These stress resultants satisfy the following equilibrium equations

$$m_{x,x} + m_{xy,y} + Q_x + q_x = 0$$

$$m_{y,y} + m_{xy,x} + Q_y + q_y = 0 \tag{9:13}$$

$$Q_{x,x} + Q_{y,y} + p_z = 0$$

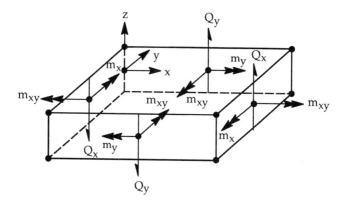

Figure 9.2   Bending Moments and Transverse Shears Per Unit Length.

where p_z is the vertical load per unit area and q_x, q_y are moment loads per unit area.

A relationship between bending moments and curvatures is found by integrating stresses through the thickness. Thus

$$\begin{Bmatrix} m_x \\ m_y \\ m_{xy} \end{Bmatrix} = \{m\} = -\int_{-t/2}^{t/2} z\{\sigma^b\}\,dz = -\int_{-t/2}^{t/2} z\begin{Bmatrix} \sigma_x^b \\ \sigma_y^b \\ \tau_{xy}^b \end{Bmatrix}\,dz \qquad (9{:}14)$$

where

$$\{\sigma^b\} = [D]\{\varepsilon^b\} = -[D]z\{\chi\} \qquad (9{:}15)$$

and $[D]$ is the two-dimensional, plane stress constitutive matrix (see Equation 2:16). Substitution then gives

$$\{m\} = \int_{-t/2}^{t/2} z^2[D]\,dz\{\chi\} = \frac{t^3}{12}\Big[D^b\Big]\{\chi\} \qquad (9{:}16)$$

where $\left[D^b \right]$ = $[D]$ for materials which are homogeneous through the thickness but not, for example, for layered composites. Membrane forces per unit length $\{N\}$ are related to membrane strains by an entirely analogous relationship

$$\{N\} = \int_{-\frac{t}{2}}^{\frac{t}{2}} \{\sigma^m\} \, dz = \int_{-\frac{t}{2}}^{\frac{t}{2}} [D] \, dz \{\varepsilon^m\} = t \left[D^m \right] \{\varepsilon^m\} \qquad (9\!:\!17)$$

Again, for materials which are homogeneous through the thickness, $\left[D^m \right] = [D] = \left[D^b \right]$.

Transverse shear forces are similarly related to transverse shear strains by integration through the thickness. Thus

$$\begin{Bmatrix} Q_x \\ Q_y \end{Bmatrix} = \{Q\} = \int_{\frac{t}{2}}^{\frac{t}{2}} \begin{Bmatrix} \tau_{xz} \\ \tau_{yz} \end{Bmatrix} dz \qquad (9\!:\!18)$$

where

$$\begin{Bmatrix} \tau_{xz} \\ \tau_{yz} \end{Bmatrix} = [G] \begin{Bmatrix} \gamma_{xz} \\ \gamma_{yz} \end{Bmatrix} = [G]\{\gamma\} \qquad (9\!:\!19)$$

and $[G]$ is a 2 x 2 matrix of transverse shear moduli. As remarked earlier, while γ_{xz} and γ_{yz} are not constant through the thickness in real plates, Reissner-Mindlin plate theory supplies constant effective values for these quantities which are related to displacements by Equation 9:7. Substitution into Equation 9:18 with an added shear effectiveness factor, k, then gives

$$\{Q\} = k \left(\int_{-\frac{t}{2}}^{\frac{t}{2}} [G] \, dz \right) \begin{Bmatrix} \gamma_{xz} \\ \gamma_{yz} \end{Bmatrix} = kt \left[G^s \right] \{\gamma\} \qquad (9\!:\!20)$$

where $\left[G^s \right] = [G]$ and $k = \frac{5}{6}$ for materials which are homogeneous through the thickness.

Taken together, Equations 9:16, 9:17, and 9:20 express the constitutive equations between stress resultants, strains, and curvatures for plates with a plane of material symmetry. The strain energy per unit area is given by

$$
\begin{aligned}
W' &= \frac{1}{2}\left(\lfloor N \rfloor \left\{ \varepsilon^m \right\} + \lfloor m \rfloor \{\chi\} + \lfloor Q \rfloor \{\gamma\} \right) \\
&= \frac{1}{2}\left(t\left\{\varepsilon^m\right\}^T \left[D^m \right]\left\{\varepsilon^m\right\} + \frac{t^3}{12}\{\chi\}^T\left[D^b\right]\{\chi\} \right. \\
&\qquad\left. + kt\{\gamma\}^T\left[G^s\right]\{\gamma\} \right)
\end{aligned}
\tag{9:21}
$$

This expression for strain energy density evidences the assumed absence of elastic coupling between membrane strains, bending curvatures, and transverse shear strains. When the Kirchhoff hypothesis is invoked, the third term, representing the density of transverse shear energy, is ignored.

In plates which lack a plane of material symmetry, coupling exists between membrane strains and bending curvatures which may not be negligible. Such coupling can very often be accounted for simply by shifting the $z = 0$ plane from the middle of the plate to a so-called *neutral plane*. This must be possible, for example, with materials which are homogeneous and isotropic with respect to x and y but not necessarily with respect to z. In more general cases, as for example when the material is anisotropic with respect to x and y and inhomogeneous with respect to z,[*] it may be necessary to treat the coupling between membrane strains and bending curvatures explicitly. For such cases the appropriate constitutive relationships can be expressed in partitioned matrix form as

$$
\left\{ \begin{array}{c} N \\ m \\ Q \end{array} \right\} = t \left[\begin{array}{c:c:c} \left[D^m\right] & t\left[D^{mb}\right] & 0 \\ \hdashline t\left[D^{mb}\right]^T & \dfrac{t^2}{12}\left[D^b\right] & 0 \\ \hdashline 0 & 0 & k\left[G^s\right] \end{array} \right] \left\{ \begin{array}{c} \varepsilon^m \\ \chi \\ \gamma \end{array} \right\}
\tag{9:22}
$$

---

[*]A plate with unidirectional stiffeners attached to one side is an example.

where $\left[D^{mb}\right]$ is the membrane-bending coupling matrix. Some finite element programs, such as MSC/NASTRAN, allow the user to offset an element's neutral plane from the plane of its connecting nodes or to supply a membrane-bending coupling matrix. Special routines may also be provided which automatically compute $\left[D^{mb}\right]$ and the other modulus matrices for layered composites and for other special cases.

9.2 KIRCHHOFF PLATE ELEMENTS

As has been noted, the plate element pioneers of the 1960s employed the Kirchhoff hypothesis in their designs. Thus, by assuming transverse shear strains to be zero, they could derive all of an element's bending properties from the assumed distribution of a single variable, the lateral displacement w. For reasons which will be explained, the resulting elements exhibit severe practical difficulties which largely render them noncompetitive with elements which do not invoke the Kirchhoff hypothesis. Today, Kirchhoff plate elements have primarily a historical interest. They also have an aesthetic interest in that they include some of the most exotic forms ever devised. Expressed in the notation we have employed, the formal relationships for Kirchhoff plate elements are developed below.

By analogy with the examples of two- and three-dimensional elasticity, the vector of curvatures $\{\chi\}$ is related to nodal displacements $\{u_i\}$ by

$$\{\chi\} = \sum_i [B_i]\{u_i\} \tag{9:23}$$

where the elements of $[B_i]$ are functions of position and the elements of $\{u_i\}$ are the degrees of freedom at node (i); for example,

$$\{u_i\} = \begin{Bmatrix} w \\ \theta_x \\ \theta_y \\ \vdots \end{Bmatrix}_i \tag{9:24}$$

where $\theta_x = \beta = w_{,y}$, $\theta_y = -\alpha = -w_{,x}$, and the dots indicate the possible inclusion of some higher order derivatives of w.

Using the terminology developed in Chapter 3, we can express the lateral displacement, w, in terms of nodal shape functions, $\lfloor N \rfloor_i$, or in terms of elementary basis functions, $\lfloor X \rfloor$, i.e.,

$$w = \sum_i \lfloor N \rfloor_i \{u_i\} = \lfloor X \rfloor \{a_j\} \tag{9:25}$$

Using the Kirchhoff hypothesis, we can express the curvatures $\{\chi\}$ directly in terms of the *second* derivatives of lateral displacement,

$$\{\chi\} = \begin{Bmatrix} \chi_x \\ \chi_y \\ \chi_{xy} \end{Bmatrix} = \begin{Bmatrix} \alpha_{,x} \\ \beta_{,y} \\ \alpha_{,y} + \beta_{,x} \end{Bmatrix} = \begin{Bmatrix} w_{,xx} \\ w_{,yy} \\ 2w_{,xy} \end{Bmatrix} \tag{9:26}$$

Substitution of w from Equation 9:25 then gives

$$\{\chi\} = \begin{bmatrix} X_{,xx} \\ X_{,yy} \\ 2X_{,xy} \end{bmatrix} \{a_j\} \tag{9:27}$$

Evaluation of the B_i matrices in Equation 9:23 requires, finally, an expression for $\{a_j\}$ in terms of the $\{u_i\}$ vectors. Clearly the number of basis functions in $\lfloor X \rfloor$ must equal the sum of the number of degrees of freedom at nodes in order to allow this determination. The actual calculation may not be as straightforward as it proved to be for membrane and solid elements where $\{u_i\}$ in Equation 9:25 contained only a single component. Assuming that the calculation can be performed, we will omit the details and concentrate instead on the selection of basis functions and nodal degrees of freedom for particular elements.

A minimum requirement for accuracy is that the interpolation of nodal displacements gives correct constant curvatures. This requirement will be satisfied if the basis functions are complete through the quadratic terms in x and y or, in case parametric interpolation is used, if nodal values of w equal to x^2, xy, and y^2 are correctly interpolated. This is a far stiffer condition than the correct interpolation of *linear* displacement fields required in two- and three-dimensional elasticity; it contributes to many of the difficulties encountered in plate element design.

Satisfaction of a constant curvature patch test requires further that constant assumed bending moments in adjacent elements be in equilibrium at nodes. We can easily extend the discussion of nodal equilibrium in Section 5.3 to show that moment equilibrium is satisfied if rotations of the normal to the middle surface are continuous along the boundaries between adjacent elements. In the case of Kirchhoff elements, this amounts to C^1 continuity of displacements (w, $w_{,x}$, and $w_{,y}$ are continuous at element boundaries). Plate bending elements are conforming if w, α, and β are continuous at element boundaries. And conforming elements which exhibit neither interpolation failure nor integration failure will pass patch tests.

9.2.1   Triangular Kirchhoff Elements

The simplest possible triangular bending element has three corner nodes and three degrees of freedom per node $\left(w, \alpha, \beta\right)$ as shown in Figure 9.3.

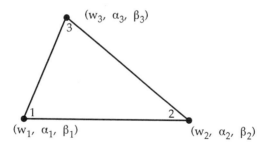

Figure 9.3   **Three-Node Element with Three Degrees of Freedom Per Node.**

The Kirchhoff hypothesis allows all of the element's bending properties to be derived from an assumed lateral displacement field, $w(x, y)$. Since there are nine external degrees of freedom, we need a field with nine independent terms. Thus we are led to consider the complete cubic polynomial

$$1, x, y, x^2, xy, y^2, x^3, x^2y, xy^2, y^3 \qquad (9:28)$$

which has ten terms, i.e., one term too many. Early finite element designers experimented with the elimination of various terms, such as xy, or combinations of terms, such as $x^2y + xy^2$, but could find no combination which preserved constant bending (the quadratic terms) and retained isotropy with respect to the element's geometry.

More generally, element designers soon discovered[6] that a completely satisfactory solution could not be achieved for *any* element with three degrees of freedom per node. Consider, for example, the four-node rectangle shown in Figure 9.4. Since we wish, as a minimum, to satisfy constant curvature patch tests, we require that the lateral displacement, w, and its slopes, α, β, be continuous along element boundaries or, in other words, that w, α, and β along any edge be determined by their nodal values at the adjacent corners. This means, for example, that along edge 1-2 of Figure 9.4, w can be, at most, a cubic function of x, and $\beta = w_{,y}$ can be, at most, a linear function of x. The six coefficients $a_0, a_1,$ etc., are determined by the six displacement components at the two adjacent nodes. Similar reasoning applies to edge 1-4.

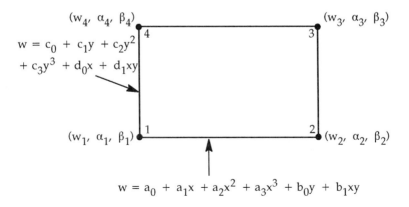

Figure 9.4   Demonstration of Continuity Requirements.

Note that the twisting curvature, $\chi_{xy} = 2w_{,xy}$, equals $2b_1$ along edge 1-2 and $2d_1$ along edge 1-4. These two values will, in general, be different because b_1 depends on the displacements at node 2 and d_1 depends on the displacements at node 4. As a result χ_{xy} *will not be unique* at the common point, node 1. As a corollary, an expression for w over the plate's surface which assures uniqueness of χ_{xy}, such as any polynomial in x and y, cannot assure slope continuity along the common edges of adjacent elements when only w and its slopes are prescribed at nodes. [6]

The argument just presented extends easily to nonrectangular corners. Three alternative courses of action present themselves to designers of Kirchhoff plate elements. They are:

1.   Accept the nonuniqueness of curvature in order to assure conformability.

2.   Accept nonconformability in order to assure uniqueness of curvature at all points within an element.

3.   Add higher order derivatives of w, such as $w_{,xy}$, as nodal degrees of freedom.

All three courses of action have been used by element designers and we will consider examples of each.

Clough and Tocher[7] achieved a conforming three-node triangle with three degrees of freedom per node by accepting the nonuniqueness of curvature. Their element, shown in Figure 9.5, consists of three subtriangles with a common node at the center of the element. In each triangle, w is expressed by a cubic polynomial with one missing term, selected to assure conformability along the exterior edge of the subtriangle. Thus, in subtriangle A, the x^2y term is deleted with the result that slope normal to the edge, $w_{,y}$, varies linearly along edge 1-2. The same term is deleted in the rotated coordinate systems of triangles B and C.

Conformability along interior edges is assured by enforcing equality between the normal slopes of adjacent elements at the midpoint of their common edge. These three constraints, for interior edges 1-4, 2-4, and 3-4 respectively, rigidly specify

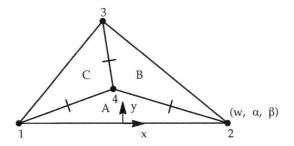

Figure 9.5   The Clough-Tocher Triangle.

the three components of displacement at node 4, the interior node, in terms of the displacements at the exterior nodes.

The Clough-Tocher triangle is a conforming bending element with a minimum degree of freedom count. It is not, however, a good performer. In Figure 1.6, for example, the element identified as NASTRAN TRIA2 is the Clough-Tocher triangle. It is seen to be excessively stiff and to require an exceedingly large number of degrees of freedom to achieve satisfactory accuracy. The reason for the slow convergence is that, while only one cubic term is eliminated in each subtriangle, these constraints act in different directions and thus will greatly impede any variation of curvature in the vicinity of the element.

The addition of higher order derivatives as degrees of freedom can eliminate troubles with nonuniqueness and nonconformability at the cost of a greatly increased number of degrees of freedom and new troubles resulting from *excessive smoothness* of the displacement field. The minimum set of higher order derivatives which makes sense for general application is the set of three second derivatives, $w,_{x^2}$, $w,_{xy}$, and $w,_{y^2}$.

A three-node triangle with second derivatives as degrees of freedom has eighteen degrees of freedom (six per node). The smallest complete polynomial in x and y which has at least this many terms is the complete 21-term quintic. Clearly we require an adjustment which could either be the elimination of three terms from the polynomial expression for w or the addition of three degrees of freedom to the element. 1968 and 1969 saw the publication of a truly remarkable number of independent papers which described either the first approach,[8,9] or the second,[10,11,12] or both.[13,14]

Along a given edge, as shown in Figure 9.6, the complete quintic polynomial allows the lateral displacement, w, to be a six-term quintic function of distance, and normal slope, $w_{,y}$, to be a five-term quartic function of distance. The coefficients in the six-term expression for w are accounted for by the values of w, $w_{,x}$, and $w_{,x^2}$ at the two adjacent nodes. The coefficients in the five-term expression for normal slope, $w_{,y}$, can be accounted for by the values of $w_{,y}$ and $w_{,xy}$ at the two ends plus one additional degree of freedom. A good choice for the additional degree of freedom is $w_{,y}$ at the midpoint of the edge. This degree of freedom plus the six degrees of freedom at each of the adjacent corner nodes account for the thirteen terms in the expression for w along edge 1-2. If the addition of nodes with only one degree of freedom is considered too awkward by the developer, he can elect instead to eliminate the b_4x^4y term in the expression for w and take analogous measures with regard to the other two edges.

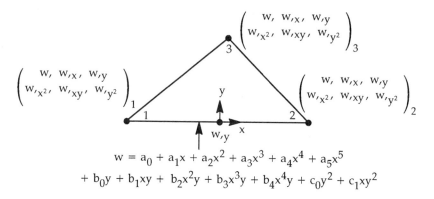

$$w = a_0 + a_1x + a_2x^2 + a_3x^3 + a_4x^4 + a_5x^5$$
$$+ b_0y + b_1xy + b_2x^2y + b_3x^3y + b_4x^4y + c_0y^2 + c_1xy^2$$

Figure 9.6   Three-Node Triangle with Six Degrees of Freedom Per Node.

Since displacements at points along the edge depend only on the degrees of freedom at the adjacent corners and on the normal slope at the center point, if that option is selected, the element satisfies compatibility. The curvatures are unique at all points within the element and, in the case of the full 21-degree-of-freedom version, have complete *cubic* dependence on position. These characteristics give the element a tremendous degree of accuracy and probably explain the enthusiasm with which it was taken up by element designers in the late sixties. Nevertheless, the element did not prove to be popular and is now mainly a curiosity.

The reason for this lack of acceptance is associated with the interpretation of second derivatives of w as nodal variables. Since they physically correspond to curvatures, how are they to be loaded and how are boundary conditions to be applied to them? Also, since they have unique values at nodes, they tend to prevent any discontinuity in curvature along element edges such as would result from applied moments or changes in material properties. As discussed in Section 4.6.3, structural analysis is very largely concerned with the discontinuities which occur in structures, such as edges, holes, corners, stiffeners, and changes in load intensity. The subdivision of a structure into finite elements normally allows such discontinuities to occur on the boundaries between elements. The enforcement of *supercontinuity* through the use of higher order nodal variables vitiates this important feature.

9.2.2 Quadrilateral Kirchhoff Elements

Credit for the earliest practical plate bending element goes to Adini and Clough[15] for their twelve degree of freedom rectangle, published in 1961. The element, sketched in Figure 9.7, has three degrees of freedom per node. Basis functions for the element were selected by noting that w should be allowed to vary cubically along each edge. Thus, the basis functions are given by

$$w = \left(1, y\right)\left(1, x, x^2, x^3\right) + \left(1, x\right)\left(1, y, y^2, y^3\right) \qquad (9:29)$$

The factors $(1, y)$ and $(1, x)$ are needed to allow independent cubic variations on opposite edges. Expansion of Equation 9:29 shows that there are twelve independent terms. The form of Equation 9:29 shows the element to be nonconforming because $\beta = w_{,y}$ is not restricted to linear dependence on x.

The question naturally arises as to whether the Adini-Clough rectangle can be extended to more general shapes. The mechanics are simple because we can substitute parametric coordinates $\left(\xi, \eta\right)$ for metric coordinates (x, y) in Equation 9:29. The real question is whether the element will still be able to

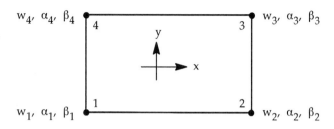

Figure 9.7 **The Adini-Clough Rectangle.**

compute constant curvatures correctly or, in other words, whether a quadratic displacement field will be correctly interpolated from its nodal values. We recall from Section 3.5 that this will be true for general quadrilateral elements with straight sides if each of the terms $\left(1, \xi, \eta, \xi^2, \xi\eta, \eta^2, \xi^2\eta, \xi\eta^2, \xi^2\eta^2\right)$ is included in the basis. This is not the case for the Adini-Clough element because, with ξ, η replacing x, y, the term $\xi^2\eta^2$ is absent from Equation 9:29. The element will, however, produce correct constant curvatures for parallelogram shapes because all that is required of the basis functions in this case is a complete set of quadratic terms.

The lack of conformability of the twelve-node Adini-Clough rectangle can be corrected by adding $w_{,xy}$ as a degree of freedom at each node. The shape functions for the resulting Bogner-Fox-Schmit rectangle[16] are formed by multiplying together beam shape functions along x and y (see Irons and Ahmad, Reference 9.17).

The chief weakness of both of these rectangular elements is that they cannot be successfully extended to general quadrilateral shapes. The earliest conforming general quadrilateral, which Irons admired as the prettiest such element, is the Fraeijs de Veubeke quadrilateral.[18] As shown in Figure 9.8, the element has sixteen degrees of freedom including $\left(w, \alpha, \beta\right)$ at corner nodes and the normal

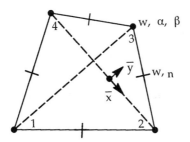

Figure 9.8   The Fraeijs de Veubeke Quadrilateral.

slope, $w_{,n}$, at the midpoints of edges. The diagonals subdivide the element into four triangles which each have a different cubic field of lateral displacement. Thus the Veubeke quadrilateral accepts nonuniqueness of curvature in order to assure conformability of w and $w_{,n}$ along exterior edges.

The displacement basis is constructed as the sum of three functions

$$w = w^a + w^b + w^c \qquad (9:30)$$

where w^a is a complete ten term cubic. The second function w^b is zero in subtriangle 1-2-4 and consists of quadratic and cubic terms in subtriangle 2-3-4,

$$w^b = \left(\bar{y}^2, \bar{y}^3, \overline{xy}^2 \right) \qquad (9:31)$$

where $\bar{x}$ is directed along diagonal 2-4. Note that displacement and slope continuity are preserved along the diagonal. In similar fashion the third function, w^c, is zero in subtriangle 1-2-3 and contains three quadratic and cubic terms in subtriangle 1-3-4 which preserve continuity along diagonal 1-3. Since w^a has ten terms and w^b, w^c each have three terms, the total number of terms matches the number of external degrees of freedom. If desired, the midside nodes can be constrained to form a twelve-degree-of-freedom element.[18]

9.2.3   Discrete Kirchhoff Elements

In the Kirchhoff elements described to this point, the Kirchhoff hypothesis has
been applied à priori to reduce the role of normal rotations to that of nodal
variables which determine the coefficients in the field assumed for w. In the
discrete Kirchhoff approach, separate fields are assumed for α and β and the
Kirchhoff constraints $\left(\alpha = w_{,x}, \; \beta = w_{,y} \right)$ are applied at discrete points, thereby
eliminating some of the nodal variables. Successful elements of this type include
both triangles[19] and quadrilaterals[20] with minimum node counts and, most
prominently, Irons' SEMILOOF quadrilateral[21] (see Figure 9.9). The nodal
degrees of freedom for the SEMILOOF element include the values of w at corners
and midsides and the values of normal slope, $w_{,n}$, at $\xi_e = \pm 1 / \sqrt{3}$ on each
edge. The SEMILOOF element is still competitive with more modern elements for
both plate and shell applications.

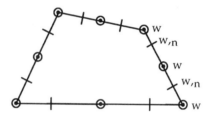

Figure 9.9   Irons' SEMILOOF Element.

9.3   MINDLIN PLATE ELEMENTS

As we have seen, elements which enforce C^1 continuity, such as some of the
Kirchhoff elements just described, are fundamentally flawed because they require
the use of second derivatives of the field variable as nodal degrees of freedom.
Kirchhoff elements which do not use second derivatives will violate C^1 continuity
either because slope continuity is not enforced on exterior edges or because the
second derivatives of the field are discontinuous at interior points (see Section 3.4
for the definition of C^1 continuity).

In this section we consider plate bending elements which do not employ the Kirchhoff hypothesis; i.e., the transverse shear strain is allowed a nonzero value. In this case the rotations of the normal (α, β) become independent variables and the C^1 continuity requirement for w translates into C^0 continuity requirements for w, α, and β. These requirements are easier to satisfy. They also open the door to the approaches to element design treated in earlier chapters.

The earliest Mindlin (or Reissner-Mindlin) element was the eight-node Ahmad[1,22] curved shell element (1969) derived by applying the Reissner-Mindlin plate assumptions to a sixteen-node elastic brick element. In Figure 9.10, the six degrees of freedom at two nodes with the same values of x and y are replaced by five degrees of freedom (u, v, w, α, β) at their midpoint. Note that this reduction eliminates the effect of vertical strain, ε_z. Linear variation of u and v with z is already implicit in the choice of nodes for the sixteen-node brick, derived from a twenty-node brick by suppressing the four nodes at the midpoints of the vertical edges.

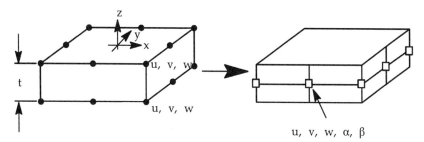

Figure 9.10 Derivation of the Ahmad Eight-Node Shell Element From a Sixteen-Node Brick Element.

An eight-node plate (or shell) element is somewhat more efficient than a thin sixteen node solid because it has 40 degrees of freedom rather than 48. A more substantial reason for preferring the plate element is that it avoids a numerical round-off problem caused by an excessively large stiffness ratio when the plate's thickness, t, is small compared to its characteristic length, L. The two stiffnesses in question are the through-the-thickness extensional stiffness, proportional to EL^2 / t, and the plate bending stiffness, proportional to Et^3 / L^2. The ratio of these two stiffnesses, proportional to L^4 / t^4, can, as noted in Section 7.4, cause excessive round-off error for values of L/t greater than a few hundred.

If we assume the geometric curvature of the Ahmad shell element to be zero, we obtain a plate element for which the Reissner-Mindlin plate theory derived in Section 9.1 applies directly. Specifically, the membrane and bending deformations can be treated separately if the material is homogeneous with respect to z. In addition, the strain energy in bending depends on rotations of the normal in a manner that is totally analogous to the dependence of membrane strain energy on in-plane translations. Thus, since the curvatures (α, β) are independent variables (independent of w, that is), the formulation of the bending part of Mindlin plate elements is identical to the formulation of two-dimensional elastic elements. As a result, except for the substitution of (α, β) for (u, v) and the substitution of $\frac{t^2}{12}\left[D^b\right]$ for $\left[D^m\right]$ in Equation 9:22, we can apply all of the theory developed for two-dimensional elastic elements in previous chapters to the bending part of Mindlin plate elements.

The formal treatment of bending curvature for Mindlin plate elements mimics that of membrane strain. Specifically,

$$\{\chi\} = \sum_i \left[B_i\right]\left\{\theta_i\right\} \qquad\qquad (9\!:\!32)$$

where

$$\left\{\theta_i\right\} = \left\{\begin{array}{c} \alpha \\ \beta \end{array}\right\} = \left\{\begin{array}{c} -\theta_y \\ \theta_x \end{array}\right\} \qquad\qquad (9\!:\!33)$$

and

$$\left[B_i\right] = \left[\begin{array}{cc} N_{i,x} & 0 \\ 0 & N_{i,y} \\ N_{i,y} & N_{i,x} \end{array}\right] \qquad\qquad (9\!:\!34)$$

where N_i is the shape function selected for $\left\{\theta_i\right\}$.

The treatment of transverse shear strains is a different story. In the defining equations for these strains in terms of displacements (Equation 9:7), the spatial derivatives of w are combined with the field values of α and β. For the important case of pure bending, the terms must cancel to produce zero transverse shear. Note however that, if the same sets of basis functions are used for w, α, and β, which would be natural if they are evaluated at the same nodes, α and β will contain higher degree terms not present in $w_{,x}$ and $w_{,y}$. These higher degree terms can cause *transverse shear locking* and, in fact, did cause locking in Ahmad's original element.

Transverse shear locking is *the* critical problem for the design of Mindlin plate elements. This problem is finessed in the design of Kirchhoff plate elements by embedding the assumption of zero shear strain in the specification of the displacement field. Transverse shear locking of Mindlin plate elements has an exact counterpart in the shear locking of two- and three-dimensional elastic elements treated in Chapter 6. In fact, we can say that Ahmad's element inherited transverse shear locking from its parent brick element. This type of locking is described, for the QUAD8 element, in Section 6.4.1 and in Table 6.4(a). It corresponds to a cubic lateral translation, $w = x^3$.

Transverse shear locking is a more serious problem for plate and shell bending elements than it is for membrane and solid elements because even uniform bending requires quadratic dependence of w on position. Thus, in the terminology developed in Chapter 5, convergence to the correct solution for bending problems requires satisfaction of patch tests in which w varies quadratically. This is more severe than the convergence condition for membrane and solid elements which requires only the satisfaction of a linear displacement patch test.

It is useful, as a prelude to the detailed discussion of transverse shear locking, to inventory the internal degrees of freedom for frequently used configurations. For this purpose, we can think of a plate element as a box with a top, bottom, and sides, as shown in Figure 9.11. Deformations in the planes of the top and bottom covers yield symmetrical membrane strains and antisymmetrical bending strains. Shear deformations of the vertical webs produce the transverse shear strains. In general there is one transverse shear mode for each vertical panel. (For the nine-

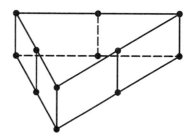

Figure 9.11 **Visualization of a Plate Element as a Box with Top, Bottom, and Vertical Webs.**

node quadrilateral we add a panel connecting the center node to one of the exterior nodes.)

Table 9.1 lists the number of internal degrees of freedom by polynomial degree for five first- and second-order plate bending elements. All of the bending modes are obtained by comparison with the degree-of-freedom inventory for membrane elements given in Tables 3.4 and 3.6. All but one of the transverse shear modes are obtained by assuming independent vertical motions at the nodes with no in-plane motion of the covers. One additional linear transverse shear mode is obtained by letting the top and bottom covers rotate in opposite directions about a vertical axis.

It is seen that the linear elements are complete through the constant bending and transverse shear strains and that the quadratic elements are complete through the linear strains. The scattered higher-order terms add significant accuracy in some cases and help performance in other ways, such as in the passage of patch tests. Actual performance, on the other hand, will be far less than that suggested by the data in Table 9.1 if transverse shear locking is not suppressed.

9.4 TRANSVERSE SHEAR LOCKING

Mindlin plate elements are differentiated by the manner in which they resolve the issue of transverse shear locking. The matter is critical because even uniform bending requires correct interpolation of a quadratic lateral displacement field.

Table 9.1

Degree-of-Freedom Inventory for Plate Bending Elements

TYPE	SPATIAL DEPENDENCE	TRIA3	QUAD4	TRIA6	QUAD8	QUAD9	NEEDED FOR COMPLETE-NESS
Rigid Body	—	3	3	3	3	3	3
Bending	Constant	3	3	3	3	3	3
	Linear	0	2	6	6	6	6
	Quadratic	0	0	0	4	4	8
	Cubic	0	0	0	0	2	10
Transverse Shear	Constant	2	2	2	2	2	2
	Linear	1	2	4	4	4	4
	Quadratic	0	0	0	2	2	6
	Cubic	0	0	0	0	1	8
TOTAL		9	12	18	24	27	50

As was shown in Section 3.5, exact representation of a quadratic field is not possible for any isoparametric element which employs its full geometric capacity to represent irregular shapes. Thus we are led to consider either restrictions on element shape or approaches not based on parametric mapping of the displacement field.

9.4.1   Eight- and Nine-Node Quadrilaterals

Ahmad's original eight-node shell element[1] employed full (3 x 3) integration. Before long (1971), Zienkiewicz, Too, and Taylor showed[23] that substantially improved results could be obtained with reduced (2 x 2) integration. We now know that most of the improvement of the Ahmad element for curved shell applications comes from the avoidance of membrane locking, but some improvement comes from the avoidance of transverse shear locking.

Reduced integration in an eight-node element provides relief from locking caused by incorrect interpolation of cubic displacement fields, as was shown in Section 6.4.1. This measure yields improvements ranging from mild to substantial in practical applications. See, for example, Table 6.5 which compares results for out-of-plane bending of a thin square plate (Figure 6.15). In transverse shear locking the ratio of spurious shear strain energy to the correct strain energy is proportional to the square of the element's length-to-thickness ratio. Thus, for very thin plates, the effect of shear locking is to lock out the mode in question. In the case of an eight-node element the locking modes are not the lowest elastic modes so that convergence with increasing numbers of elements is rapid.

An equally important issue for the eight-node quadrilateral is the question of shape sensitivity. As demonstrated in Section 3.5, an eight-node element cannot correctly interpolate $w = x^2$ when it has a general quadrilateral shape with straight sides. Thus, regardless of the relative intensity of the resulting locking, the element cannot pass constant curvature patch tests. One remedy, which supplies the required $\xi^2\eta^2$ term to the displacement basis, is to add a ninth interior node. This is not an unmixed blessing, as we shall see. Another remedy is to add a term to each of the element's shape functions (see Equation 8:54 in Section 8.3) that will ensure correct interpolation of quadratic fields.[24]

Solutions for the lateral deflection of a slender cantilever plate are compared in Figure 9.12. Note that the end loading on this structure, which has repeatedly been analyzed as a cantilever beam in previous chapters, is now normal to the plane of the paper. The width-to-thickness ratio is 2.0. The modifiers F, R, and RM applied to the names of the elements in Figure 9.12 designate full integration, reduced integration, and reduced integration with shape functions modified as in Section 8.3. It is observed that the substitution of reduced integration for full integration makes a substantial improvement. The subsequent modification of the shape functions reduces the already small error by nearly a factor of two.

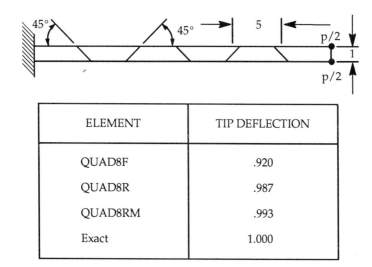

ELEMENT	TIP DEFLECTION
QUAD8F	.920
QUAD8R	.987
QUAD8RM	.993
Exact	1.000

Figure 9.12 **Solutions for Tip Loading of a Slender Cantilever Plate. Thickness = 0.5.**

We have not as yet specified the location of integration points for bending curvatures. The natural tendency is to use the same points as are used for transverse shear. If reduced (2 x 2) integration is used with an eight-node element, the element will have one noncommunicating spurious mode which is similar in appearance to the type 2 mode described for membrane action in Figure 7.1 with α, β replacing u, v. A lateral displacement $w = a\xi^2 - b\eta^2$ will accompany the mode to make the transverse shear strains zero at the Gauss points.

The treatment of bending curvature is less critical than the treatment of transverse shear because the required polynomial dependence of the rotations of the normal will, in nearly all cases, be one degree lower than the required polynomial dependence of lateral displacement. Consequently, it makes only a small difference if full integration is used in place of reduced integration for the calculation of bending curvatures.

As has been noted, the QUAD9 is the lowest order isoparametric quadrilateral which can satisfy constant curvature patch tests for general straight-sided shapes.[*] This fact provides a powerful incentive to substitute QUAD9s for QUAD8s. Another reason is that a little noticed QUAD8 locking mode inherited from the twenty-node brick involves the missing $\xi^2\eta^2$ term and is not corrected by reduced integration[**] (see Table 6.4). Whatever the reason, many developers of plate and shell elements turned to nine-node configurations in the 1980s.

The locking problem for cubic lateral displacement does not disappear when nine nodes are substituted for eight. Unfortunately the obvious solution, 2 x 2 integration of transverse shear, produces a type 1 communicating spurious mode, $w = \left(1 - 3\xi^2\right)\left(1 - 3\eta^2\right)$ (see Figure 7.4). Since $w_{,\xi}$ and $w_{,\eta}$ vanish at Gauss points, zero transverse shear strains result if α and β are zero. This spurious mode can be avoided only by abandoning reduced integration or by sticking with eight nodes. Hughes and Cohen[25] took the novel approach in their 1978 Heterosis element of using eight nodes for w and nine nodes for α, β. The integration schemes are, naturally enough, 2 x 2 for transverse shear and 3 x 3 for bending curvature. The element cannot pass constant curvature patch tests with straight-sided elements, but it does appear[26] to offer improved accuracy over the eight-node element with uniform 2 x 2 integration.[***]

---

[*]The issue of distortion due to offset of an edge node toward one of the adjacent corners can be resolved (see Figure 8.10 and related discussion).

[**]The MSC/NASTRAN QUAD8 element does not display this locking mode, possibly because it adds residual bending flexibility to transverse shear (see Section 9.5).

[***]The results published in Reference 9.26 for the eight-node element are considerably poorer than those published for the MSC/NASTRAN QUAD8.[27] The reasons for the differences are not known.

If the designer prescribes nine nodes for all displacement components, he must also use full integration for both transverse shear and bending curvature to avoid communicating spurious modes. Locking of cubic displacement fields then presents a problem which the designer can avoid by some of the techniques described in previous chapters. Some authors[28,29,30] use lower-order substitute strain fields which they evaluate by selective underintegration or by the Hellinger-Reissner variational principle (see Sections 7.6 and 7.7). Taking a different turn, Belytschko and his associates derive a successful nine-node shell element[31] by separating strain states into low order modes which they evaluate by 2 x 2 integration and high order modes whose stiffnesses are approximated analytically (see Section 7.8). Finally, in their assumed natural strain (ANS) shell element, Park and Stanley[32] evaluate each component of transverse shear strain at six strategic points on lines of constant ξ or η (see Figure 8.16), and interpolate them to 3 x 3 Gauss points. The elements derived by these various approaches differ somewhat in performance, but they all have acceptable accuracy as, indeed, does the QUAD8 element with reduced integration.

9.4.2   The Six-Node Triangle

A six-node triangle with straight edges can correctly interpolate any quadratic displacement field. It follows that a standard isoparametric TRIA6 element will satisfy a constant curvature patch test as long as the edges are straight.[*] Three integration points are sufficient to avoid spurious modes and to treat any linear variation of transverse shear strain.

It is possible to go farther and to ensure correct transverse shear strains for a cubic displacement field. If s is distance along one of the edges, then $w_{,s}$ is correct for cubic w at points $\xi_e = \pm \frac{1}{\sqrt{3}}$ along the edge (see Section 8.4.3). The tangential component of transverse shear strain, $\gamma_{sz} = w_{,s} - \alpha_s$, where α_s is the component of normal rotation in the plane of the edge, will also be correct at these points. The specification of transverse shear strain might then be completed by interpolating

---

[*]See Section 8.3 and Figure 8.10 for modifications to accommodate edge node offset toward one of the adjacent corners.

the edge values to Gauss points, for example, by the method described in Section 8.4.1.

Some recent research efforts report improvements of this sort in the performance of the six-node triangle. One approach[33] eliminates cubic shear locking by adding a cubic correction to w which cancels the quadratic terms in α and β. Another approach[34] uses a mixed formulation in which values of the transverse shear *stress* at the six special edge points, $\xi_e = \pm 1 / \sqrt{3}$, are used to enforce linear variation of transverse shear. The latter approach is related to the discrete Kirchhoff triangle (DKT) element.[19]

While advanced treatment of transverse shear strain can improve performance for flat plates, little can be gained for shell analysis where membrane locking effectively limits a six-node triangle's accuracy to constant bending moments.

9.4.3   The Four-Node Quadrilateral

In the case of eight- and nine-node quadrilaterals and the six-node triangle, standard isoparametric mapping provides correct interpolation of a quadratic displacement field for elements with straight edges. The only difficulty appears in connection with the eight-node element, where a special modification to the shape functions is necessary to provide correct interpolation for general quadrilateral shapes. When we come to consider four-node quadrilaterals and three-node triangles the difficulties multiply because there are only a few *points* at which one or the other of the first derivatives of a quadratic displacement field are correct. Accordingly, the correct evaluation of transverse shear strain can utilize only the values computed at these points.

In 1977, Hughes, Taylor, and Kanoknukulchai proposed a quadrilateral element[35] which uses single-point quadrature for transverse shear. Since the internal degree-of-freedom count includes four values of transverse shear (see Table 9.1), the element has two spurious modes. One mode is a global spurious mode, $w = \xi\eta$, $\alpha = \beta = 0$. The other mode is a noncommunicating mode which involves equal and opposite in-plane rotations of the top and bottom surfaces,

$w = 0$, $\alpha = y$, $\beta = -x$. The global spurious mode severely limits the element's practical application. It cannot, for example, resist twisting moments.

In MacNeal's 1978 version[36] of his QUAD4 element, γ_{xz} is evaluated at $\xi = 0$, $\eta = \pm 1 / \sqrt{3}$ and γ_{yz} is evaluated at $\xi = \pm 1 / \sqrt{3}$, $\eta = 0$ (see Figure 9.13). These values are then transferred to the adjacent 2 x 2 Gauss points for numerical integration. The element has the required minimum number of transverse shear strain evaluations and does not, therefore, suffer from spurious modes. The rationale for the selection of points along $\xi = 0$ for the evaluation of γ_{xz} is taken from one-dimensional locking theory (see Table 7.1). Clearly, there are difficulties for nonrectangular shapes. For example, in the case of parallelogram shape, the components of transverse shear strain should at least be oriented parallel to the edges. Such difficulties mount in practical applications of the element and, not surprisingly, trapezoidal shapes give particularly bad results. To illustrate the point, let us calculate the transverse shear strains for the isosceles trapezoid shown in Figure 9.14 by following the procedure described in Section 6.6.2.

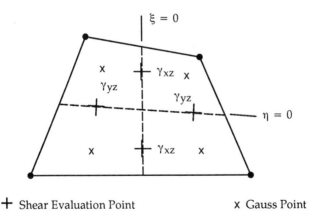

+ Shear Evaluation Point　　　　　　　　x Gauss Point

Figure 9.13　Location of Transverse Shear Evaluation Points for MacNeal's 1978 QUAD4 Element.

By inspection, the relationships between (x, y) and (ξ, η) are

$$x = \Lambda c \xi (1 - \alpha \eta) \quad , \quad y = c \eta \qquad (9:35)$$

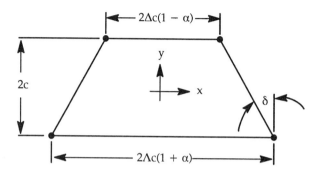

Figure 9.14   Four-Node Isosceles Trapezoid, $\alpha\Lambda = \tan\delta$.

from which it is easily deduced that

$$[J]^{-1} = \begin{bmatrix} \xi'_x & \eta'_x \\ \xi'_y & \eta'_y \end{bmatrix} = \frac{1}{c}\left[\begin{array}{c|c} \dfrac{1}{\Lambda(1-\alpha\eta)} & 0 \\ \hline \dfrac{\alpha\xi}{1-\alpha\eta} & 1 \end{array}\right] \tag{9:36}$$

The alias for a quadratic normal displacement, $w = x^2$, is derived as follows using the alias rules in Table 6.1.

$$w = x^2 = \Lambda^2 c^2 \xi^2 (1-\alpha\eta)^2 \Rightarrow w^a = \Lambda^2 c^2 \left(1 - 2\alpha\eta + \alpha^2\right) \tag{9:37}$$

The rotations of the normal are simply

$$\alpha = w'_x = 2x = 2\Lambda c \xi(1-\alpha\eta) \quad , \quad \beta = 0 \tag{9:38}$$

which are correctly interpolated. The aliases for the transverse shear strains are

$$\gamma^a_{xz} = w^a_{,x} - \alpha = w^a_{,\xi}\,\xi'_x + w^a_{,\eta}\,\eta'_x - \alpha$$
$$= -2\Lambda c \xi(1-\alpha\eta) \tag{9:39}$$

$$\gamma^a_{yz} = w^a_{,y} - \beta = w^a_{,\xi}\,\xi'_y + w^a_{,\eta}\,\eta'_y - \beta$$
$$= -2\alpha\,\Lambda^2 c = -2\Lambda c \tan\delta \tag{9:40}$$

It is seen that γ_{xz} is correctly evaluated at $\xi = 0$ but that γ_{yz} has an incorrect constant value *everywhere* within the element. This fact invalidates both the Hughes 1977 element[35] and the MacNeal 1978 element.[36] Bending strain, on the other hand, is correctly computed within the element. Its value is

$$\varepsilon^b_x = -z\chi_x = -z\alpha_{,x} = -2z \tag{9:41}$$

Thus, comparing Equations 9:40 and 9:41,

$$\frac{\gamma^a_{yz}}{\varepsilon^b_x} = \frac{\Lambda c \tan \delta}{z} \tag{9:42}$$

This ratio can be expected to be large for at least some elements in the finite element model of a thin, irregularly-shaped plate ($\frac{t}{\Lambda c}$ small, $\tan \delta$ not insignificant). As a result, transverse shear locking is a serious threat to the integrity of four-node Mindlin plate elements.

So back to the drawing boards. Before long, both Hughes and MacNeal replaced their treatments of transverse shear with an assumed strain approach [see Hughes and Tezduyar[37] (1981) and MacNeal[38] (1982)]. The basic concept is described in Section 8.4.1. At the center of each of the element's four sides we evaluate the component of transverse shear in the vertical plane of the side by the formula

$$\gamma_{sz} = w_{,s} - \alpha_s = \frac{w_2 - w_1}{\ell_e} - \frac{\left(\alpha_2 + \alpha_1\right)}{2} \tag{9:43}$$

where ℓ_e is the length of the edge, α is measured in the plane of the edge, and subscripts 1 and 2 refer to the nodes at the ends of the edge. Equation 9:43 is correct for any field in which w varies quadratically and α varies linearly.

The methods used by Hughes and MacNeal differ in the way that transverse shear strains are interpolated from the edges to 2 x 2 Gauss points. The method used does not matter for the case of constant curvature because the transverse shears are zero for that case. As a result, any method will satisfy constant curvature patch tests. A word of caution is in order regarding the implementation of such patch tests: all external loads should be applied to α and β. A twisting moment applied as a vertical couple will, for example, produce large local transverse shear strains. Some of us are so used to the Kirchhoff hypothesis that we tend to ignore transverse shear flexibility, even when it's important.

To proceed, we note that the (contravariant) transverse shear strains transform like displacement vectors in the plane of the element (they use a projection rule). Thus

$$\gamma_{sz} = \gamma_{xz} \cos \delta_e + \gamma_{yz} \sin \delta_e \tag{9:44}$$

where δ_e is the angle that the edge makes with the x axis.

In his revised QUAD4 element,[38] MacNeal assumes the following fields for γ_{xz} and γ_{yz}

$$\begin{Bmatrix} \gamma_{xz} \\ \gamma_{yz} \end{Bmatrix} = \begin{bmatrix} 1 & y & 0 & 0 \\ 0 & 0 & 1 & x \end{bmatrix} \begin{Bmatrix} a_1 \\ a_2 \\ a_3 \\ a_4 \end{Bmatrix} \tag{9:45}$$

so that, for side 1,

$$\gamma_{sz} = \lfloor \cos \delta_1, \, y \cos \delta_1, \, \sin \delta_1, \, x \sin \delta_1 \rfloor \begin{Bmatrix} a_1 \\ a_2 \\ a_3 \\ a_4 \end{Bmatrix} \tag{9:46}$$

Repeating for the other three sides, we obtain four equations which can be solved for a_1, a_2, a_3, a_4 in terms of transverse edge shears and then, from Equation 9:43, in terms of nodal displacements. Finally, Equation 9:45 provides the values of the transverse shear strains at Gauss points.

In their T1 element,[37] Hughes and Tezduyar define a skewed field of transverse shear strains $\left(\gamma_{sz}, \gamma_{tz} \right)$ where γ_{sz} is tangent to lines of constant η and γ_{tz} is tangent to lines of constant ξ (see Figure 9.15). At each edge, Equation 9:43 evaluates either γ_{sz} or γ_{tz} in terms of nodal displacements.

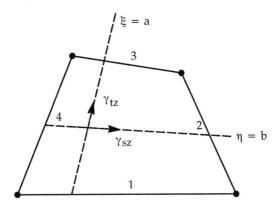

Figure 9.15   Covariant Transverse Shear Components.

The transverse shear strains at the centers of edges are transferred to the adjacent corners where they form a contravariant vector pair (projection rule). That vector is then transformed into a covariant vector (parallelogram rule) and is interpolated to Gauss points with standard isoparametric shape functions. The final step is to convert the covariant shear strain vectors at Gauss points into Cartesian components. Note that the approach of Hughes and Tezduyar avoids solution of a 4 x 4 matrix. The two approaches give identical results for rectangular elements and differ only slightly for other shapes. In 1985 Bathe and Dvorkin[39] proposed an element that is very similar to the Hughes T1 element.

Four node elements constructed in the manner described above will clearly satisfy constant curvature patch tests. They are also competitive with higher order elements. Table 9.2 compares the (revised) MSC/NASTRAN QUAD4 with the MSC/NASTRAN QUAD8 for the problem illustrated in Figure 9.12. Both elements include minor modifications from the theories which have been described. In the table, *rectangular elements* are those obtained when the 45° angles in Figure 9.12 are replaced by 90° angles.

Table 9.2

**Performance of MSC/NASTRAN QUAD4 and QUAD8 Elements
for Tip Loading of a Slender Cantilever Plate**

	QUAD4	QUAD8
Rectangular Elements	.986	.991
Trapezoidal Elements	.968	.998
Exact	1.000	1.000

9.4.4   The Three-Node Triangle

In a three-node triangle, there is no point at which linear interpolation of a general quadratic displacement field gives correct values for both components of transverse shear strain. We are, therefore, again forced to consider alternatives to continuous field interpolation of nodal displacements. As in the case of the four-node quadrilateral, Equation 9:43 gives, for quadratic w, correct tangential components of transverse shear strain at the midpoints of the edges.

All that remains is to interpolate the tangential edge values of transverse shear strains to integration points. Hughes and Taylor[40] follow the method of Hughes and Tezduyar[37] described in the previous section. In their method (see Figure 9.16), the edge shears are transferred to adjacent corners where the two components form a skewed vector pair in a local ξ, η system. After transformation from contravariant to covariant components, the nodal values are linearly interpolated to integration points.

In his method, MacNeal[41] assumes that the transverse shear field consists of two uniform distributions, γ_{xz}^0 and γ_{yz}^0, and a third circulating component, γ_{tz}^0, which resists relative rotation of the top and bottom surfaces.* Then, assuming an

---

*It may help to visualize the element as a box (see Figure 9.11).

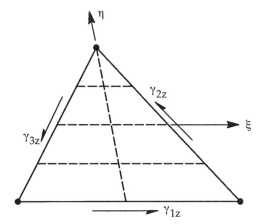

Figure 9.16 **The Hughes-Taylor Three-Node Triangle.**

x-axis that coincides with side 1, he obtains the following relationship between the edge components and the field components of transverse shear.

$$
\begin{Bmatrix} \gamma_{1z} \\ \gamma_{2z} \\ \gamma_{3z} \end{Bmatrix} = \begin{bmatrix} 1 & 0 & 1 \\ \cos \delta_2 & \sin \delta_2 & 1 \\ \cos \delta_3 & \sin \delta_3 & 1 \end{bmatrix} \begin{Bmatrix} \gamma_{xz}^0 \\ \gamma_{yz}^0 \\ \gamma_{tz}^0 \end{Bmatrix}
\qquad (9{:}47)
$$

where δ_2 and δ_3 are the angles that the x-axis makes with sides 2 and 3 respectively. After inversion of Equation 9.47, the values of γ_{xz}^0 and γ_{yz}^0 are used at a single integration point. We can assign a stiffness to γ_{tz}^0 fairly arbitrarily since its main purpose is to eliminate a noncommunicating spurious mode. MacNeal uses it as a free parameter to improve test results. Again, as noted in connection with the four-node quadrilateral, the method used for interpolation of transverse shear strains does not affect the satisfaction of a constant curvature patch test.

Another successful three-node triangle is the discrete Kirchhoff (DKT) triangle.[19] It defines separate fields for w, α, and β and uses the Kirchhoff conditions, $w,_x = \alpha$, $w,_y = \beta$, at discrete points to eliminate coefficients in the assumed displacement fields. In this way, twenty-one original coefficients reduce to the nine standard nodal degrees of freedom. Note that transverse shear flexibility is assumed to be zero in DKT elements, which renders them inappropriate for the analysis of thick plates. The reader should consult References 9.42 and 9.43 for additional discussion of discrete Kirchhoff elements.

9.5 RESIDUAL BENDING FLEXIBILITY

With the special interpolation procedures noted in Sections 9.4.3 and 9.4.4, the competence of the linear triangle and the bilinear quadrilateral rises to the level of complete quadratic normal displacement. That would seem a fair accomplishment but it is possible to go farther. For four-node rectangles, at least, we can achieve accurate results for some cubic displacement patterns by augmenting the transverse shear flexibility with a term known as *residual bending flexibility*.[36]

The concept of residual bending flexibility is most easily explained in the context of a two-node beam element. Since integration in one dimension can be exact, the beam elements found in commercial finite element codes generally give exact results, at least for prismatic shapes. For our purpose, however, it is instructive to consider a two-node beam element (Figure 9.17) formed by separate linear interpolation of lateral displacement and normal rotation, with reduced (one point) integration of transverse shear.

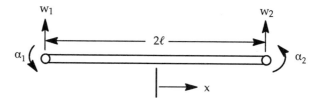

Figure 9.17   Two-Node Beam Element.

The two-dimensional rectangular plate elements described in Section 9.4.3 reduce to this beam element for bending in one direction. A beam element with single-point evaluation of transverse shear has been proposed[35] and is appropriate for use in conjunction with four-node plate elements which make similar assumptions. Table 9.3 compares exact expressions for lateral displacement, w, normal rotation, α, curvature, χ, and transverse shear, γ, with their linearly interpolated values for the case of a general *cubic lateral* displacement. Note that x^2 and x^3 are replaced by their aliases, ℓ^2 and $\ell^2 x$, in the linearly interpolated field values for w and α. Note also that χ and γ are correct for $w_3 = 0$, provided that γ is measured at the center of the beam. Both χ and γ are incorrect if $w_3 \neq 0$.

Table 9.3

**Comparison of Linear Interpolation with Exact Interpolation
of Field Values for the Two-Node Beam Element in Figure 9.17**

VARIABLE	EXACT	LINEAR INTERPOLATION
w	$w_0 + w_1 x + w_2 x^2$ $+ w_3 x^3 + \gamma_0 x$	$w_0 + w_1 x + w_2 \ell^2$ $+ w_3 \ell^2 x + \gamma_0 x$
α	$w_1 + 2 w_2 x + 3 w_3 x^2$	$w_1 + 2 w_2 x + 3 w_3 \ell^2$
χ	$2 w_2 + 6 w_3 x$	$2 w_2$
γ	$w_{,x} - \alpha = \gamma_0$	$\gamma_0 - 2 w_2 x - 2 w_3 \ell^2$

To proceed further, let us compute the strain energy

$$W = \frac{1}{2} \int_{-\ell}^{\ell} \left(EI\chi^2 + kAG\gamma^2 \right) dx \tag{9:48}$$

for the exact and linearly interpolated fields. The result for the exact field is

$$W_e = EI\ell \left(4 w_2^2 + 12\ell^2 w_3^2 \right) + kAG\ell\gamma_0^2 \tag{9:49}$$

and for the linearly interpolated fields, with center point evaluation of γ,

$$W_i = EI\ell\left(4w_2^2\right) + kAG\ell\left(\gamma_0 - 2w_3\ell^2\right)^2 \qquad (9:50)$$

The point we wish to examine is whether the latter expression can be modified to be identical to the first because, in that case, the stiffness matrix for the beam element, as seen at its nodes, will be exact. For this purpose we replace the shear effectiveness factor, k, in Equation 9:50 by k^* and see if we can select k^* so that $W_i = W_e$ for all values of γ_0 and w_3. Fortunately, γ_0 and w_3 are related by the moment equilibrium equation (see Figure 9:18),

$$M_{,x} + V = 0 \qquad (9:51)$$

Figure 9.18   Moment Equilibrium for a Beam.

Substituting $M = EI\chi$ and $V = kAG\gamma_0$ we obtain

$$EI\chi_{,x} + kAG\gamma_0 = 0 \qquad (9:52)$$

or, by substituting the exact value for χ from Table 9.3

$$6EIw_3 + kAG\gamma_0 = 0 \qquad (9:53)$$

Thus

$$w_3 = -\frac{kAG}{6EI}\gamma_0 \qquad (9:54)$$

Substitution into the strain energy expressions then gives, with k replaced by k^* in Equation 9:50,

$$W_e = 4EI\ell w_2^2 + kAG\ell\gamma_0^2\left(1 + \frac{kAG\ell^2}{3EI}\right) \qquad (9:55)$$

and

$$W_i = 4EI\ell w_2^2 + k^* AG\ell\gamma_0^2 \left(1 + \frac{kAG\ell^2}{3EI}\right)^2 \qquad (9:56)$$

The strain energy expressions will be identical if

$$k^* = k\left(1 + \frac{kAG\ell^2}{3EI}\right)^{-1} \qquad (9:57)$$

or, in other words, if the modified transverse shear flexibility

$$\frac{1}{k^* AG} = \frac{1}{kAG} + \frac{\ell^2}{3EI} \qquad (9:58)$$

Thus we see that the exact strain energy and consequently an exact end point stiffness matrix are achieved by the addition of the term $\ell^2 / 3EI$ to the transverse shear flexibility. This term is the residual bending flexibility, so called because it is derived from and completes the beam's bending flexibility. It is seen that the value of the residual bending flexibility can exceed that of the transverse shear flexibility for large enough ℓ. The ratio of residual bending flexibility to transverse shear flexibility is, for a rectangular cross section of depth, t,

$$\frac{kAG\ell^2}{3EI} = \frac{5}{6}\left(\frac{\ell}{t}\right)^2 \frac{1+\upsilon}{2} \qquad (9:59)$$

With $\upsilon = .3$ this value exceeds unity for $\ell / t > 1.36$ and increases as $\left(\ell / t\right)^2$.

The use of this concept in the computerized analysis of beams and frameworks is quite old; it goes back at least to 1953.[44]

From a slightly different perspective, the comparison of the exact energy with that for the uncorrected beam element reveals the extent to which locking afflicts the element's ability to represent a cubic displacement field. Thus, from Equations 9:55 and 9:56, the strain energy is seen to be too large by the factor $1 + kAG\ell^2 / 3EI$. This factor becomes very large for shallow beams and, by implication, for thin three- and four-node plate elements, effectively eliminating the cubic content of the lateral displacement in such elements.

The extent to which residual flexibility can improve performance is illustrated in Table 9.4 for the case of an end-loaded cantilever beam.

Table 9.4

Tip Displacement for a Thin Cantilever Beam, L/t = 100

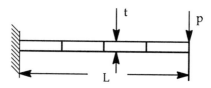

Number of Elements	1	2	4	8	16
Without Residual Flexibility*	.750	.938	.984	.996	.999
With Residual Flexibility*	1.000	1.000	1.000	1.000	1.000

*Both elements employ single point integration of transverse shear.

The implementation of residual bending flexibility in plate elements[36] is entirely heuristic. We could, for example, simply replace the transverse shear effectiveness factor, k, by the value of k^* given in Equation 9:57. This would give correct results for cubic bending parallel to an edge of a rectangular field of elements used, for example, to model tip loading of a rectangular cantilever plate. It would have other beneficial effects such as the softening of transverse shear locking for other shapes and the avoidance of numerical precision difficulties for extremely thin plates ($L / t > O\left(10^4\right)$).

There are, however, situations in which the addition of residual bending flexibility is inappropriate as, for example, when the element is subjected to twist.* In this

---

*Here, again, it helps to visualize the element as a rectangular box.

case, since w = xy is included in the element's set of basis functions, the twisting curvature χ_{xy} and the transverse shear strains are correct and the modification to k is not needed to get the correct strain energy. Fortunately we can distinguish between transverse shear strain due to bending and transverse shear strain due to twist because a rectangular element has two independent values of γ_{xz}, located at $\eta_a = 1/\sqrt{3}$, $\eta_b = -1/\sqrt{3}$.

Thus if we label these two values γ_{xz}^a and γ_{xz}^b, we can form the following constitutive relationship, similar to Equation 9:20.

$$\left\{ \begin{array}{c} Q_x^a \\ Q_x^b \end{array} \right\} = \left[K^{sx} \right] \left\{ \begin{array}{c} \gamma_{xz}^a \\ \gamma_{xz}^b \end{array} \right\} \tag{9:60}$$

where Q_x^a and Q_x^b are transverse shear forces per unit width at points a and b. The constitutive matrix

$$\left[K^{sx} \right] = \left[Z^{sx} + Z^{bx} \right]^{-1} \tag{9:61}$$

where $\left[Z^{sx} \right]$ is the transverse shear flexibility matrix

$$\left[Z^{sx} \right] = \frac{1}{ktG} \begin{bmatrix} 1 & 0 \\ 0 & 1 \end{bmatrix} \tag{9:62}$$

and $\left[Z^{bx} \right]$ is the residual bending flexibility matrix

$$\left[Z^{bx} \right] = \frac{\Delta x^2}{2Et^3} \begin{bmatrix} 1 + a & 1 - a \\ \hline 1 - a & 1 + a \end{bmatrix} \tag{9:63}$$

where Δx is the length of the element in the x direction and a is a small parameter, a $\ll$ 1, selected to serve other concerns. One concern is that $\left[Z^{bx} \right]$ is singular for a = 0 which opens the door to locking and round-off troubles for very thin plates. Another is that, while twist is the main reason that γ_{xz}^a can differ from γ_{xz}^b, differential bending can, as illustrated in Figure 9.19, also contribute to the

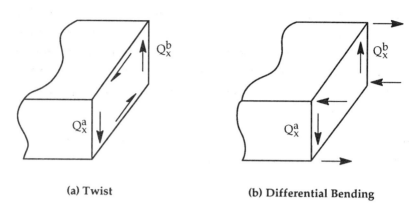

(a) Twist   (b) Differential Bending

Figure 9.19   Two Mechanisms Which Cause γ_x^b to Differ From γ_x^a.

difference particularly in the case of short cantilever plates. Residual bending
flexibility should be included for differential bending but not for twist. The actual
value given to the parameter a in MSC/NASTRAN is

$$a = \frac{.04}{.04 + .96\left(\dfrac{\Delta x}{\Delta y}\right)^2} \qquad (9:64)$$

This value was selected to optimize test results for the bending of square plates
with either clamped or simply-supported edges.

In practical applications the correct interpolation of quadratic displacement fields
for general element shapes is a more important objective than getting improved
results for cubic displacement fields. Nevertheless, residual bending flexibility
has a small but positive benefit. Hughes has shown,[45] for example, that the
MSC/NASTRAN QUAD4 element slightly outperforms his T1 element (which
does not have residual bending flexibility) for simply supported and clamped
rectangular plates. The advantage is largest for coarse element fields and decays
to insignificance for finely divided fields.

The concept of residual bending flexibility also has application to eight-node
quadrilateral elements[46] and three-node triangular elements.[41] In the latter
case, the residual bending flexibility is applied to the integration point values of

transverse shear strain. Thus, referring to the components defined in Equation 9:47, the constitutive relationship is

$$
\begin{Bmatrix} Q_x \\ Q_y \\ Q_t \end{Bmatrix} = \begin{bmatrix} Z^s + Z^b \end{bmatrix}^{-1} \begin{Bmatrix} \gamma_{xz}^0 \\ \gamma_{yz}^0 \\ \gamma_{tz}^0 \end{Bmatrix} \tag{9:65}
$$

where

$$
\begin{bmatrix} Z^s \end{bmatrix} = \frac{1}{KtG} \begin{bmatrix} 1 & 0 & 0 \\ 0 & 1 & 0 \\ 0 & 0 & 1 \end{bmatrix} \tag{9:66}
$$

and

$$
\begin{bmatrix} Z^b \end{bmatrix} = \begin{bmatrix} Z_{11} & Z_{12} & 0 \\ Z_{12} & Z_{22} & 0 \\ 0 & 0 & Z_{33} \end{bmatrix} \tag{9:67}
$$

The values of Z_{11}, Z_{12}, and Z_{22} are selected to give correct results for cubic bending in directions perpendicular to the three sides of the triangle. The value of Z_{33} is, as discussed in Section 9.4.3, unimportant and can be selected to improve particular test results.

The chief effect of residual bending flexibility in practical examples is to soften the response to applied load. This is illustrated in Figure 1.6 where it is seen that the MSC/NASTRAN TRIA3, which has residual bending flexibility, greatly outperforms the NASTRAN TRIA2, which does not. As will be recalled, the NASTRAN TRIA2 element is the Clough-Tocher triangle[7] which, while based on cubic interpolation, includes constraints which effectively lock out the cubic components of lateral displacement.

REFERENCES

9.1   S. Ahmad, "Curved Finite Elements in the Analysis of Solid, Shell, and Plate Structures," Ph.D. Thesis, University of Wales, Swansea, 1969.

9.2   B. M. Irons, "Engineering Application of Numerical Integration in Stiffness Methods," *J. AIAA*, 14, pp 2035-7, 1966.

9.3   J. N. Reddy, "A Simple Higher-Order Theory for Laminated Composite Plates," *J. Appl. Mech.*, ASME 51, pp 745-52, 1984.

9.4   E. Reissner, "The Effect of Transverse Shear Deformation on the Bending of Elastic Plates," *J. Appl. Mech.*, ASME 12, pp 69-76, 1945.

9.5   R. D. Mindlin, "Influence of Rotary Inertia and Shear on Flexural Motions of Isotropic Elastic Plates," *J. Appl. Mech*, ASME 18, pp 31-58, 1951.

9.6   B. M. Irons and J. K. Draper, "Inadequacy of Nodal Connections in a Stiffness Solution for Plate Bending," *J. AIAA*, 3, p. 5, 1965.

9.7   R. W. Clough and J. L. Tocher, "Finite Element Stiffness Matrices for Analysis of Plates in Bending," Proc. Conf. Matrix Methods in Struct. Mech., Air Force Inst. of Tech., Wright-Patterson AFB, Ohio, 1965.

9.8   G. A. Butlin and R. Ford, "A Compatible Plate Bending Element," Univ. of Leicester Eng. Dept. Report, 68-15, 1968.

9.9   G. R. Cowper, E. Kosko, G. M. Lindberg, and M. D. Olson, "Formulation of a New Triangular Plate Bending Element," Trans. Canad. Aero-Space Inst., 1, pp 86-90, 1968 (see also N.R.C. Aero Report LR514, 1968).

9.10   B. M. Irons, "A Conforming Quartic Triangular Element for Plate Bending," *Intl. J. Numer. Methods Eng.*, 1, pp 29-46, 1969.

9.11   W. Bosshard, "Ein Neues Vollverträgliches Endliches Element für Plattenbiegung," Mt. Assoc. Bridge Struct. Eng. Bulletin, 28, pp 27-40, 1968.

9.12   W. Visser, "The Finite Element Method in Deformation and Heat Conduction Problems," Dr. W. Dissertation, T.H., Delft, 1968.

9.13   K. Bell, "A Refined Triangular Plate Bending Element," *Intl. J. Numer. Methods Eng.*, 1, pp 101-22, 1969.

9.14   J. H. Argyris, I. Fried, and D. W. Scharpf, "The TUBA Family of Plate Elements for the Matrix Displacement Method," *The Aeronautical J. R. Ae. S.*, 72, pp 701-9, 1968.

9.15   A. Adini and R. W. Clough, "Analysis of Plate Bending by the Finite Element Method," and Report to Nat. Sci. Found./USA, G.7337, 1961.

9.16   F. K. Bogner, R. L. Fox, and L. A. Schmit, "The Generation of Inter-Element Compatible Stiffness and Mass Matrices by the Use of Interpolation Formulas," Proc. Conf. Matrix Methods in Struct. Mech., Air Force Inst. of Tech., Wright-Patterson AFB, Ohio, pp 397-444, 1966.

9.17   B. M. Irons and S. Ahmad, *Techniques of Finite Elements*, Ellis Horwood, Chichester, p. 268, 1980.

9.18   B. Fraeijs de Veubeke, "A Conforming Finite Element for Plate Bending," *Intl. J. Solids Struct.*, 4, pp 95-108, 1968.

9.19   J. A. Stricklin, W. Haisler, P. Tisdale, and R. Gunderson, "A Rapidly Converging Triangular Plate Element," *J. AIAA*, 7, pp 180-1, 1969.

9.20   J. L. Batoz and M. Ben Tahar, "Formulation et Evaluation d'un Nouvel Elément Quadrilatéral à 12 D.L. pour la Flexion des Plaques Minces," Département de Génie Mécanique, Université de Technologie, Compiègne, France.

9.21   B. M. Irons, "The Semiloof Shell Element," *Finite Elem. for Thin Shells & Curved Members*, 11, pp 197-222, 1976.

9.22   S. Ahmad, B. M. Irons, and O. C. Zienkiewicz, "Analysis of Thick and Thin Shell Structures by Curved Elements," *Intl. J. Numer. Methods Eng.*, 2, pp 419-51, 1970.

9.23   O. C. Zienkiewicz, J. Too, and R. L. Taylor, "Reduced Integration Technique in General Analysis of Plates and Shells," *Intl. J. Numer. Methods Eng.*, 3, pp 275-90, 1971.

9.24   R. H. MacNeal and R. L. Harder, "Eight Nodes or Nine?," *Intl. J. Numer. Methods in Eng.*, 33, pp 1049-58, 1992.

9.25   T. J. R. Hughes and M. Cohen, "The 'Heterosis' Finite Element for Plate Bending," *Comput. Struct.*, 9, pp 445-50, 1978.

9.26   T. J. R. Hughes, *The Finite Element Method*, Prentice-Hall, Englewood Cliffs, NJ, pp 338-42, 1987.

9.27   R. H. MacNeal and R. L. Harder, "A Proposed Standard Set of Problems to Test Finite Element Accuracy," *Finite Elem. Analysis & Design*, 1, pp 3-20, 1985.

9.28   H. C. Huang and E. Hinton, "A New Nine-Node Degenerated Shell Element with Enhanced Membrane and Shear Interpolation," *Int. J. Numer. Methods Eng.*, 22, pp 73-92, 1986.

9.29   J. Jang and P. M. Pinsky, "An Assumed Co-Variant Strain Based 9-Node Shell Element," *Intl. J. Numer. Methods Eng.*, 24, pp 2389-411, 1987.

9.30   J. J. Rhiu and S. W. Lee, "A New Efficient Mixed Formulation for Thin Shell Finite Element Models," *Intl. J. Numer. Methods Eng.*, 24, pp 581-604, 1987.

9.31   T. Belytschko, J. S.-J. Ong, and W. K. Liu, "A Consistent Control of Spurious Singular Modes in the Nine-Node Lagrange Element for the Laplace and Mindlin Plate Equations," *Comput. Methods Appl. Mech. Engrg,* 44, pp 269-95, 1984.

9.32   K. C. Park and G. M. Stanley, "A Curved C^0 Shell Element Based on Assumed Natural-Coordinate Strains," *J. Appl. Mech.*, 53, pp 278-90, 1986.

9.33   M. H. Verwoerd and A. W. M. Kok, "A Shear Locking Free Six-Node Mindlin Plate Bending Element," *Comput. Struct.*, 36, pp 547-51, 1990.

9.34   O. C. Zienkiewicz, R. L. Taylor, P. Papadopoulos, and E. Oñate, "Plate Bending Elements with Discrete Constraints: New Triangular Elements," *Comput. Struct.*, 35, pp 502-22, 1990.

9.35   T. J. R. Hughes, R. L. Taylor, and W. Kanoknukulchai, "A Simple and Efficient Element for Plate Bending," *Intl. J. Numer. Methods Eng.*, 11, pp 1529-43, 1977.

9.36   R. H. MacNeal, "A Simple Quadrilateral Shell Element," *Comput. Struct.*, 8, pp 175-83, 1978.

9.37   T. J. R. Hughes and T. E. Tezduyar, "Finite Elements Based Upon Mindlin Plate Theory with Particular Reference to the Four-Node Bilinear Isoparametric Element," *J. Appl. Mech.*, pp. 587-96, 1981.

9.38   R. H. MacNeal, "Derivation of Element Stiffness Matrices by Assumed Strain Distributions," *Nucl. Eng. Design,* 70, pp 3-12, 1982.

9.39   K. J. Bathe and E. N. Dvorkin, "A Four-Node Plate Bending Element Based on Mindlin/Reissner Plate Theory and a Mixed Interpolation," *Intl. J. Numer. Methods Eng.*, 21, pp 367-83, 1985.

9.40   T. J. R. Hughes and R. L. Taylor, "The Linear Triangular Bending Element," *Mathematics of Finite Elem. and Appl. IV, MAFELAP 1981*, pp 127-42, 1982.

9.41   R. H. MacNeal, "The TRIA3 Plate Element," MacNeal-Schwendler Corp. Memo RHM-37, 1976.

9.42   J. L. Batoz, K. J. Bathe, and L. W. Ho, "A Study of Three-Node Triangular Plate Bending Elements," *Intl. J. Numer. Methods Eng.*, 15, pp 1771-812, 1980.

9.43   J. L. Batoz, "An Explicit Formulation for an Efficient Triangular Plate-Bending Element," *Intl. J. Numer. Methods Eng.*, 18, pp 1077-89, 1982.

9.44   W. T. Russell and R. H. MacNeal, "An Improved Electrical Analogy for the Analysis of Beams in Bending," *J. Applied Mech.*, 1953.

9.45   T. J. R. Hughes, *The Finite Element Method*, Prentice-Hall, Englewood Cliffs, NJ, p. 363, 1987.

9.46   R. H. MacNeal, "Specifications for the QUAD8 Quadrilateral Curved Shell Element," MacNeal-Schwendler Corp. Memo RHM-46B, 1980.

10
Shell Elements

The treatment of curved shells appears last in the study of finite elements and for good reason. Curved shell elements include all of the features of two-dimensional elastic elements and plate bending elements, plus new complexities arising from the curved geometry. Shell elements are considered to be the most difficult of all elements and are the constant subject of advanced research. Still, we will encounter some very simple shell elements tucked in among those of the greatest sophistication.

10.1 SHELL THEORY

Like plate bending theory, curved shell theory is an abstraction of three-dimensional elasticity with simplifying assumptions. The midplane of plate bending theory becomes a curved surface, and strain components parallel to the midsurface are assumed to vary (approximately linearly) with distance from the midsurface. Unlike plate bending theory, however, shell theory requires simultaneous treatment of membrane strains and bending strains because the shell's curvature inherently couples them. Distinctions are made, in practical applications, between cases where membrane strains predominate and cases where bending strains predominate.

At bottom, the assumptions made in shell theory are the same as those made in plate theory with due allowance for the effects of curvature. The reader is invited to consult a standard text on shell analysis.[1,2,3] As we shall see, finite element designers have tended to develop shell theory from finite element concepts without reference to earlier theoretical work.

10.1.1 The Degenerated Shell Element Procedure

The earliest curved shell element appears to be S. Utku's fifteen-degree-of-freedom triangular element,[4] 1967, which is derived within the framework of linear shell theory. Most later curved shell elements, however, trace their origin to S. Ahmad's "degenerate" isoparametric shell element,[5] 1969.

The design concept for Ahmad's element was made possible by Irons' extension, three years earlier, of the isoparametric method to curved solid elements of all orders.[6] Ahmad derived his element by applying simplifying assumptions to a twenty-node brick. As shown in Figure 10.1, he first eliminated the four midside nodes along edges pointing in the ζ direction to form a sixteen-node brick. The top and bottom surfaces, $\zeta = \pm 1$, coincide with the surfaces of the shell. He then replaced each pair of top and bottom surface nodes by a single node at $\zeta = 0$. The degrees of freedom at this shell node include three components of translation and either two or three components of rotation.

The procedure just described, and the subsequent calculations required to form a strain-displacement matrix, etc., are sometimes called the *degenerated shell element procedure*.[7]

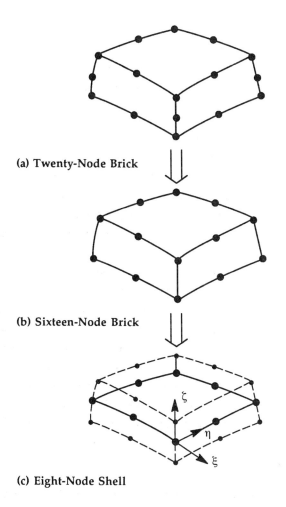

(a) Twenty-Node Brick

(b) Sixteen-Node Brick

(c) Eight-Node Shell

Figure 10.1 **Degeneration of a Twenty-Node Brick Element into an Eight-Node Shell Element.**

While the assignment of six degrees of freedom to the shell node will equal the number assigned to the top and bottom surface nodes, they cannot exactly duplicate the functions of the surface nodes. For one thing, the difference of displacements in the ζ direction, sometimes called the *fiber* direction, is not duplicated (see Figure 10.1). This is consistent with the assumption in plate theory (and also in shell theory) that strain in the direction of the normal to the plate (shell) can be ignored. Another more serious difference is that rotation about the ζ axis at the shell node has no counterpart in the solid element. This *drilling freedom*, as it was called in Section 8.2, clearly cannot produce strain in the solid element and should not, in the degenerated shell element, if the ζ direction coincides with the normal to the shell.

At this point the element designer is faced with a variety of options. To make sure that rotation about the fiber direction does not enter the calculation of strains, he can orient the z-axis of the displacement coordinate system at the shell node in the ζ direction and constrain θ_z. This is the approach used in Ahmad's original element.[5] It requires the user, or some upstream computer code,[8] to supply the ζ direction at shell nodes as extra input. (The top and bottom surface nodes don't really exist.) Alternatively, the element designer can define the ζ direction as the local normal to the $\xi\eta$ plane[9] (the shell midsurface) and then constrain rotation about the normal. This runs into the difficulty that the normal direction may not be the same for all of the elements joined at a node. As we have emphasized, one of the strengths of finite element analysis is its ability to cope with structural discontinuities such as a change in slope. But even when the slope of a shell is continuous, the polynomial representation used in finite element analysis will create slope discontinuities between elements. Consider, for example, a circular arc modeled by a set of parabolic arcs representing three-node beam elements. For 30° arc segments, the slope discontinuity at inter-element nodes is about 0.53°, too large to be safely ignored.

It is possible to ensure that normal rotation does not enter each element's strain calculations without defining a unique normal rotation at each node. For example, we can begin with a unique vector of rotations at node (i), defined in a Cartesian system, x, y, z, and transform it into a vector for each element which has one component, $\theta_{\bar{z}}$, oriented in the normal direction for that element. Thus

$$\begin{Bmatrix} \theta_{\bar{x}} \\ \theta_{\bar{y}} \\ \theta_{\bar{z}} \end{Bmatrix}_i = \begin{bmatrix} E^i \end{bmatrix} \begin{Bmatrix} \theta_x \\ \theta_y \\ \theta_z \end{Bmatrix}_i \qquad (10:1)$$

Next we multiply the transformed vector by an incomplete unit matrix to eliminate the $\theta_{\bar{z}}$ component,

$$\begin{Bmatrix} \theta^c_{\bar{x}} \\ \theta^c_{\bar{y}} \\ \theta^c_{\bar{z}} \end{Bmatrix}_i = \begin{bmatrix} 1 & 0 & 0 \\ 0 & 1 & 0 \\ 0 & 0 & 0 \end{bmatrix} \begin{Bmatrix} \theta_{\bar{x}} \\ \theta_{\bar{y}} \\ \theta_{\bar{z}} \end{Bmatrix}_i \qquad (10:2)$$

and transform back to the original coordinate system to get

$$\begin{Bmatrix} \theta^c_x \\ \theta^c_y \\ \theta^c_z \end{Bmatrix}_i = \begin{bmatrix} E^i \end{bmatrix}^{-1} \begin{bmatrix} 1 & 0 & 0 \\ 0 & 1 & 0 \\ 0 & 0 & 0 \end{bmatrix} \begin{bmatrix} E^i \end{bmatrix} \begin{Bmatrix} \theta_x \\ \theta_y \\ \theta_z \end{Bmatrix}_i \qquad (10:3)$$

The corrected rotation vector now gives a null result for rotation about the normal to the element at the node. Each element connected to the node can have a different corrected rotation vector. The MSC/NASTRAN QUAD8 element[10,11] uses a variation of this procedure.

As a final alternative, the element designer can choose to use the uncorrected nodal rotation vector and rely on subsequent strain calculations to ignore contributions from normal rotation.[12]

Before proceeding, let us consider for a moment the consequences of noncoincident element normals at a node. It will not be known before the elements are assembled whether the normals coincide. If they do, the nodal stiffness matrix will be locally singular with respect to one of the three components of nodal rotation. Thus it is left to the user, or to some automatic singularity detector (such as AUTØSPC in MSC/NASTRAN), to determine whether the stiffness matrix is singular (or nearly singular) and to apply an appropriate constraint to one of the rotational degrees of freedom. As an aside, these considerations imply difficulties in the use of normal rotation to improve performance in the manner proposed in Section 8.2 (Drilling Freedoms).

Turning next to a consideration of the interior of the element, we note that the midsurface, $\zeta = 0$, forms a convenient reference for the definition of strain components. This surface and all surfaces for which $\zeta =$ constant are called *laminas* or *laminae*. In general, since the shell's thickness can vary, the laminas are not necessarily parallel to the midsurface or to each other. This fact is, however, of little consequence and it will be assumed that all strain components are oriented in planes parallel or perpendicular to the midsurface regardless of the thickness variation.

The location of points in the midsurface is derived from the location of nodes by the usual parametric method

$$\left\{\begin{matrix} x \\ y \\ z \end{matrix}\right\} = \sum_i N_i'(\xi, \eta) \left\{\begin{matrix} x \\ y \\ z \end{matrix}\right\}_i \tag{10:4}$$

where (x, y, z) are Cartesian components in an *element* coordinate system which may be different for each element. The shape function for position, $N_i'(\xi, \eta)$, may be different from the shape function for displacement, $N_i(\xi, \eta)$.

The location of points off the midsurface can likewise be derived from the location of points on the nodal fibers. Thus

$$\left\{\begin{matrix} x \\ y \\ z \end{matrix}\right\} = \sum_i N_i'(\xi, \eta) \left(\left\{\begin{matrix} x \\ y \\ z \end{matrix}\right\}_i + \zeta \frac{t}{2} \left\{\begin{matrix} e_{\bar{z}x} \\ e_{\bar{z}y} \\ e_{\bar{z}z} \end{matrix}\right\}_i \right) \tag{10:5}$$

where t is the thickness of the shell and $\left(e_{\bar{z}x}, e_{\bar{z}y}, e_{\bar{z}z} \right)$ are direction cosines of the fiber direction in the element coordinate system, i.e., the third row of the $\left[E^i \right]$ matrix defined in Equation 10:1. The second term in Equation 10:5 represents an offset from the node in the fiber direction. From earlier discussion, the fiber direction may, or may not, be the same for all elements connected at a node. The value of thickness appearing in Equation 10:5 can either be the nodal value or the local value.

Components of strain are defined in a *local* Cartesian system, $\bar{x}, \bar{y}, \bar{z}$, with $\bar{z}$ normal to the midsurface. Note that the direction of $\bar{z}$ does not necessarily coincide with the interpolated fiber direction. The transformation from (x, y, z) to $(\bar{x}, \bar{y}, \bar{z})$ is given by the $[E]$ matrix (see Equation 10:1), where $[E]$ takes the value $[E^i]$ at node (i) and the value $[E^g]$ at integration point (g).

In standard texts on shell analysis,[1] the components of strain are usually defined in curvilinear coordinates. It is, however, much easier to use Cartesian components in the degenerated shell element procedure. The relationship between Cartesian components of strain in the shell and the local components of displacement $(\bar{u}, \bar{v}, \bar{w})$ is simply

$$\varepsilon_{\bar{x}} = \bar{u}_{,\bar{x}} \qquad\qquad \gamma_{\overline{xz}} = \bar{u}_{,\bar{z}} + \bar{w}_{,\bar{x}}$$

$$\varepsilon_{\bar{y}} = \bar{v}_{,\bar{y}} \qquad\qquad \gamma_{\overline{xz}} = \bar{v}_{,\bar{z}} + \bar{w}_{,\bar{y}} \qquad\qquad (10:6)$$

$$\gamma_{\overline{xy}} = \bar{u}_{,\bar{y}} + \bar{v}_{,\bar{x}}$$

We require, finally, a relationship between the local components of displacement $(\bar{u}, \bar{v}, \bar{w})$ and the nodal components of displacement $(u, v, w, \theta_x, \theta_y, \theta_z)$. In first-order plate bending theory, displacements are assumed to vary linearly with distance from the midplane. The same assumption, with distance computed in the direction of the local normal, can be applied to a curved plate. Thus

$$\bar{u} = \bar{u}_o - \frac{\zeta t}{2} \alpha$$

$$\bar{v} = \bar{v}_o - \frac{\zeta t}{2} \beta \qquad\qquad (10:7)$$

$$\bar{w} = \bar{w}_o$$

where α and β are rotations of the normal and the subscript (o) indicates a value at the midsurface.

Since $\alpha = -\theta_{\bar{y}}$ and $\beta = \theta_{\bar{x}}$, an equivalent expression for Equation 10:7 is

$$
\left\{ \begin{array}{c} \bar{u} \\ \bar{v} \\ \bar{w} \end{array} \right\} = \left\{ \begin{array}{c} \bar{u}_o \\ \bar{v}_o \\ \bar{w}_o \end{array} \right\} + \frac{\zeta t}{2} \left[\begin{array}{ccc} 0 & 1 & 0 \\ -1 & 0 & 0 \\ 0 & 0 & 0 \end{array} \right] \left\{ \begin{array}{c} \theta_{\bar{x}} \\ \theta_{\bar{y}} \\ \theta_{\bar{z}} \end{array} \right\} \tag{10:8}
$$

We can relate the local displacement components to nodal variables in different ways. One way is, first, to interpolate the nodal variables to an integration point (g) and to rotate them into the local coordinate system. Then, since

$$
\left\{ \begin{array}{c} u_o \\ v_o \\ w_o \end{array} \right\}_g = \sum_i N_i(\xi, \eta) \left\{ \begin{array}{c} u_o \\ v_o \\ w_o \end{array} \right\}_i \quad ; \quad \left\{ \begin{array}{c} \theta_x \\ \theta_y \\ \theta_z \end{array} \right\}_g = \sum_i N_i(\xi, \eta) \left\{ \begin{array}{c} \theta_x \\ \theta_y \\ \theta_z \end{array} \right\}_i \tag{10:9}
$$

we obtain

$$
\left\{ \begin{array}{c} \bar{u} \\ \bar{v} \\ \bar{w} \end{array} \right\}_g = \left[E^g \right] \sum_i N_i \left\{ \begin{array}{c} u_o \\ v_o \\ w_o \end{array} \right\}_i + \frac{\zeta t}{2} \left[\begin{array}{ccc} 0 & 1 & 0 \\ -1 & 0 & 0 \\ 0 & 0 & 0 \end{array} \right] \left[E^g \right] \sum_i N_i \left\{ \begin{array}{c} \theta_x \\ \theta_y \\ \theta_z \end{array} \right\}_i \tag{10:10}
$$

We note from Equation 10:8 that rotation about the local normal, $\theta_{\bar{z}}$, produces no displacement and hence no strain. This does not guarantee, however, that rotations about the normals at node points will produce no strain. To ensure this outcome, we can first evaluate Equation 10:10 at node points, transform it to element coordinates, interpolate it to integration points, and then transform it to local coordinates. The result is

$$
\left\{ \begin{array}{c} \bar{u} \\ \bar{v} \\ \bar{w} \end{array} \right\}_g = \left[E^g \right] \sum_i N_i \left(\left\{ \begin{array}{c} u_o \\ v_o \\ w_o \end{array} \right\}_i + \frac{\zeta t}{2} \left[\bar{E}^i \right] \left\{ \begin{array}{c} \theta_x \\ \theta_y \\ \theta_z \end{array} \right\}_i \right) \tag{10:11}
$$

where

$$\left[\,\overline{E}^{\,i}\,\right] = \left[\,E^{\,i}\,\right]^{-1} \begin{bmatrix} 0 & 1 & 0 \\ -1 & 0 & 0 \\ 0 & 0 & 0 \end{bmatrix} \left[\,E^{\,i}\,\right] \tag{10:12}$$

This is the procedure used in the MSC/NASTRAN element[10,11] and, except for notational differences, in the Ahmad element.[5]

Comparing Equations 10:10 and 10:11 we note that, while Equation 10:11 requires the evaluation of $[E]$ at integration points and at nodes, Equation 10:10 requires the evaluation of $[E]$ at integration points only. This distinction is significant in some contexts, such as when using hierarchical degrees of freedom which have no discrete locations.

While the variation of displacements with distance from the midsurface is assumed to be linear, the variation of strains will not be linear unless the shell is locally flat. Observe, for example, that

$$\varepsilon_{\overline{x}} = \overline{u}_{,\overline{x}} = \overline{u}_{,\xi}\,\xi_{,\overline{x}} + \overline{u}_{,\eta}\,\eta_{,\overline{x}} + \overline{u}_{,\zeta}\,\zeta_{,\overline{x}} \tag{10:13}$$

includes an explicit linear dependence on ζ in the expressions for $\overline{u}_{,\xi}$ and $\overline{u}_{,\eta}$ (Equation 10:10 or 10:11) and an implied dependence on ζ in $\xi_{,\overline{x}}$, $\eta_{,\overline{x}}$, and $\zeta_{,\overline{x}}$. As a result, $\varepsilon_{\overline{x}}$ is not a linear function of ζ. The dependence of $x_{,\xi}$, etc., on ζ is illustrated in Figure 10.2 for curvature in one direction. The dimension of a patch of laminar surface is seen to be proportional to $1 + z/R$ where R is the radius of curvature.

In plate bending theory, the linear dependence of strains on distance from the midplane is used to define the curvatures $\chi_x = -\varepsilon_{x,z}$, etc., (see Equation 9:9) and also to effect the separation of membrane and bending strain energies. In curved shell theory, such a separation is not strictly possible unless we assume t/R to be negligibly small. If, for t/R finite, we take the definition of elastic curvature to be

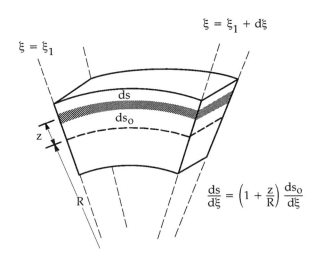

Figure 10.2   Dependence of Arc Length on Distance from the Midsurface.

$$\{\chi\} = \begin{Bmatrix} \chi_{\bar{x}} \\ \chi_{\bar{y}} \\ \chi_{\overline{xy}} \end{Bmatrix} = -\lim_{\zeta \to 0} \frac{2}{t} \frac{\partial}{\partial \zeta} \begin{Bmatrix} \varepsilon_{\bar{x}} \\ \varepsilon_{\bar{y}} \\ \gamma_{\overline{xy}} \end{Bmatrix} \qquad (10{:}14)$$

then the state of strain is approximated by

$$\{\varepsilon\} = \{\varepsilon^m\} - \frac{\zeta t}{2}\{\chi\} \qquad (10{:}15)$$

where $\{\varepsilon^m\}$ is the laminar strain vector at $\zeta = 0$. Strains can, of course, be computed directly at all points in the cross section without introducing the concept of elastic curvature. The strain energy due to laminar strains is obtained in the usual way as the volume integral

$$W_\varepsilon = \tfrac{1}{2} \int_{V_e} \{\sigma\}^T \{\varepsilon\} dV_e = \tfrac{1}{2} \int_{V_e} \{\varepsilon\}^T [D]\{\varepsilon\} dV_e = \tfrac{1}{2} \int_{V_e'} J\{\varepsilon\}^T [D]\{\varepsilon\} d\xi\, d\eta\, d\zeta$$

$$= \tfrac{1}{2} \sum_g w_g \int_{-1}^{1} \left(J\{\varepsilon\}^T [D]\{\varepsilon\} \right)_g d\zeta \qquad (10{:}16)$$

where w_g is the weighting factor at integration point (g) on the midsurface. The integral with respect to ζ in Equation 10:16 can be put in closed form by assuming, for example, that the Jacobian determinant J depends linearly on ζ and that $\{\varepsilon\}$ is given by Equation 10:15. Assumptions regarding the dependence of J on ζ are dangerous, however, and can lead to violation of rigid body properties (self-straining). See the discussion in Section 10.5.1. A similar expression for strain energy due to transverse shear is

$$W_\gamma = \tfrac{k}{2} \sum_g w_g \int_{-1}^{1} \left(J\{\gamma\}^T \left[G^s \right] \{\gamma\} \right) d\zeta \tag{10:17}$$

where $\{\gamma\}^T = \left\lfloor \gamma_{\overline{xz}}, \gamma_{\overline{yz}} \right\rfloor$, k is the shear effectiveness factor, and $\left[G^s \right]$ is the modulus matrix for transverse shear (see Section 9.1).

The remaining operations (calculation of nodal forces and stiffness) are standard. Stress and moment resultants, illustrated in Figure 9.2 for the case of flat plate bending, can be computed for curved shells but they do not satisfy equilibrium equations that are as simple as Equation 9:13.

For thick shells, the dependence of J on ζ and the nonlinear part of the dependence of $\{\varepsilon\}$ on ζ can be significant. One of the more important effects of geometric curvature is that deformations within the midsurface will cause elastic curvature. As a result, the neutral surface (the surface wherein deformations produce zero elastic curvature) is shifted away from the midsurface. In very thin shells, such that t/R << 1, these effects are negligible, and it becomes reasonable to employ a simpler formulation [12] in which $\{\chi\}$ depends only on rotational displacements (see Section 9.5.1).

Looking back over the shell "theory" presented here, the reader must be struck by the number and variety of ad hoc assumptions made to accommodate the standard procedures of finite element formulation. One of the more arbitrary assumptions is that displacement varies linearly in the ζ direction. This, of course, is done to conform to the nature of the degrees of freedom assumed at nodes. As we have seen, this assumption gives rise to strains that, in the presence of curvature, vary nonlinearly with ζ. Is this reasonable? Might it not be equally reasonable to assume that strains vary linearly and that displacements vary nonlinearly with ζ? Few finite element developers have bothered to test the merits of such assumptions.

10.1.2 The Relative Importance of Bending and Extensional Deformations

Curved shells differ from flat plates in that they can carry normal loads through extensional deformation and not just through bending. This is an important feature in the design of many types of structures, such as pressure vessels and automobile panels. Curvature stiffens these structures and gives them greater strength. It is important for the design engineer and also for the finite element analyst to know whether loads are taken primarily by membrane action (extension) or primarily by bending. Where bending action predominates, the part will be weaker and the stresses will tend to be higher and to vary more rapidly; as one consequence, the finite element model may need a finer mesh.

As a start toward the classification of shell response into bending dominant and membrane dominant categories, consider the shallow roof truss shown in Figure 10.3.

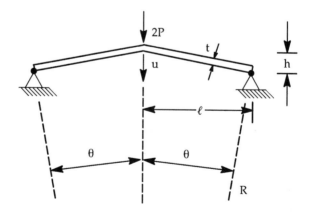

Figure 10.3   Shallow Roof Truss with Apex Load.

If we assume that the apex can carry bending moment from one member to the other, we see that the applied load is resisted by both bending action and extensional action. Thus

$$P = \left(K_b + K_e\right)u \tag{10:18}$$

For $h / \ell \ll 1$, the bending stiffness is

$$K_b = \frac{3EI}{\ell^3} = \frac{Et^3 w}{4\ell^3} \tag{10:19}$$

where t and w are the depth and width of the (assumed) rectangular cross section. Likewise, the effective stiffness due to extension is

$$K_e = \left(\frac{h}{\ell}\right)^2 \frac{EA}{\ell} = \frac{Eh^2 tw}{\ell^3} \tag{10:20}$$

The ratio of extensional stiffness to bending stiffness is

$$\frac{K_e}{K_b} = \frac{4h^2}{t^2} \tag{10:21}$$

For $h / \ell \ll 1$, the height of the truss, h, can be related as follows to the radius, R, of a circle passing through its three nodes

$$h \approx \frac{1}{2} R\theta^2 \approx \frac{1}{2} R\left(\frac{\ell}{R}\right)^2 \tag{10:22}$$

where θ is half the included angle of the truss. Substitution into Equation 10:21 gives

$$\frac{K_e}{K_b} = \frac{\ell^4}{R^2 t^2} = \left(\frac{\ell}{\ell_c}\right)^4 \tag{10:23}$$

where

$$\ell_c = \sqrt{Rt} \tag{10:24}$$

is a *characteristic length*. Clearly, if $\ell > \ell_c$ the extensional stiffness predominates and if $\ell < \ell_c$ the bending stiffness predominates. The fourth power ensures that the dominance of one term or the other increases rapidly as ℓ / ℓ_c deviates from unity.

The relevance of this example to shell analysis is that the truss shown in Figure 10.3 behaves like a flat-faceted finite element model of a shallow shell. Consequently we should expect that bending action will predominant within the characteristic length, $\ell_c = \sqrt{Rt}$, of structural discontinuities, such as concentrated point and line loads.

As an example to confirm this expectation, consider the infinitely long cylindrical shell with a concentrated radial line load, shown in Figure 10.4(a). An analytical solution for the radial displacement (Reference 10.1, p. 282) is

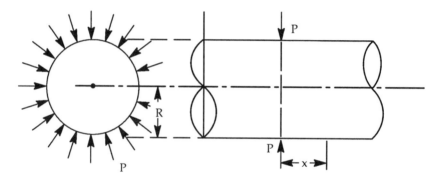

(a) Shell with Applied Load

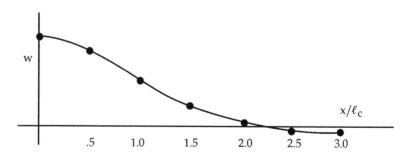

Figure 10.4   **Radial Displacement of an Infinitely Long Cylindrical Shell Under a Concentrated Radial Line Load.**

$$w = w_o e^{-kx / R} \sin\left(\frac{\pi}{4} + \frac{kx}{R}\right) \tag{10:25}$$

where

$$k = \left(3\left(1 - v^2\right)R^2 / t^2\right)^{\frac{1}{4}} \tag{10:26}$$

In the notation developed for the shallow roof truss, the characteristic length in the x direction is

$$\ell_c = \frac{R}{k} = \frac{\sqrt{Rt}}{\left(3\left(1 - v^2\right)\right)^{\frac{1}{4}}} \tag{10:27}$$

The solution is graphed in Figure 10.4(b). We see that the bending solution decays to insignificance at a small multiple of the characteristic length. This can be a very short distance for thin shells. Assume, for example, that $R/t = 100$ and $v = 0.3$ in Equation 10:27. Then, from Equation 10:27, $\ell_c = .0778R$. Clearly a finite element mesh spacing of this size or smaller is required to capture the bending behavior near the applied load.

Configurations also exist where the bending behavior of curved shells predominates at long distances from concentrated loads or other discontinuities. The most important class of such examples is characterized by the presence of free edges.[*] A thin cylindrical shell with open ends can, for example, be squashed by any nonuniform radial load distribution without inducing extensional deformations. This result follows from the definition of membrane strain in cylindrical coordinates

$$\varepsilon_x = u_{,x}$$

$$\varepsilon_\phi = \frac{1}{R}\left(v_{,\phi} + w\right) \tag{10:28}$$

$$\gamma_{x\phi} = v_{,x} + \frac{1}{R} u_{,\phi}$$

---

[*]Flugge (Reference 10.1, p. 86) also describes pure bending deformations for closed axisymmetric shells with re-entrant (convex inward) corners.

where (u, v, w) are longitudinal, circumferential, and radial displacements respectively.

All three components remain zero if

$$u = f_3(\phi)$$

$$v = -\frac{x}{R} u_{,\phi} + f_4(\phi) \qquad (10:29)$$

$$w = -v_{,\phi}$$

Equation 10:29 constitutes the general *inextensional solution* for a cylindrical shell. One particular solution is

$$u = 0 \quad , \quad v = \cos m\phi \quad , \quad w = m \sin m\phi \qquad (10:30)$$

Since m is an arbitrary integer, solutions of the form of Equation 10:30 can be summed to fit any arbitrary circumferential deformation mode.

Another interesting particular solution is

$$u = \sin m\phi \quad , \quad v = \frac{-mx}{R} \cos m\phi \quad , \quad w = \frac{-m^2 x}{R} \sin m\phi \quad (10:31)$$

This solution illustrates that pure bending solutions still exist if a rigid diaphragm is placed at x = 0 such that v(0) = w(0) = 0 with u(0) unrestrained. The solution for m = 1 represents rigid body rotation.

General inextensional solutions for spherical shells with free edges have also been published (Reference 10.1, p. 85). One such solution forms the basis for the Morley spherical shell test problem[13] (Figure 10.14). We will develop inextensional solutions for shallow shell caps in Section 10.4.1 and use them to study membrane locking.

From the standpoint of engineering applications, inextensional solutions exist in the neighborhood of any free edge such as, for example, the perimeter of a small hole in an otherwise closed shell. In general, bending behavior dominates near free edges.

10.2 THE USE OF FLAT PLATE ELEMENTS AND SOLID ELEMENTS IN SHELL ANALYSIS

The procedure described in Section 10.1.1 uses a solid twenty-node brick as a starting point for development of an eight-node curved shell element. The advantages of a shell element over a solid element are a reduction in the number of degrees of freedom, reduced sensitivity to round-off error through the elimination of strain in the normal direction, and greater attention to the effects of curvature. Situations exist, however, where solid elements have the advantage. One example is the analysis of arch dams and other thick shells where the assumption that displacements vary linearly in the normal direction may be suspect. In this case, an assembly of brick elements may be preferable. Another example is the modeling of the intersection of shell branches, as shown in Figure 10.5.

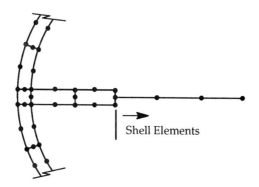

Shell Elements

Figure 10.5   Model of Shell Intersection.

Figure 10.5 also illustrates a transition from solid elements to shell elements at a point sufficiently removed from the intersection. The transition is customarily treated by constraints (MPCs) between the degrees of freedom at a shell node and the degrees of freedom at the pair of top and bottom surface nodes of the solid. It is also possible to design a transition element which includes some shell nodes and some solid node pairs. Such elements relieve the user of the need to supply constraints.

Solid elements can perform satisfactorily in moderately thin shell applications if remedies are provided for their various locking disorders. Section 10.3 contains comparative data on the performance of solid elements and shell elements in such applications.

The shell theory presented in Section 10.1 requires the existence of at least one node per element edge to define curvature in the midsurface. This would appear to rule out shell elements with corner nodes only, although it is indeed possible to use the edge nodes to define geometry without assigning degrees of freedom to them. As discussed in Section 3.5, a *superparametric* formulation such as this would have difficulty in representing even a constant membrane strain condition.

In point of fact, however, *flat* elements which have only corner nodes frequently perform quite well in curved shell applications. Their formulation is identical to that of flat plate elements provided that membrane action and bending action are both included. The coupling between membrane action and bending action occurs mainly through differences between element orientations at nodes,[*] as illustrated in Figure 10.6.

The only essential requirement for the use of flat plate elements in three-dimensional applications relates to their internal coordinate systems. Since the normal to a plate is treated differently than the other two directions, each element must have an *element* coordinate system that is fixed with respect to the element's geometry and independent of the coordinate systems used at nodes. The *nodal* coordinate system can, in most finite element programs, be specified by the user to have a different orientation at each node. The chief uses of a nodal coordinate system are to display displacement output and to provide directions for the application of loads, boundary conditions, and constraints. In MSC/NASTRAN the collection of coordinate directions at all nodes is called the *global* coordinate system. The code for each element includes a

---

[*]Coupling can also be defined within a plate element if material properties are not symmetrical with respect to the midplane (see Equation 9:22).

subroutine which computes the transformation between nodal degrees of freedom in element coordinates, $\{u_e\}$, and nodal degrees of freedom in global coordinates, $\{u_g\}$. Thus, at each node,

$$\{u_e\} = \left[T_{eg}\right]\{u_g\} \qquad (10:32)$$

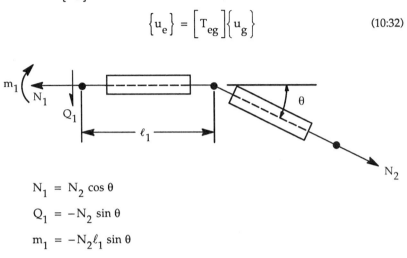

$$N_1 = N_2 \cos\theta$$

$$Q_1 = -N_2 \sin\theta$$

$$m_1 = -N_2 \ell_1 \sin\theta$$

Figure 10.6   Coupling Between Membrane Action and Bending Action Due to a Change in Element Direction.

The three-node, triangular plate element can, without any modification, be used in curved shell applications. The four-node, quadrilateral plate element requires additional work because its four nodes may not necessarily lie in a plane. The MSC/NASTRAN QUAD4 element accounts for this fact by transferring forces, in a manner that is statically correct, from the corners of the element, which remains flat, to the nearby nodes. Stated differently, the displacements at the element's corners are related to the displacements at nodes in such a way that rigid body conditions are satisfied. Details are described in Section 10.2.1. Although this approach appears crude in comparison to treatment of the four-node quadrilateral as a straight-sided but twisted shell with anticlastic curvature, it works well in practice and is a good deal simpler to implement.

10.2.1 Modification to QUAD4 for Nonplanar Nodes

Figure 10.7 illustrates the connections of a four-node flat plate element to nodes which do not lie in a plane. The xy plane of the element is parallel to diagonals 1-3 and 2-4 in Figure 10.7 and lies midway between them. The corners of the element are connected to the nodes by vertical links of equal length, h.

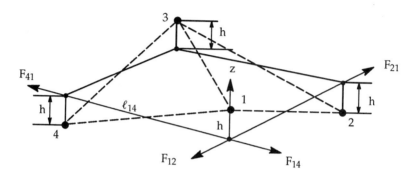

Figure 10.7   A Four-Node Flat Plate Element with Offset Nodes.

The stiffness matrix of the element as seen at the nodes, $\left[K^n \right]$, is computed from the stiffness of the element in its plane, $\left[K^p \right]$, by the transformation

$$\left[K^n \right] = [S]^T \left[K^p \right] [S] \tag{10:33}$$

where $[S]$ relates displacements in the plane to nodal displacements

$$\left\{ u^p \right\} = [S] \left\{ u^n \right\} \tag{10:34}$$

and $[S]^T$ relates nodal forces and moments to forces and moments in the plane

$$\left\{ F^n \right\} = [S]^T \left\{ F^p \right\} \tag{10:35}$$

The task is to derive $[S]$ or $[S]^T$. We select $[S]^T$ because it is easier in this application to think in terms of forces and moments. As a starting point we note that the vector $\left\{ F^p \right\}$ at node i includes the terms

$$\left\{ F^p \right\}_i = \left\lfloor F_x, F_y, F_z, M_x, M_y, 0 \right\rfloor_i \qquad (10{:}36)$$

The last term, M_z, is zero because a flat plate generates no moments about its normal. Transfer of this vector to the node induces additional moments

$$\Delta M_x^n = \pm h\, F_y \ , \ \Delta M_y^n = \mp h\, F_x \qquad (10{:}37)$$

where the choice of sign depends on whether the node is above or below the xy plane. This is just the result we get if we replace the vertical links by little rigid beams. A moment's thought shows, however, that it is not a very good result. It will, for example, induce bending moments in a shell problem that is otherwise membrane dominant; and, since membrane stiffness is usually much larger than bending stiffness, the bending deformations are likely to be quite large.

In order to avoid the inadvertent excitation of bending moments, the moments described by Equation 10:37 should be replaced by vertical couples. To do this we first decompose the forces at element corner (i) into forces parallel to the edges. Then, for the case of edge 1-4 in Figure 10.7, we apply vertical forces to nodes 1 and 4.

$$\Delta F_{1z} = -\Delta F_{4z} = \frac{h}{\ell_{14}} \left(F_{14} - F_{41} \right) \qquad (10{:}38)$$

Repeating for all edges and collecting results, we can easily generate the entire $[S]^T$ matrix.

There is, however, one remaining difficulty which was discovered, in the case of the MSC/NASTRAN QUAD4, in the validation test of a twisted cantilever beam (Figure 10.8). Portions of two adjacent elements in the beam model are shown in Figure 10.9. Due to the twist, the plane of each element makes an angle α with respect to a straight line between nodes A and B. As a result, the transfer of bending moment along edge A-B from element 1 to element 2 at node A requires a bending moment about the vertical axis,

$$M_z = \left(M_1 - M_2 \right) \sin \alpha \qquad (10{:}39)$$

Fixed End

Length = 12.0, Width = 1.1, Depth = 0.32, Twist = 90°

Figure 10.8   Twisted Beam Problem.

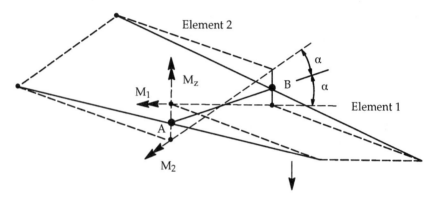

Figure 10.9   Model of a Twisted Ribbon.

Since nothing restrains this moment, the finite element model will collapse under applied load. (In the actual test, the recorded tip displacements were 200 times too large.)

The remedy adopted for this problem was to equilibrate M_1 by moment about the node line A-B, and a couple normal to z,

$$F_A = -F_B = \frac{\sin \alpha}{\ell_{AB}} M_1 \tag{10:40}$$

This modification completes the specification for the $[S]^T$ matrix.

10.2.2 The "Simplest" Shell Elements

Without doubt, the treatment of rotation about the normal to a shell is a delicate matter for the designer and for the user of shell elements. As we have just seen, an inadvertently placed moment about the normal can cause large unwanted deformations. At the least, the question of whether or not to apply constraints so as to remove singularities or near singularities associated with normal rotation is a bother which must be addressed at execution time by the user or by an automatic singularity suppression routine. Ahmad[5] finessed this issue by suppressing rotation in a user-specified fiber direction at each node. His approach has the drawback that it does not accommodate legitimate changes in shell slope.

A very simple triangular plate bending element presented by Morley[14] in 1971 avoids the question of normal rotation in an entirely different manner. In his element, shown in Figure 10.10, the degrees of freedom include normal displacements at the corners and rotations about each of the three edges at their midpoints. The element is converted into a shell element by combining it with a constant strain membrane triangle.

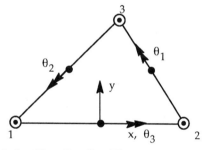

Figure 10.10   The Morley Plate Bending Element.

Morley's element has six degrees of freedom which are just sufficient to represent the rigid body modes and the three components of constant elastic curvature. Morley's element is clearly a Kirchhoff element because the additional degrees of freedom needed to allow transverse shear strains are not available.

The normal displacement function for w is the complete quadratic

$$w = a + bx + cy + dx^2 + exy + fy^2 \qquad (10:41)$$

The coefficients $a, b, \cdots, f$ are linear combinations of the nodal displacements $\left(w_1, w_2, w_3, \theta_1, \theta_2, \theta_3 \right)$ which can be evaluated by collocation. Then, since, from Equation 9:26,

$$\chi_x = w_{,xx} = 2d \; ; \; \chi_y = w_{,yy} = 2f \; ; \; \chi_{xy} = 2w_{,xy} = 2e \quad (10:42)$$

the constant curvatures can be related directly to nodal displacements to form $[B]$ in

$$\{\chi\} = [B]\{u_i\} \qquad (10:43)$$

Finally, the element's stiffness matrix is, using the notation of Equation 9:21,

$$[K] = \frac{At^3}{12} [B]^T \left[D^b \right][B] \qquad (10:44)$$

where A is the area of the triangle.

Morley's element is obviously nonconforming. For example, the three nodal displacements on edge 1-2 (i.e., w_1, w_2, and θ_3) are unable to represent a quadratic variation of w or a linear variation of $w_{,y}$ along edge 1-2. As a result, the Morley element cannot pass constant curvature patch tests.

When Morley elements are joined together to model a shell, the rotational degrees of freedom accommodate easily to changes in slope between elements, and there is no problem with nearly singular degrees of freedom. These are the chief virtues of the Morley element.

Recently some of the ideas contained in the Morley element have been used by Phaal and Calladine[15] to design shell elements with *no* rotational degrees of freedom. Their basic SHB element, shown in Figure 10.11, consists of two rigid triangles joined together by a rotational spring along diagonal 1-2. This element has only one elastic mode, bending about the hinge 1-2. This limitation is overcome in practice by overlapping elements as shown in Figure 10.12. The central triangle in Figure 10.12 consists of three overlapping rigid triangles which are joined to the outer triangles by rotational springs. In this respect, the central triangle in Figure 10.12 resembles the Morley triangle.

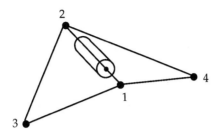

Figure 10.11   The Phaal-Calladine *Simple Hinged Bending* **(SHB) Element.**

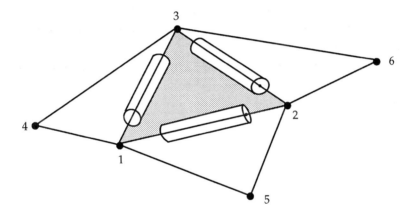

Figure 10.12   Overlapping SHB Elements.

While Phaal and Calladine present data showing good agreement with results obtained with more conventional elements, the selection of the overlapping patterns appears to require considerable ingenuity. Furthermore, the partial overlapping of elements violates the first basic assumption or rule about finite elements stated in Section 2.2:

> *"Each finite element fills a well-defined region of space and represents all of the relevant physics within the space."*

The Phaal-Calladine elements may appeal to adventurous users. For finite element purists, they are beyond the pale.

10.3 COMPARISON OF PERFORMANCE IN CURVED SHELL APPLICATIONS

The claim is made in Section 10.2 that solid elements and lowest order flat plate elements are frequently competitive with curved shell elements in curved shell applications. To test that claim, we consider here the performance of all three types of elements—solid, flat plate, and curved shell—for two problems which have become de facto benchmark standards. [16]

The first problem is the Scordelis-Lo roof, Figure 10.13, which we have already used in Section 1.3 to illustrate differences in the performance of elements. It is a moderately thin shell, $R/t = 100$; and, since two of its edges are free, it must possess inextensional bending modes. Results show, however, that bending action and membrane action are about equally important in the Scordelis-Lo roof problem. Symmetry allows us to model one-fourth of the roof. The quantity which we will use to compare the performance of elements is the vertical deflection at the center of the free edge.

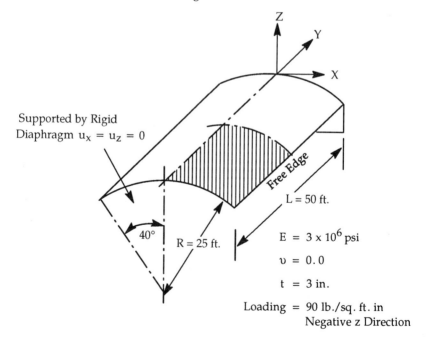

Figure 10.13 Shell Roof Under Gravity Load (the Scordelis-Lo Roof).

Table 10.1 shows results obtained in the Scordelis-Lo roof problem with ten different elements. Comparisons are made for equal numbers of "node spaces" along the edges of the mesh. This is done to compare meshes on an approximately equal basis with respect to the number of degrees of freedom; for example, meshes of $p = 2$ elements have two "node spaces" per element and meshes of $p = 1$ elements have one "node space" per element. The meshes for quadrilateral elements are rectangular arrays with equal numbers of elements in the two directions. The meshes for triangular elements are formed by dividing each quadrilateral element into two triangles. The results depend slightly on which diagonal is used to subdivide the rectangles, but this fact is not accounted for in Table 10.1.

The results for the different elements are seen to vary widely. The best results are achieved by the QUAD8 curved shell element with reduced (2 x 2) integration, followed closely by the HEXA20 solid element with reduced integration. The same elements with *full* integration give the worst results. Score another victory for reduced integration, similar to the victories illustrated in Figure 9.12 and Table 6.5 for the case of flat plate bending!

The four-node plate elements QUAD4 and QUADR also perform well, certainly much better than the three-node triangles TRIA2 and TRIA3. The difference between TRIA2 and TRIA3 is that TRIA2 incorporates the Clough-Tocher bending triangle[19] while TRIA3 uses the assumed strain formulation of transverse shear described in Section 9.4.4 and the residual bending flexibility described by Equation 9:67. We see that the latter formulation has the better performance for the Scordelis-Lo roof problem.

The second problem is the Morley hemispherical shell, shown in Figure 10.14. The 18° hole at the apex, which does not appear in the original description of the problem,[13] has been added to allow a regular mesh of quadrilateral elements without the addition of triangles at the apex. The loading consists of alternating radial forces at 90° intervals on the free edge. For this loading the hole at the apex does not greatly affect the results (it increases the strain energy by less than 2%). The quantity used to compare performance will be the radial displacement at load points. This quantity is a measure of the strain energy.

Table 10.1

Results for Scordelis-Lo Roof

(Vertical Deflection at Midpoint of Free Edge Normalized to 0.3024)

NUMBER OF NODE SPACES ALONG EDGE OF MESH	2	4	6	8	10	12
SOLID ELEMENTS						
HEXA8 (Reduced)	1.320	1.028	1.012	1.005	——	——
HEXA20 (Full)	0.092	0.258	0.589	0.812	——	——
HEXA20 (Reduced)	1.046	0.967	1.003	0.999	——	——
FLAT PLATE ELEMENTS						
TRIA2$^{(1)}$	0.828	0.676	0.788	0.860	0.903	——
TRIA3$^{(2)}$	1.127	0.769	0.858	0.923	0.964	——
QUAD4 (Reduced)	1.376	1.050	1.018	1.008	1.004	——
QUADR$^{(3)}$	1.379	1.055	1.020	1.009	1.004	1.002
CURVED SHELL ELEMENTS						
QUAD8 (Full)	0.201	0.486	0.689	0.841	0.919	0.956
QUAD8 (Reduced)	1.021	0.984	1.002	0.997	0.996	——
TRIA6$^{(4)}$	0.881	0.893	1.009	1.034	1.006	——

Notes: (1) Constant strain membrane + Clough-Tocher bending (Section 9.2.1).

      (2) See Section 9.4.4.

      (3) See Section 8.2.

      (4) See Section 10.5.3.

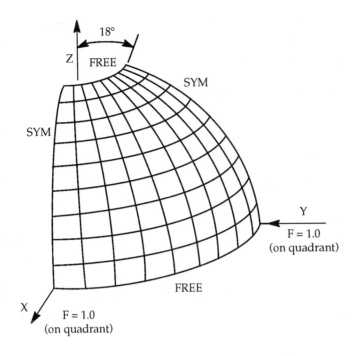

Radius = 10.0, Thickness = .04, E = 6.825 x 10^7,

$\upsilon = 0.3$, Mesh = N x N (on Quadrant)

Loading: Concentrated Forces as Shown

Figure 10.14 Spherical Shell Problem.

The Morley shell is rather thin, R/t = 250, and it possesses inextensional bending modes which are strongly excited by the loading. The radial component of displacement in the inextensional solutions is [18]

$$w_n = A(n + \cos \phi) \tan^n \frac{\phi}{2} \cos n\theta \qquad (10:45)$$

where ϕ is measured along a meridian and θ is measured along a polar circle. At the equator, $\phi = \pi/2$,

$$w_n\left(\frac{\pi}{2}\right) = An \cos n\theta \qquad (10:46)$$

The loading applied in the Morley shell problem excites harmonics $n = 2, 6, 10, \cdots$. Test results indicate that bending deformations of harmonic $n = 2$ are dominant in the response.

Table 10.2 compares results obtained in the Morley hemispherical shell problem. Again the meshes for quadrilateral elements are regular arrays of tapered elements (see Figure 10.14) and the meshes for triangular elements are obtained by splitting each quadrilateral along a diagonal.

The results for the Morley hemispherical shell vary widely—even more widely than the results for the Scordelis-Lo roof. It comes as no surprise that the worst results are those obtained using full integration in second order elements, HEXA20 and QUAD8. What *is* surprising is that the best results are achieved by the simplest elements—TRIA2, TRIA3, and QUAD4. The range of performance is enormous. For example, with eight "node spaces," the error varies from 1% (or less) to 98%. Clearly a very virulent form of locking is evident here, much more so than in the Scordelis-Lo roof problem or in the flat plate problems examined earlier.

Other smaller but noteworthy points are evident. For example, the addition of drilling freedoms to the QUADR element degrades its performance relative to QUAD4. The reason, as noted earlier, is the coupling of normal rotation to bending deformations through geometric curvature. It is also seen that while reduced integration improves the performance of the solid elements HEXA8 and HEXA20, it does not make them competitive with the flat plate elements or with the better curved shell elements. The best curved shell element appears to be the nine-node "assumed natural strain" element, 9ANS, of Park and Stanley.[9] We will defer speculation about the reasons for the superiority of 9ANS until later.

The otherwise good results for the TRIA6 element are spoiled by its dependence on restraint of rotation about the normal to the shell. In the case of QUAD8 and the flat plate elements, the difference in results obtained with full restraint and restraint based on a relatively mild automatic singularity criterion did not vary significantly. In the case of TRIA6, the result obtained with the mild automatic singularity criterion appears to converge to an answer that is about 12% too soft.

Table 10.2

Results for Morley Hemispherical Shell with Hole

(Displacement Under Load Normalized to 0.0940)

NUMBER OF NODE SPACES ALONG EDGE OF MESH	2	4	6	8	10	12	16
SOLID ELEMENTS							
HEXA8 (Reduced)	——	0.039	——	0.730	——	0.955	——
HEXA20 (Full)	——	0.001	——	0.021	——	0.097	——
HEXA20 (Reduced)	——	0.162	——	0.776	——	0.972	——
FLAT PLATE ELEMENTS							
TRIA2	0.942	1.006	1.013	1.013	1.013	1.012	——
TRIA3	1.059	1.046	1.029	1.017	1.009	1.004	——
QUAD4 (Reduced)	0.972	1.024	1.013	1.005	1.001	0.998	——
QUADR	0.202	0.447	0.858	0.962	0.985	0.992	0.996
CURVED SHELL ELEMENTS							
QUAD8 (Full)	0.001	0.006	0.029	0.069	0.134	0.210	0.373
QUAD8 (Reduced)[3]	1.124	0.194	0.639	0.895	0.985	1.005	1.008
9ANS[1]	——	0.80	——	0.97	——	——	0.99
TRIA6 (Best Case)[2]	0.840	0.927	0.976	0.990	0.995	——	——
TRIA6 (Worst Case)[2]	0.847	0.976	1.059	1.096	1.115	——	——

Notes: (1) See Reference 10.9.
   (2) The best case has constrained normal rotation $\left(\theta_r\right)$; the worst case has free normal rotation. Results also depend slightly (2%) on which diagonal is used to split quadrilaterals into triangles.
   (3) Results for QUAD8R in Reference 10.20.

Table 10.3 attempts to summarize the detailed results of Tables 10.1 and 10.2 through the assignment of value judgments (Excellent, Good, Fair, Poor). It also passes judgment on results obtained with the simple hinged bending (SHB) element described in Section 10.2.2. No one element gives excellent results in both problems. The QUAD4 has the best overall score (*Good* in the Scordelis-Lo roof, *Excellent* in the Morley hemisphere). Recall, however, the deficiencies recorded elsewhere of QUAD4 as a membrane element.

All in all, the test results vary widely and exhibit differences which are not easily explained. A variety of subtle effects is at work which makes the study of shell elements an interesting and rewarding experience for those with enough patience and curiosity.

10.4 MEMBRANE LOCKING

The extremely poor results we observed for QUAD8 and HEXA20 with full integration in the Morley hemispherical shell problem (Table 10.2) are caused by *membrane locking*. It was also observed that the simpler flat plate elements are immune to this disorder, which implies that it relates to the way curvature is handled in curved shell elements. The earliest explanations of membrane locking[21,22] employed the equations of a curved three-node beam element to show that the excessive stiffness comes from incorrect interpolation of a cubic term in the extensional strain. Here we will develop, with only slightly more effort, the equations for a shallow shell cap and apply them directly to curved shell elements.

10.4.1 Inextensional Solutions for a Shallow Shell

Figure 10.15 shows a portion of the middle surface of a shallow shell which makes angles γ_x, γ_y at point p with the x-, y-axes of a fixed Cartesian coordinate system. The local axes $\bar{x}$, $\bar{y}$ are tangent to the midsurface at point p and are located, respectively, in the xz and yz planes. The membrane strains at point p are computed in the $\bar{x}$, $\bar{y}$ plane. Note that $\bar{x}$ and $\bar{y}$ are not (quite) orthogonal but this is immaterial for our purpose which is to define a valid set of membrane strains (in this case slightly skewed) and to relate them to displacements u, v, w in the x, y, z system.

Table 10.3

Summary Comparison of Results for Shell Problems

ELEMENT	INTEGRATION METHOD	SCORDELIS-LO ROOF	MORLEY HEMISPHERE
SOLID ELEMENTS			
HEXA8	Reduced	Good	Fair
HEXA20	Full	Poor	Poor
HEXA20	Reduced	Excellent	Fair
FLAT PLATE ELEMENTS			
TRIA2	(1)	Poor	Excellent
TRIA3	(2)	Fair	Excellent
QUAD4	Reduced	Good	Excellent
QUADR	(3)	Good	Good
SHB	(4)	Fair	Good
CURVED SHELL ELEMENTS			
QUAD8	Full	Poor	Poor
QUAD8	Reduced	Excellent	Fair
TRIA6	(5)	Good	Fair
9ANS	(6)	———	Good

Notes: (1) Constant strain membrane + Clough-Tocher bending (Section 9.2.1).

      (2) See Section 9.4.4.

      (3) See Section 8.2.

      (4) See Section 10.2.2.

      (5) See Section 10.5.3.

      (6) See Reference 10.9.

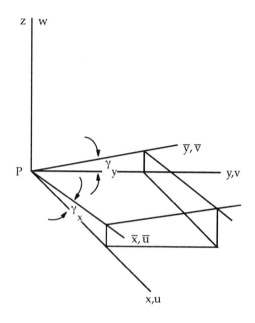

x, y, z: Fixed Cartesian coordinate axes.

$\bar{x}, \bar{y}$:   Axes which are locally tangent to the midsurface;
        $\bar{x}$ in xz plane, $\bar{y}$ in yz plane.

Figure 10.15   Coordinates of a Shallow Shell.

The definition of skewed membrane strain components in the $\bar{x}$, $\bar{y}$ system is

$$\varepsilon_{\bar{x}} = \bar{u}_{,\bar{x}} \ , \ \varepsilon_{\bar{y}} = \bar{v}_{,\bar{y}} \ , \ \bar{\gamma} = \bar{u}_{,\bar{y}} + \bar{v}_{,\bar{x}} \tag{10:47}$$

Since, from Figure 10.15,

$$x_{\bar{x}} = \cos \gamma_x \ , \ y_{\bar{y}} = \cos \gamma_y \ , \ x_{,\bar{y}} = 0 \ , \ y_{,\bar{x}} = 0 \tag{10:48}$$

and

$$\bar{u} = u \cos \gamma_x + w \sin \gamma_x \ , \ \bar{v} = v \cos \gamma_y + w \sin \gamma_y \tag{10:49}$$

we can immediately write the membrane strains in the form

$$\varepsilon_{\overline{x}} = u_{,x} \cos^2 \gamma_x + w_{,x} \sin \gamma_x \cos \gamma_x$$

$$\varepsilon_{\overline{y}} = v_{,y} \cos^2 \gamma_y + w_{,y} \sin \gamma_y \cos \gamma_y$$

$$\overline{\gamma} = \left(u_{,y} + v_{,x} \right) \cos \gamma_x \cos \gamma_y + w_{,x} \sin \gamma_y \cos \gamma_x + w_{,y} \sin \gamma_x \cos \gamma_y$$

$$(10:50)$$

These expressions are valid for large angles $\left(\gamma_x, \gamma_y < 90° \right)$. Setting them to zero gives conditions which the inextensional solutions must satisfy. These conditions are, after dividing by common factors,

$$u_{,x} + w_{,x} \tan \gamma_x = 0$$

$$v_{,y} + w_{,y} \tan \gamma_y = 0 \qquad (10:51)$$

$$u_{,y} + v_{,x} + w_{,x} \tan \gamma_y + w_{,y} \tan \gamma_x = 0$$

In order to simplify the search for inextensional solutions, let us assume constant curvature in the x and y directions so that the angles γ_x and γ_y increase linearly along the x- and y-axes. Assume further that these angles are small enough that $\tan \gamma_x$ can be replaced by γ_x and $\tan \gamma_y$ can be replaced by γ_y without significant error. Equation 10:51 then becomes

$$\varepsilon_{\overline{x}} = u_{,x} + \frac{x}{R_x} w_{,x} = 0$$

$$\varepsilon_{\overline{y}} = v_{,y} + \frac{y}{R_y} w_{,y} = 0 \qquad (10:52)$$

$$\overline{\gamma} = u_{,y} + v_{,x} + \frac{y}{R_y} w_{,x} + \frac{x}{R_x} w_{,y} = 0$$

where R_x, R_y are the radii of curvature in the x and y directions respectively. We see that, for a shallow shell, the effect of curvature is to add small terms proportional to the spatial derivatives of w to the membrane strain components.

The form of Equation 10:52 admits polynomial solutions in x and y. We consider two cases: $R_y \to \infty$ which corresponds to a shallow cylindrical shell, and $R_x = R_y = R$ which corresponds to a shallow spherical shell. A general solution for the cylindrical case can be found by assuming $w = x^n y^m$ and finding what, if any, polynomial forms for u and v then satisfy Equation 10:52. Table 10.4 lists the general solutions found in this way. The solution for $n = 1$, $m = 0$ is a rigid body mode. General solutions for the spherical case are more difficult to construct. Table 10.5 lists the two lowest order inextensional solutions for a shallow spherical shell. The second solution is, in fact, the first solution rotated by 45° about the z-axis.

Table 10.4

Inextensional Solutions for Shallow Cylindrical Shells, $n \geq 1$

w	u	v
x^n	$-\dfrac{nx^{n+1}}{(n + 1)R_x}$	0
$x^n y$	$-\dfrac{nx^{n+1}y}{(n + 1)R_x}$	$-\dfrac{x^{n+2}}{(n + 1)(n + 2)R_x}$

Table 10.5

Lowest Order Inextensional Solutions for Shallow Spherical Shells

w	u	v
$x^2 - y^2$	$-\dfrac{2x^3}{3R}$	$\dfrac{2y^3}{3R}$
xy	$-\dfrac{y}{2R}\left(x^2 + \dfrac{y^2}{3}\right)$	$-\dfrac{x}{2R}\left(y^2 + \dfrac{x^2}{3}\right)$

The important thing to note about the inextensional solutions listed in Tables 10.4 and 10.5 is that the polynomial degree for u and v is one degree higher than the polynomial degree for w. For example, simple uniform bending of a cylindrical shell, $w = x^2$, requires u to be proportional to x^3. From the study of aliasing in Section 6.1, we know that an eight- or nine-node isoparametric element cannot correctly interpolate this cubic term. The result is membrane locking.

10.4.2 Membrane Locking of Eight- and Nine-Node Rectangular Elements

The inextensional solutions listed in Tables 10.4 and 10.5 provide a convenient starting point for the definition of membrane locking in rectangular elements. We begin with the substitutions $x = a\xi$, $y = b\eta$ and note the following aliases for eight- and nine-node elements (see Section 6.1).

$$\xi^3 \Rightarrow \xi \ , \ \xi^3\eta \Rightarrow \xi\eta \ , \ \xi^4 \Rightarrow \xi^2 \qquad (10:53)$$

Making these substitutions in Tables 10.4 and 10.5 and using Equation 10:52, we obtain the expressions for membrane strains recorded in Table 10.6. (Only the displacement components with aliases need be considered in the derivation since the correct displacements produce no strains.)

It is evident from Table 10.6 that the membrane strains for quadratic normal displacement fields vanish at the 2 x 2 Gauss points, $\xi = \pm 1/\sqrt{3}$, $\eta = \pm 1/\sqrt{3}$. Reduced integration will, therefore, provide relief for the membrane locking of eight-node elements and this was seen to be true for the QUAD8 results listed in Table 10.2. The spurious modes associated with reduced integration make serious difficulties for nine-node elements, but other equivalent measures—assumed strains (Section 7.7), spurious mode stabilization (Section 7.8), bubble functions (Section 8.1)—can be taken which do not suffer from spurious modes. The only cubic lateral displacement state listed in Table 10.6, $w = \xi^2\eta$, produces a membrane shear strain which does not vanish at the 2 x 2 Gauss points.

We can also use the results in Table 10.6 to construct a parameter with which to estimate the severity of membrane locking. To conform to the definition of locking

Table 10.6
Spurious Membrane Strains in Eight- and Nine-Node Rectangular Elements

$$x = a\xi, \quad y = b\eta, \quad \Lambda = a/b$$

(a) <u>Shallow Cylindrical Geometry</u>

w	u	v	$\varepsilon_{\bar{x}}$	$\varepsilon_{\bar{y}}$	$\bar{\gamma}$
ξ^2	$-\dfrac{2a\xi^3}{3R_x}$	0	$\dfrac{2}{R_x}\left(\xi^2 - \dfrac{1}{3}\right)$	0	0
$\xi\eta$	$\dfrac{a\xi^2\eta}{2R_x}$	$-\dfrac{\Lambda a\xi^3}{6R_x}$	0	0	$\dfrac{\Lambda}{2R_x}\left(\xi^2 - \dfrac{1}{3}\right)$
$\xi^2\eta$	$-\dfrac{2a\xi^3\eta}{3R_x}$	$-\dfrac{\Lambda a\xi^4}{12R_x}$	$\dfrac{2\eta}{R_x}\left(\xi^2 - \dfrac{1}{3}\right)$	0	$\dfrac{\Lambda\xi}{2R_x}\left(\xi^2 - \dfrac{5}{6}\right)$

(b) <u>Shallow Spherical Geometry</u>

w	u	v	$\varepsilon_{\bar{x}}$	$\varepsilon_{\bar{y}}$	$\bar{\gamma}$
$\Lambda^2\xi^2 - \eta^2$	$-\dfrac{2\Lambda^2 a\xi^3}{3R}$	$\dfrac{2b\eta^3}{3R}$	$\dfrac{2\Lambda^2}{R}\left(\xi^2 - \dfrac{1}{3}\right)$	$-\dfrac{2}{R}\left(\eta^2 - \dfrac{1}{3}\right)$	0
$\xi\eta$	$-\dfrac{a\eta}{2R}\left(\xi^2 + \dfrac{\Lambda^2\eta^2}{3}\right)$	$-\dfrac{a\xi}{2\Lambda R}\left(\eta^2 + \dfrac{\Lambda^2\xi^2}{3}\right)$	0	0	$\dfrac{\Lambda^3}{3R}\left(\eta^2 - \dfrac{1}{3}\right) + \dfrac{\Lambda}{2R}\left(\xi^2 - \dfrac{1}{3}\right)$

parameters in other locking modes, the locking parameter in this example should measure the ratio of spurious membrane strains to desired bending strains. For the first example listed in Table 10.6(a) $\left(w = \xi^2 \right)$, the elastic curvature $\chi_x = w_{,xx} = 2 / a^2$. The location for bending strain which best measures the bending energy is at $z = t / 2\sqrt{3}$ where the bending strain is $\varepsilon_b = z\chi_x = t / \sqrt{3}\, a^2$. The ratio of spurious membrane strain to bending strain is, therefore,

$$\frac{\varepsilon_{\overline{x}}}{\varepsilon_b} = \frac{a^2}{R_x t} 2\sqrt{3}\left(\xi^2 - \frac{1}{3} \right) \tag{10:54}$$

Dropping the subscript on R_x, we take the membrane locking parameter to be a^2 / Rt. Other equivalent expressions are

$$P_{m\ell} = \frac{a^2}{Rt} = \left(\frac{a}{\ell_c} \right)^2 = \left(\frac{a}{R} \right)^2 \frac{R}{t} \tag{10:55}$$

where ℓ_c is the characteristic length defined in Section 10.1.2. Note that, in contrast to the locking problems treated in earlier chapters, the parameter for membrane locking is the *square* of a length ratio. As a result, the onset of membrane locking with increasing element size is more abrupt than for other forms of locking. For the Morley hemispherical shell problem $R / t = 250$ and $a / R = \pi / 2n$, where n is the number of node spaces in the quarter model of the shell. Thus, for this example, $P_{m\ell} = 250\,\pi^2 / 4n^2$. The severity of membrane locking for the Morley hemispherical shell is evidenced by the fact that n must be at least 25 to drive $P_{m\ell}$ below 1.0.

It is also possible to use the formulas in Table 10.6 to predict results in Table 10.2 for QUAD8 and HEXA20 with full integration. For example, we can use the formulas listed for $w = \Lambda^2 \xi^2 - \eta^2$ in Table 10.6(b) (Shallow Spherical Geometry) with $\Lambda = 1.0$ to predict the ratio of spurious membrane energy to bending energy. The result is

$$\frac{W_m}{W_b} = \frac{16}{15} P_{m\ell}^2 \tag{10:56}$$

The ratio of the displacement under load, which is a direct measure of the strain energy, to the correct displacement is

$$\frac{u}{u_c} = \frac{W_b}{W_b + W_m} = \frac{1}{\frac{W_m}{W_b} + 1} \tag{10:57}$$

That ratio, computed from Equations 10:55, 10:56, and 10:57, is compared in Table 10.7 with the results for QUAD8 and HEXA20 with full integration. It is seen that the theory has the correct trend but underpredicts the magnitude of the displacements, particularly for QUAD8. The explanation for the excessively stiff prediction may be that the finite element model has more than just one deformation mode and that the response in other modes will be substantial when the mode in question is very stiff.

Table 10.7

Comparison of Theoretically Predicted and Observed Displacements Under Load in the Morley Hemispherical Shell Problem

n	THEORY	QUAD8	HEXA20
4	.00063	.006	.001
8	.0100	.069	.021
12	.0485	.210	.097
16	.1387	.373	——

10.4.3 Membrane Locking of Eight- and Nine-Node Trapezoidal Elements

We have just seen that membrane locking of rectangular eight- and nine-node isoparametric shell elements can be avoided for constant bending states by the use of reduced integration or equivalent measures. Nevertheless, the experimental results for a bending dominant spherical shell problem (Table 10.2) still show relatively slow convergence for the QUAD8 with reduced integration. A possible explanation is that the taper of the elements in the meridianal direction (see Figure 10.14) causes locking which is not relieved by reduced integration. We will explore that possibility here by analyzing the effect of taper on membrane locking. Figure 10.16 depicts the projection of a tapered shallow shell element onto the xy plane. The relationships between x, y and the parametric coordinates, ξ, η, are

$$x = a\xi(1 - \alpha\eta) \quad , \quad y = b\eta \tag{10:58}$$

or

$$\xi = \frac{x}{a(1 - \alpha y / b)} \quad , \quad \eta = y / b \tag{10:59}$$

The inverse Jacobian matrix, which we will require for the calculation of strains, is

$$[J]^{-1} = \begin{bmatrix} \xi'_x & \eta'_x \\ \xi'_y & \eta'_y \end{bmatrix} = \begin{bmatrix} \dfrac{1}{a(1 - \alpha\eta)} & 0 \\ \dfrac{\alpha\xi}{b(1 - \alpha\eta)} & \dfrac{1}{b} \end{bmatrix} \tag{10:60}$$

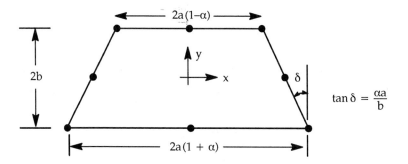

$$\tan\delta = \frac{\alpha a}{b}$$

Figure 10.16   Eight-Node Tapered Shell Element.

Assuming the shell to be spherical we take, from Table 10.5, the inextensional (bending) solution

$$w = x^2 - y^2 \quad , \quad u = -\frac{2x^3}{3R} \quad , \quad v = \frac{2y^3}{3R} \tag{10:61}$$

and determine the resulting spurious membrane strains due to the element's inability to interpolate cubic and higher order terms correctly. From Equation 10:58

$$u = -\frac{2a^3}{3R}\xi^3(1 - \alpha\eta)^3 = -\frac{2a^3}{3R}\xi^3\left(1 - 3\alpha\eta + 3\alpha^2\eta^2 - \alpha^3\eta^3\right)$$

$$v = \frac{2b^3}{3R}\eta^3 \qquad\qquad (10:62)$$

$$w = a^2\xi^2(1 - \alpha\eta)^2 - b^2\eta^2 = a^2\xi^2\left(1 - 2\alpha\eta + \alpha^2\eta^2\right) - b^2\eta^2$$

All functions of ξ, η except those in the element's basis ($1, \xi, \eta, \xi^2, \xi\eta, \eta^2$, $\xi^2\eta, \xi\eta^2$, plus $\xi^2\eta^2$ for nine-node elements) are incorrectly interpolated. Consequently, we see that all terms in u and v are incorrectly interpolated. In order to simplify the calculation of strains, we exclude terms proportional to α^2 and α^3. If we assume, further, that strains are evaluated at 2 x 2 Gauss points then, as we have seen in Section 10.4.2, the term proportional to ξ^3 in u and the term proportional to η^3 in v produce no error even though they are incorrectly interpolated. We are left with the term proportional to $\xi^3\eta$ in u which has the alias $\xi^3\eta \Rightarrow \xi\eta$ and which produces the error in u

$$E(u) = \frac{2a^3\alpha}{R}\left(\xi\eta - \xi^3\eta\right) \qquad\qquad (10:63)$$

Expressions for the membrane strains in a shallow spherical shell are provided by Equation 10:52. We note that, since the only retained error is in u, the nonzero strains are

$$\varepsilon_{\overline{x}} = E(u)_{,x} = E(u)_{,\xi}\,\xi_{,x} + E(u)_{,\eta}\,\eta_{,x}$$
$$\gamma = E(u)_{,y} = E(u)_{,\xi}\,\xi_{,y} + E(u)_{,\eta}\,\eta_{,y} \qquad (10:64)$$

where

$$E(u)_{,\xi} = \frac{2a^3\alpha}{R}\eta\left(1 - 3\xi^2\right) \quad , \quad E(u)_{,\eta} = \frac{2a^3\alpha}{R}\xi\left(1 - \xi^2\right) \quad (10:65)$$

Then, since $E(u)_{,\xi}$ vanishes at 2×2 Gauss points and $\eta_{,x}$ is zero (see Equation 10:60), we are left with

$$\gamma = E(u)_{,\eta} \, \eta_{,y} = \frac{2a^3\alpha}{bR} \, \xi\left(1 - \xi^2\right) \tag{10:66}$$

as the only membrane strain component which does not vanish at 2×2 Gauss points. This term causes membrane locking for tapered elements even when reduced integration is used.

To measure the magnitude of the locking effect we can compare γ to a typical bending strain component; take, for example, the bending strain in the x direction at a distance $t / 2\sqrt{3}$ from the midsurface, which is,

$$\varepsilon_b = \frac{t}{2\sqrt{3}} \chi_x = \frac{t}{2\sqrt{3}} w_{,xx} = \frac{t}{\sqrt{3}} \tag{10:67}$$

Then, if we substitute $\xi = 1 / \sqrt{3}$ in Equation 10:66,

$$\frac{\gamma}{\varepsilon_b} = \frac{4a^3\alpha}{3\,bRt} = \frac{4\alpha a}{3b}\left(\frac{a^2}{Rt}\right) = \frac{4}{3} \tan \delta \left(\frac{a}{\ell_c}\right)^2 \tag{10:68}$$

where $\tan \delta = \alpha a / b$ in Figure 10.16 and $\ell_c = \sqrt{Rt}$ is the characteristic length for shell bending.

Equation 10:68 shows that small values of the taper angle, δ, can cause significant membrane locking of an eight- or nine-node element when the element's size is greater than the characteristic length. Analysis using the inextensional deformation of a cylindrical shell leads to precisely the same result if $w = x^2$.

Equation 10:68 may somewhat overstate the severity of membrane locking for shells with polar symmetry. Tapered element shapes occur naturally in such cases and since ξ measures the polar angle, the normal displacement field is more likely to vary as ξ^n than as x^n. This is true for the inextensional spherical shell solutions (Equation 10:45) which, accordingly, will probably produce a less severe locking condition.

10.5 CURVED SHELL ELEMENTS

Previous sections of this chapter include a good deal of information about curved shell elements including theoretical considerations (Section 10.1), performance comparisons (Section 10.3), and an analysis of membrane locking (Section 10.4). In this section we continue discussion of the formal relationships needed to construct stiffness matrices and describe salient features of representative elements.

10.5.1 The Strain-Displacement Matrix and Other Details

The discussion of coordinate systems in Section 10.1 introduced an *element* coordinate system, x, y, z, whose axes are fixed throughout the element, and a *local* Cartesian coordinate system, $\bar{x}, \bar{y}, \bar{z}$, whose axes are tangent or perpendicular to the midsurface of the element. Here we complete that discussion by examining available choices for these two coordinate systems and the transformation between them.

The xy plane of the element coordinate system is typically chosen to lie near the curved surface of the element. In triangular elements this can be achieved by making the xy plane pass through the corner nodes. A good choice for the xy plane of a quadrilateral element is the plane which is parallel to the diagonals joining corner nodes and midway between them (see Figure 10.7). In order to avoid dependence of the stiffness matrix on node sequencing, the direction of the x- and y-axes should be chosen so that they only change by multiples of 90° when the node sequence changes. A scheme for doing this is described in Figure 7.12. In the case of triangles, we can make no better choice than to take the x-axis parallel to a line joining two of the corner nodes.

The $\bar{x}$- and $\bar{y}$-axes of the local coordinate system must be tangent to the element's curved midsurface and should also be selected to avoid induced anisotropy. A popular scheme for accomplishing this is to select $\bar{x}$ and $\bar{y}$ so that they make equal angles with the local directions of ξ and η (see Figure 10.17).

Construction of these axes requires first the direction of the ξ and η axes with respect to the element coordinate system.

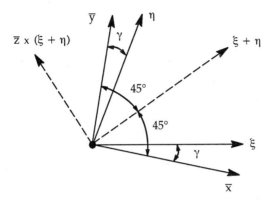

Figure 10.17   Construction of Orthogonal Local Axes near ξ and η.

$$\tilde{e}_\xi = \tilde{i}\, x_{,\xi} + \tilde{j}\, y_{,\xi} + \tilde{k}\, z_{,\xi}$$

$$\tilde{e}_\eta = \tilde{i}\, x_{,\eta} + \tilde{j}\, y_{,\eta} + \tilde{k}\, z_{,\eta}$$

(10:69)

Since the $\bar{z}$ axis is perpendicular to the ξ and η directions, its direction is

$$\tilde{e}_{\bar{z}} = \frac{\tilde{e}_\xi \times \tilde{e}_\eta}{\left| \tilde{e}_\xi \times \tilde{e}_\eta \right|}$$

(10:70)

As shown in Figure 10.17, the $\bar{x}$ and $\bar{y}$ axes make $\pm 45°$ angles with the direction $\tilde{e}_\xi + \tilde{e}_\eta$. The direction of the $\bar{y}$ axis can be constructed as follows.

$$\tilde{e}_{\bar{y}} = \frac{1}{\sqrt{2}} \left(\frac{\tilde{e}_{\bar{z}} \times \left(\tilde{e}_\xi + \tilde{e}_\eta \right)}{\left| e_{\bar{z}} \times \left(\tilde{e}_\xi + \tilde{e}_\eta \right) \right|} + \frac{\tilde{e}_\xi + \tilde{e}_\eta}{\left| \tilde{e}_\xi \times \tilde{e}_\eta \right|} \right)$$

(10:71)

As an alternative choice, we may wish to keep the $\bar{x}, \bar{y}$ axes close to the x, y axes of the element coordinate system. To begin construction of these axes, we note that $\bar{x}, \bar{y}$ in Figure 10.15 lie in the midplane and are close to x, y but are not orthogonal. Relabeling them $\bar{\bar{x}}$ and $\bar{\bar{y}}$, we can construct their directions as follows.

$$\tilde{e}_{\overline{\overline{x}}} = \frac{\tilde{j} \times \tilde{e}_{\overline{z}}}{\left| \tilde{j} \times e_{\overline{z}} \right|} \qquad \tilde{e}_{\overline{\overline{y}}} = \frac{\tilde{e}_{\overline{z}} \times \tilde{i}}{\left| \tilde{e}_{\overline{z}} \times \tilde{i} \right|} \tag{10:72}$$

These vectors can then be used in place of $\tilde{e}_\xi$ and $\tilde{e}_\eta$ in Equation 10:71 to form $\tilde{e}_{\overline{y}}$. Whichever option is selected, the components of $\tilde{e}_{\overline{x}}$, $\tilde{e}_{\overline{y}}$, and $\tilde{e}_{\overline{z}}$ are the elements of the transformation matrix

$$\left\{ \begin{array}{c} d\overline{x} \\ d\overline{y} \\ d\overline{z} \end{array} \right\}_o = [E] \left\{ \begin{array}{c} dx \\ dy \\ dz \end{array} \right\}_o = \left[\begin{array}{ccc} e_{\overline{x}x} & e_{\overline{x}y} & e_{\overline{x}z} \\ e_{\overline{y}x} & e_{\overline{y}y} & e_{\overline{y}z} \\ e_{\overline{z}x} & e_{\overline{z}y} & e_{\overline{z}z} \end{array} \right] \left\{ \begin{array}{c} dx \\ dy \\ dz \end{array} \right\}_o \tag{10:73}$$

where the modifier (o) indicates evaluation in the midsurface.

Turning our attention to the formation of the strain displacement matrix, we recall that Section 10.1.1 provided two methods for interpolating displacements from node points to integration points. In the first method, similar to that employed in the original Ahmad formulation,[5]

$$\left\{ \begin{array}{c} \overline{u} \\ \overline{v} \\ \overline{w} \end{array} \right\}_g = \left[E^g \right] \sum_i N_i \left(\left\{ \begin{array}{c} u_o \\ v_o \\ w_o \end{array} \right\}_i + \frac{\zeta t}{2} \left[\overline{E}^i \right] \left\{ \begin{array}{c} \theta_x \\ \theta_y \\ \theta_z \end{array} \right\}_i \right) \tag{10:74}$$

where $\left[E^g \right]$ is the value of $[E]$ at integration (Gauss) points and

$$\left[\overline{E}^i \right] = \left[E^i \right]^{-1} \left[\begin{array}{ccc} 0 & 1 & 0 \\ -1 & 0 & 0 \\ 0 & 0 & 0 \end{array} \right] \left[E^i \right] \tag{10:75}$$

The effect of $\left[\overline{E}^i \right]$ is, from the right, to transform $\{\theta\}_i$ into the local coordinate system at node i, to extract from it the angles α, β that increment displacements parallel to the midsurface, and to return the result to the element coordinate system. In the original Ahmad formulation $\left[E^i \right]$ was supplied as input. In some later formulations[9,11] $\left[E^i \right]$ is derived from the geometric properties of the midsurface.

While Equation 10:74 can be differentiated to form expressions for the strains in any lamina, it does not provide a direct means to evaluate the elastic curvatures, $\chi_{\bar{x}}, \chi_{\bar{y}}, \chi_{\overline{xy}}$. The latter step requires approximations and those approximations are fraught with difficulties, not the least of which is the possibility that rigid body properties will be violated. We will return to this issue, but first we will examine the second method for interpolating displacements given by Equation 10:9, repeated below.

$$\left\{ \begin{array}{c} u_o \\ v_o \\ w_o \end{array} \right\}_g = \sum_i N_i \left\{ \begin{array}{c} u_o \\ v_o \\ w_o \end{array} \right\}_i \quad ; \quad \left\{ \begin{array}{c} \theta_x \\ \theta_y \\ \theta_z \end{array} \right\}_g = \sum_i N_i \left\{ \begin{array}{c} \theta_x \\ \theta_y \\ \theta_z \end{array} \right\}_i \tag{10:76}$$

where subscript (o) indicates location in the midsurface. Since translations and rotations are available *separately* at the integration point, we can use them to derive separate expressions for strains and bending curvatures in the midsurface. By so doing we make the assumptions of plate theory apply to curved shells or, in other words, we ignore the local effects of curvature on strain. Kebari and Cassell[12] (1991) have shown that this approach gives good results for thin shells. To proceed, we require the displacements in local coordinates

$$\{\bar{u}\}_g = \left\{ \begin{array}{c} \bar{u} \\ \bar{v} \\ \bar{w} \end{array} \right\}_g = \left[E^g \right] \sum_i N_i \{u_o\}_i \quad ; \quad \{\bar{\theta}\}_g = \left[E^g \right] \sum_i N_i \{\theta\}_i$$

$$\tag{10:77}$$

and the definitions of membrane strain and elastic curvature in the plane tangent to the midsurface

$$\{\varepsilon^m\} = \left\{ \begin{array}{c} \varepsilon_{\bar{x}} \\ \varepsilon_{\bar{y}} \\ \gamma_{\overline{xy}} \end{array} \right\} = \left\{ \begin{array}{c} \bar{u}_{,\bar{x}} \\ \bar{v}_{,\bar{y}} \\ \bar{u}_{,\bar{y}} + \bar{v}_{,\bar{x}} \end{array} \right\} \quad ; \quad \{\chi\} = \left\{ \begin{array}{c} \chi_{\bar{x}} \\ \chi_{\bar{y}} \\ \chi_{\overline{xy}} \end{array} \right\} = \left\{ \begin{array}{c} -\theta_{\bar{y},\bar{x}} \\ \theta_{\bar{x},\bar{y}} \\ -\theta_{\bar{y},\bar{y}} + \theta_{\bar{x},\bar{x}} \end{array} \right\}$$

$$\tag{10:78}$$

Substituting from Equation 10:77 into Equation 10:78 we find that the expression for membrane strains can be written as

$$\left\{\epsilon^m\right\} = \begin{bmatrix} \dfrac{d}{d\bar{x}} & 0 & 0 \\[2mm] 0 & \dfrac{d}{d\bar{y}} & 0 \\[2mm] \dfrac{d}{d\bar{y}} & \dfrac{d}{d\bar{x}} & 0 \end{bmatrix} \left[E^g\right] \sum_i N_i \left\{u_o\right\}_i \qquad (10:79)$$

The only function of position in this expression is $N_i(\xi, \eta)$ which, being a scalar, can be interchanged with $\left[E^g\right]$ which is constant for the plane tangent to the midsurface at point (g). As a result

$$\left\{\epsilon^m\right\} = \sum_i \left[B_i^{mu}\right]\left\{u_o\right\}_i \qquad (10:80)$$

where the strain displacement matrix is

$$\left[B_i^{mu}\right] = \sum_i \begin{bmatrix} N_{i,\bar{x}} & 0 & 0 \\[2mm] 0 & N_{i,\bar{y}} & 0 \\[2mm] N_{i,\bar{y}} & N_{i,\bar{x}} & 0 \end{bmatrix} \left[E^g\right] \qquad (10:81)$$

Differentiation of N_i with respect to $\bar{x}$ and $\bar{y}$ requires application of the chain rule, viz.,

$$N_{i,\bar{x}} = \left(N_{i,\xi}\,\xi_{,x} + N_{i,\eta}\,\eta_{,x}\right)x_{,\bar{x}} + \left(N_{i,\xi}\,\xi_{,y} + N_{i,\eta}\,\eta_{,y}\right)y_{,\bar{x}} \qquad (10:82)$$

This expression, together with the expression for $N_{i,\bar{y}}$ can conveniently be put in the vector form

$$\left\{\begin{matrix} N_{i,\bar{x}} \\[2mm] N_{i,\bar{y}} \end{matrix}\right\} = \left[E^{(2)g}\right]^{-1,T}\left[J^{(2)}\right]^{-1,T}\left\{\begin{matrix} N_{i,\xi} \\[2mm] N_{i,\eta} \end{matrix}\right\} \qquad (10:83)$$

where

$$\left[E^{(2)g}\right] = \begin{bmatrix} \bar{x}_{,x} & \bar{x}_{,y} \\[2mm] \bar{y}_{,x} & \bar{y}_{,y} \end{bmatrix}_g \qquad (10:84)$$

expresses the two-dimensional transformation from element coordinates to local coordinates obtained by setting $d\bar{z} = 0$ and eliminating dz in $[E^g]$, the three-dimensional transformation given by Equation 10:73. $[J^{(2)}]$ is the two-dimensional Jacobian matrix evaluated in the midsurface. Hence

$$\left[J^{(2)}\right]^{-1,T} = \begin{bmatrix} \xi_{,x} & \eta_{,x} \\ \xi_{,y} & \eta_{,y} \end{bmatrix} = \frac{1}{J^{(2)}} \begin{bmatrix} y_{,\eta} & -y_{,\xi} \\ -x_{,\eta} & x_{,\xi} \end{bmatrix} \tag{10:85}$$

where $J^{(2)} = x_{,\xi} \, y_{,\eta} - x_{,\eta} \, y_{,\xi}$ is the two-dimensional Jacobian determinant. Equation 10:4 provides values for $x_{,\xi}$, etc.

Due to the use of the same interpolation rules for $\{u\}$ and $\{\theta\}$, the curvature-rotation matrix derives from the strain-displacement matrix by a simple interchange of matrix terms. Thus

$$\{\chi\} = \sum_i \left[B_i^{b\theta}\right]\{\theta\}_i \tag{10:86}$$

where

$$\left[B_i^{b\theta}\right] = \begin{bmatrix} 0 & -N_{i,\bar{x}} & 0 \\ N_{i,\bar{y}} & 0 & 0 \\ N_{i,\bar{x}} & -N_{i,\bar{y}} & 0 \end{bmatrix} \left[E^g\right] \tag{10:87}$$

and $N_{i,\bar{x}}$, $N_{i,\bar{y}}$ are given by Equation 10:83.

For transverse shear strains, we begin with their definition

$$\{\gamma^s\} = \begin{Bmatrix} \gamma_{\overline{xz}} \\ \gamma_{\overline{yz}} \end{Bmatrix} = \begin{Bmatrix} \bar{w}_{,\bar{x}} + \theta_{\bar{y}} \\ \bar{w}_{,\bar{y}} - \theta_{\bar{x}} \end{Bmatrix} \tag{10:88}$$

and substitute for $\bar{w}$, $\theta_{\bar{x}}$, and $\theta_{\bar{y}}$ from Equation 10:77 The result is

$$\{\gamma^s\} = \sum_i \left(\left[B_i^{su}\right]\{u_o\}_i + \left[B_i^{s\theta}\right]\{\theta\}_i \right) \tag{10:89}$$

where

$$
\left[B_i^{su} \right] = \begin{bmatrix} 0 & 0 & N_{i,\bar{x}} \\ 0 & 0 & N_{i,\bar{y}} \end{bmatrix} \left[E^g \right]
$$

(10:90)

$$
\left[B_i^{s\theta} \right] = \begin{bmatrix} 0 & N_i & 0 \\ -N_i & 0 & 0 \end{bmatrix} \left[E^g \right]
$$

Given the strain-displacement and curvature-displacement matrices as derived above, we complete the computation of the element's stiffness matrix by invoking flat plate theory as represented by the expression for strain energy per unit area in Equation 9:21. In the required integration over the element's surface, we use the two-dimensional Jacobian determinant defined in the midsurface (see Equation 10:85).

Returning now to the interpolation of displacements by the first method (Equation 10:74), we must decide where and how to compute strains. One way is to compute strains in several laminas and find $\{\varepsilon^m\}$ and $\{\chi\}$ by approximate numerical integration. The computation of strain in any particular lamina, $\zeta = \zeta_\ell$, roughly follows the procedure described above for $\{\varepsilon^m\}$ with extra terms to represent the dependence of $\{\varepsilon^\ell\}$ on $\{\theta\}_i$. Note however that the Jacobian transformation matrix, $[J^{(2)}]$, which now depends on ζ (see Equation 10:5), is different for each lamina. (The rotational transformation matrices, $[E]$ and $[E^{(2)}]$, are independent of ζ because the laminas are assumed to be parallel to the midsurface.) If the laminar strains are properly computed, the calculation of $\{\varepsilon^m\}$ and $\{\chi\}$ presents no difficulty. We can, for example, assume $\{\varepsilon^\ell\} = \{\varepsilon^m\} - \frac{\zeta t}{2}\{\chi\}$ and extract $\{\varepsilon^m\}$ and $\{\chi\}$ by a best fit procedure. Alternatively we can elect to compute laminar stresses and evaluate their resultants $\{N\}$ and $\{m\}$.

The method just described is appropriate for problems with material nonlinearity, where the relationship between stress and strain is not known in advance. It is, on the other hand, relatively expensive for linear analysis so that most developers have opted for explicit evaluation of $\{\varepsilon^m\}$ and $\{\chi\}$. In doing so, the exact dependence of all quantities on ζ must be accounted for or else, as Irons[23] was the first to observe, the rigid body criterion will be violated.

In order to illustrate the nature of the difficulties, we will outline a method for computing $\{\chi\}$ without first computing laminar strains. The method begins with Equation 10:14 of Section 10.1.1, repeated below.

$$\{\chi\} = -\lim_{\zeta \to 0} \frac{2}{t} \frac{\partial}{\partial \zeta} \begin{Bmatrix} \varepsilon_{\overline{x}} \\ \varepsilon_{\overline{y}} \\ \gamma_{\overline{xy}} \end{Bmatrix} = -\lim_{\zeta \to 0} \frac{2}{t} \frac{\partial}{\partial \zeta} \begin{Bmatrix} \overline{u}_{,\overline{x}} \\ \overline{v}_{,\overline{y}} \\ \overline{u}_{,\overline{y}} + \overline{v}_{,\overline{x}} \end{Bmatrix} \tag{10:91}$$

The displacement derivatives on the right are terms in the matrix

$$\left[\overline{u}_{,\overline{x}} \right] = \begin{bmatrix} \overline{u}_{,\overline{x}} & \overline{u}_{,\overline{y}} & \overline{u}_{,\overline{z}} \\ \overline{v}_{,\overline{x}} & \overline{v}_{,\overline{y}} & \overline{v}_{,\overline{z}} \\ \overline{w}_{,\overline{x}} & \overline{w}_{,\overline{y}} & \overline{w}_{,\overline{z}} \end{bmatrix} \tag{10:92}$$

which is formed as follows.

$$\left[\overline{u}_{,\overline{x}} \right] = \left[\overline{u}_{,u} \right] \left[u_{,\xi} \right] \left[\xi_{,x} \right] \left[x_{,\overline{x}} \right] = \left[E^g \right] \left[u_{,\xi} \right] \left[\overline{J} \right] \left[E^g \right]^{-1} \tag{10:93}$$

where $\left[\overline{J} \right] = [J]^{-1,T}$ and $[J]$ is the three-dimensional Jacobian matrix. Equation 10:91 requires the derivative of $\left[\overline{u}_{,\overline{x}} \right]$ with respect to ζ,

$$\frac{\partial}{\partial \zeta} \left[\overline{u}_{,\overline{x}} \right] = \left[E^g \right] \left(\left(\frac{\partial}{\partial \zeta} \left[u_{,\xi} \right] \right) \left[\overline{J} \right] + \left[u_{,\xi} \right] \frac{\partial}{\partial \zeta} \left[\overline{J} \right] \right) \left[E^g \right]^{-1} \tag{10:94}$$

where we note the identity

$$\frac{\partial}{\partial \zeta} \left[\overline{J} \right] = \left[\overline{J}' \right] = -\left[\overline{J} \right] \left(\frac{\partial}{\partial \zeta} [J]^T \right) \left[\overline{J} \right] \tag{10:95}$$

The transpose of the Jacobian matrix, formed from the derivatives of Equation 10:5 with $N_i' = N_i$, is

$$[J]^T = \sum_i \left(\begin{Bmatrix} x \\ y \\ z \end{Bmatrix}_i \left[N_{i,\xi} \ , \ N_{i,\eta} \ , \ 0 \right] + \frac{t}{2} \begin{Bmatrix} e_{\overline{z}x} \\ e_{\overline{z}y} \\ e_{\overline{z}z} \end{Bmatrix} \left[\zeta N_{i,\xi} \ , \ \zeta N_{i,\eta} \ , \ N_i \right] \right) \tag{10:96}$$

Consequently,

$$\left[\bar{J}'\right] = \frac{\partial}{\partial\zeta}\left[\bar{J}\right] = -\frac{t}{2}\left[\bar{J}\right]\sum_i\begin{Bmatrix}e_{\bar{z}x}\\e_{\bar{z}y}\\e_{\bar{z}z}\end{Bmatrix}\left\lfloor N_{i,\xi} \quad,\quad N_{i,\eta} \quad,\quad 0\right\rfloor\left[\bar{J}\right] \qquad (10\text{:}97)$$

In like manner, the matrix $\left[u_{,\xi}\right]$, computed from Equation 10:74, is

$$\left[u_{,\xi}\right] = \sum_i\left(\{u_o\}_i\left\lfloor N_{i,\xi} \quad,\quad N_{i,\eta} \quad,\quad 0\right\rfloor + \frac{t}{2}\left[\bar{E}^i\right]\{\theta_i\}\left\lfloor \zeta N_{i,\xi} \quad,\quad \zeta N_{i,\eta} \quad,\quad N_i\right\rfloor\right)$$

$$(10\text{:}98)$$

so that

$$\frac{\partial}{\partial\zeta}\left[u_{,\xi}\right] = \frac{t}{2}\sum_i\left[\bar{E}^i\right]\{\theta_i\}\left\lfloor N_{i,\xi} \quad,\quad N_{i,\eta} \quad,\quad 0\right\rfloor \qquad (10\text{:}99)$$

The remaining calculations needed to evaluate $\{\chi\}$ are nothing but straightforward, tedious algebra. The final result is

$$\{\chi\} = \sum_i\left(\left[B_i^{bu}\right]\{u_o\}_i + \left[B_i^{b\theta}\right]\{\theta\}_i\right) \qquad (10\text{:}100)$$

where

$$\left[B_i^{bu}\right] = -\begin{bmatrix}F_1 & 0 & 0\\0 & F_2 & 0\\F_2 & F_1 & 0\end{bmatrix}\left[E^g\right] \qquad \left[B_i^{b\theta}\right] = -\begin{bmatrix}D_1 & 0 & 0\\0 & D_2 & 0\\D_2 & D_1 & 0\end{bmatrix}\left[E^g\right]\left[\bar{E}^i\right]$$

$$\begin{Bmatrix}D_1\\D_2\\0\end{Bmatrix} = \left[E^g\right]\left[\bar{J}'\right]^T\begin{Bmatrix}0\\0\\N_i\end{Bmatrix} + \begin{Bmatrix}B_1\\B_2\\0\end{Bmatrix} \qquad \begin{Bmatrix}B_1\\B_2\\0\end{Bmatrix} = \left[E^g\right]\left[\bar{J}\right]^T\begin{Bmatrix}N_{i,\xi}\\N_{i,\eta}\\0\end{Bmatrix}$$

$$\begin{Bmatrix}F_1\\F_2\\0\end{Bmatrix} = \frac{2}{t}\left[E^g\right]\left[\bar{J}'\right]^T\begin{Bmatrix}N_{i,\xi}\\N_{i,\eta}\\0\end{Bmatrix} \qquad (10\text{:}101)$$

These equations, together with similar expressions for membrane strain and transverse shear, are implemented in the MSC/NASTRAN QUAD8 element.[10,11] Note that, unlike the result shown in Equation 10:86, the elastic curvature contains terms proportional to $\left\{u_o\right\}_i$. Fair enough, since it is expected that geometric curvature will change the location of the neutral surface. What is most remarkable, however, is that $\left[B_i^{bu}\right]$ *does not vanish as t goes to zero!* The factor 2/t in front of F_1, F_2 just cancels the factor t/2 in front of $\left[\bar{J}'\right]$ in Equation 10:97. As pointed out by Kebari and Cassell,[12] the membrane-bending coupling so introduced, while necessary to avoid violation of the rigid body property, or self straining, is a costly and unnecessary complication given that the simpler process represented by Equation 10:86 is available.

10.5.2 Eight- and Nine-Node Shell Elements

Reference 10.24 contains a list of 350 papers about shell elements written between 1965 and 1988. Here we will attempt to classify the distinguishing features of a few representative eight- and nine-node curved shell elements (which are the types most frequently described in the literature).

A good first question to ask is why developers continue to concentrate on eight- and nine-node shell elements given their mediocre performance compared to the simpler flat elements, as summarized in Table 10.3. Finite element users are aware of the difference and will opt for the simpler elements nine times out of ten. A plausible answer to the question is that eight- and nine-node curved shell elements often display superior performance (compare, for example, the results for QUAD8 and QUAD4 with reduced integration in the Scordelis-Lo roof problem, Table 10.1) and that there is hope for the future. The real reasons, we suspect, are the challenge presented by the persistently poor and poorly understood test scores registered in bending dominant problems (Table 10.2) and the opportunity to display mathematical ingenuity in a complex area.

The papers on curved shell elements cover many aspects of their design, such as the kinematical issues treated in Sections 10.1.1 and 10.5.1, questions of computational efficiency, and special applications such as nonlinear analysis. But more space is spent by far on ways to master their locking disorders, especially membrane locking. It makes sense, therefore, to classify the eight- and nine-node curved shell elements on the basis of the means selected to suppress locking.

In the case of eight-node shell elements, uniform, reduced (2 x 2) integration has been the standard method for suppressing locking (of all types) since Zienkiewicz, Too, and Taylor[25] demonstrated the resulting performance improvements in 1971. The accompanying spurious modes (one for membrane, one for bending) are noncommunicating and relatively benign. The MSC/NASTRAN QUAD8 element[10,11] suppresses these modes by adding integration points selectively (see Figure 7.14). On balance, the effect of the extra integration points is to degrade performance. Other modifications which have a minor effect on the performance of the eight-node shell element include residual bending flexibility (Section 9.5) and alteration of the shape functions to simulate a ninth internal node (Section 8.3).

The 1980s were the decade of the nine-node shell element. As noted in Section 9.4.1, the extra internal node allows constant curvature patch tests to be satisfied under more general conditions and unlocks a biquadratic, flat plate bending mode $\left(w = \xi^2\eta^2\right)$. The presence of the ninth node also eliminates uniform reduced integration as a remedy for locking disorders because all of the additional five spurious modes (one each for $u, v, w, \theta_x, \theta_y$) are global communicating modes (see Figure 7.4). This fact is largely responsible for the development of the substitutes for uniform reduced integration described in Chapters 7 and 8. Developers have applied all of these remedies, and more besides, to the nine-node shell element.

The Hughes and Cohen Heterosis element[26] (1978) marks the transition of developer preference from eight nodes to nine. Their design specifies nine nodes but, for translational displacements (u, v, w) it rigidly slaves the ninth interior node to the exterior nodes through the values of their parametric shape functions at the center. With these constraints, the design can then use 2 x 2 integration to eliminate locking for membrane and transverse shear strains. It uses full (3 x 3) integration for bending curvatures where locking is not an issue. Thus, in its critical aspects, the Heterosis element is more like an eight-node element than a nine-node element.

The combined remedies for locking and spurious modes in nine-node elements include the following classifications: selective underintegration, assumed strain hybrid formulations, mode stabilization, and direct assumed strain formulations.

While selective underintegration[10,27] is the simplest remedy, the fact that it causes violation of constant strain patch tests (see Section 7.6.3) provided inspiration for the other methods.

The assumed strain hybrid method comes in a variety of forms. The basic idea (see Section 7.7) is to specify a separate strain field with enough carefully selected low order terms to simultaneously avoid locking and spurious modes. In the case of the nine-node shell element, a reasonable set of basis vectors for the assumed (natural) strain components is

$$
\begin{aligned}
\varepsilon_\xi\,,\chi_\xi & \quad:\quad \left\lfloor 1,\xi,\eta,\xi\eta,\eta^2,\xi\eta^2 \right\rfloor \\[4pt]
\varepsilon_\eta\,,\chi_\eta & \quad:\quad \left\lfloor 1,\xi,\eta,\xi\eta,\xi^2,\xi^2\eta \right\rfloor \\[4pt]
\gamma_{\xi\eta}\,,\chi_{\xi\eta} & \quad:\quad \left\lfloor 1,\xi,\eta,\xi\eta \right\rfloor \\[4pt]
\gamma_{\xi\zeta} & \quad:\quad \left\lfloor 1,\xi,\eta,\xi\eta,\xi\eta^2 \right\rfloor \\[4pt]
\gamma_{\eta\zeta} & \quad:\quad \left\lfloor 1,\xi,\eta,\xi\eta,\xi^2\eta \right\rfloor
\end{aligned}
\qquad (10\!:\!102)
$$

The strain components in the local coordinate system $(\bar{x},\bar{y},\bar{z})$ can also use these basis vectors. Note that all of the basis vectors carefully avoid quadratic variation in the direction of the strain component or, in the case of shear strains, in the plane of the strain component. The number of terms is also sufficient to suppress spurious modes.

Subsequent calculations in the assumed strain hybrid formulation extract the coefficients of the assumed strain field from the strain field produced by the nodal displacements. In elements derived by mixed variational methods[28,29] such as the Hellinger-Reissner variational principle (Section 7.7.2), the assumed strain field is matched to the strain field derived from nodal displacements by a least squares fit. Other nine-node element developers[30,31] have used constrained variational principles[32] to derive the equations for their elements.

The basic idea in mode stabilization[33,34] (Section 7.8) is to separate the strain states into a low order set which is evaluated at reduced integration points and a less important high order set which vanishes at reduced integration points and which is orthogonal to the low order set in an integral sense. The designer is then free to choose an approximate method for evaluating stiffness due to the high order set. However, in a recent paper, Belytschko, Wong, and Stolarski[35] seek improved accuracy by using the Hellinger-Reissner variational principle to evaluate the coefficients of the high order set, thereby uniting two of our classifications for remedying the disorders of nine-node elements.

Finally, we must include the nine-node assumed natural strain (ANS) element developed by Park and Stanley.[9] As will be recalled, strains are evaluated at special points on six lines of constant ξ or η (see Figure 8.16). Of the elements which have been discussed, the ANS element is the only one which *requires* a ninth interior node. The treatment of the element in Section 8.4.3 for the plane case extends easily to a curved shell. The strain components require no element-to-local coordinate transformation because they are initially formed in the ξ, η *natural* (skewed) system. The only required transformation is from ξ, η to $\bar{x}, \bar{y}$ and then only if the developer insists on an orthogonal system for the stress-strain relationship. The ANS element performs well in locking tests. Its chief drawback is that it does not pass constant curvature patch tests.

Table 10.8 compares performance of representative eight- and nine-node shell elements in the Morley hemispherical shell problem. Two of the elements, the 9ANS element and the nine-node $\gamma-\psi$ element, use the form of the problem illustrated in Figure 10.18 while the remaining five use the form illustrated in Figure 10.14. Neither version is clearly more appropriate than the other. While the irregular mesh pattern of Figure 10.18 emphasizes skewness, the pattern of Figure 10.14 emphasizes taper.

The difference in the performance of the elements is not large in comparison with the differences illustrated in Table 10.2. In particular the modified shape functions used by the QUAD8RM (Section 8.3), appear to have little effect. Nor does there appear to be a decisive difference between the eight- and nine-node elements. Park and Stanley's 9ANS element is seen to have the best performance of the

Table 10.8

**More Results for Eight- and Nine-Node Elements
in the Morley Hemispherical Shell Problem**

(See Figure 10.14)

NUMBER OF NODE SPACES ALONG EDGE OF MESH	REFERENCE	4	8	12	16
QUAD8 (Reduced)	20	0.194	0.895	1.005	1.008
QUAD8RM	20	0.240	0.905	1.005	1.005
SHELM9	29	0.34	0.97	0.99	1.00
Nine-Node $\gamma-\psi$	35	0.074	0.828	0.988	1.001
LAG9	34	0.312	0.952	——	0.998
9ANS	9	0.80	0.97	——	0.99
SEMILOOF	23	0.474	0.932	0.992	——

group. A possible reason is that the method used for evaluating strains avoids the membrane locking problem described in Table 10.6(a) for $w = \xi^2\eta$. Irons' SEMILOOF element,[23] which employs a discrete Kirchhoff formulation, is thrown in for good measure. It is seen that this older element is still quite competitive.

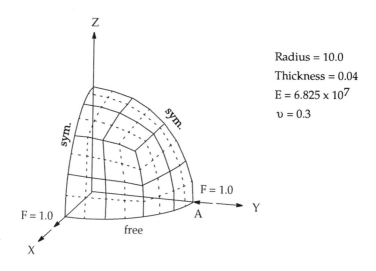

Radius = 10.0
Thickness = 0.04
$E = 6.825 \times 10^7$
$\upsilon = 0.3$

Figure 10.18   Alternate Form of the Morley Hemispherical Shell Problem.

10.5.3 A Flat-Faceted Six-Node Shell Element

The six-node triangle is the pariah of shell elements. While the six-node flat plate element has enjoyed a certain amount of recent attention (see Section 9.4.2), only a few papers[36,37,38] describe six node curved shell elements.

The reason for this lack of interest becomes apparent when one examines the element's membrane locking problem. From the discussion in Section 10.4, we know that constant curvature in the ξ direction of a quadratic element[*] induces a spurious membrane strain, $\varepsilon_{\bar{x}}$, which is proportional to $\xi^2 - 1/3$. Thus we can avoid locking by locating integration points at $\xi = \pm 1/\sqrt{3}$. As applied to a straight-sided triangle (Figure 10.19), this means that, for bending in a direction normal to edge 1-2, we can avoid locking by placing integration points at points on lines $\xi_3 = \pm 1/\sqrt{3}$ in Figure 10.19. In like manner, membrane locking for bending normal to one of the other two edges can be avoided by placing

---

[*]Here we are talking about one-dimensional response which will be the same for a line element, a triangle, or a rectangle with the same number of nodes in the selected direction.

integration points along special lines parallel to that edge. Note, however, that these three sets of lines do not intersect at common points. As a result, no matter where we place the integration points, we cannot avoid locking for all three constant bending states; nor does the selection of a compromise set of integration points, such as the standard three-point set shown in Figure 10.19, provide a satisfactory solution.

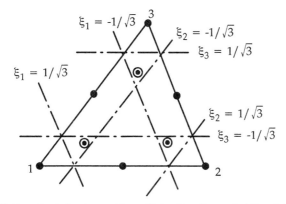

$\odot$ Standard Optimum Three-Point Set, $\xi_i = -2/3, +1/3$

Figure 10.19   Locations of Integration Points to Avoid Locking of a Six-Node Triangle in Particular Directions.

When we examine alternative approaches, such as the assumed strain hybrid formulations of Section 7.7, we find an essential difficulty in the fact that $\xi_i^2 - 1/3$, with ξ_i measured from a point halfway between an edge and the opposite corner, does not integrate to zero over the surface of the element in either metric or parametric space. Consequently a spurious strain state of this form will induce a constant assumed strain state.

Relief from membrane locking might be achieved by employing cubic in-plane displacement states (u, v) as bubble functions. To the author's knowledge, this approach has not been tried.

A remedy for membrane locking which does work is the direct assumed strain formulation described in Section 8.4.1. Here we evaluate tangential strains at the midpoints of nine straight line segments joining nodes (see Figure 8.12) and collocate them with an assumed linearly varying strain field.

The method extends easily to curved shell geometry. In effect, each of the triangles in Figure 8.12 is assumed to be a flat facet for the purpose of computing tangential strains along its edges. The assumed membrane strain field is still linear in ξ and η over the whole element. We can also assume the surface of the element to be faceted for the purpose of computing bending curvatures and transverse shear strains but, while such consistency would be intellectually satisfying, it is not necessary. The MSC/NASTRAN TRIA6 element,[21] in fact, uses the older isoparametric formulation as described by Equation 10:100 and 10:101 for bending curvatures and uses similar expressions for transverse shear.

Performance of the MSC/NASTRAN TRIA6 element, as illustrated in Tables 10.1, 10.2, and 10.3, is satisfactory. Its chief difficulty, which has not been explained, is an excessive tendency toward flexibility when rotation about the normal to the shell is not restrained (see Table 10.2).

10.5.4 Higher Order Curved Shell Elements

In view of the membrane locking difficulties of eight- and nine-node shell elements, it is appropriate to look at the next higher order, $p = 3$, elements. The relationship between $p = 2$ and $p = 3$ shell elements is much like the relationship between $p = 1$ and $p = 2$ membrane elements. The lower order element in each pair encounters locking difficulties in constant strain (or curvature) states which are relieved by going to the next higher order.

Rhiu, Russell, and Lee[39,40] (1989, 1990) have developed third order degenerated shell elements with sixteen nodes (Figure 10.20). The elements use an assumed strain field to eliminate locking and the Hellinger-Reissner variational principle to minimize error. The assumed strain field is separated into low order states and high order states. The low order strain states are taken equal to corresponding states provided by the displacement field and evaluated at reduced integration (3×3) points. The high order states are evaluated at (4×4) integration points.[*] Their basis vector contains terms x^2y^3 and x^3y^2. The separation of strain states into low and high orders has much in common with the mode stabilization method of Belytschko, et al.,[35] but Rhiu, Russell, and Lee do not require the higher and lower order states to be orthogonal.

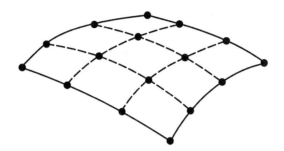

Figure 10.20   A Sixteen Node Lagrange Shell Element.

Performance of the sixteen-node element is, as should be expected, better than that of eight- and nine-node elements. Table 10.9 compares performance in the Morley hemispherical shell problem (Figure 10.14) for the better of two variations of the sixteen-node element with that for the QUAD8 with uniform reduced integration. Note that, in the case of the sixteen-node element, six node spaces correspond to two elements, etc.

Table 10.9

Results for the Morley Hemispherical Shell

NUMBER OF NODE SPACES ALONG EDGE OF MESH	REFERENCE	6	8	9	12
QUAD8 (Reduced)	20	0.639	0.895	——	1.005
M16A	40	0.95	——	1.00	1.00

---

[*]Recall that all Lagrange elements have communicating spurious modes when reduced integration is employed (Section 7.3).

Arbitrarily high order hierarchic shell elements are beginning to appear in the technical literature. In perhaps the earliest paper on the subject, Szabó and Sahrmann[41] (1988) employ solid brick p elements, designed along the lines described in Section 4.5, to analyze shell problems. They take p = 1 in the thickness direction to simulate shell behavior and prescribe a zero value of Poisson's ratio to avoid dilatation locking (see Section 6.2.2). The basis vectors span the Lagrange product space $\xi^a \eta^b \zeta^c$ with $0 \le a \le p$, $0 \le b \le p$, $c = 0, 1$. Full integration is used. Their analysis of the Scordelis-Lo roof (Figure 10.13) demonstrates that reasonable convergence of displacement is achieved for a one element model with p = 6 and that comparable convergence is achieved for a four element model with p = 4.

Surana and Sorem[42] (1991) take an approach which is similar to that of Szabó and Sahrmann and, in fact, achieve the same results for the Scordelis-Lo roof problem. They also study the effect of different values of p in the thickness direction.

The work of Leino and Pitkäranta[43] (1992) begins to unite the conventional development of shell elements, as described in this chapter, with the high order hierarchical approach. Although their element is based on the curvilinear coordinates of a particular example (a cylindrical shell), they use it effectively to study convergence characteristics in the presence of membrane locking. The basis vectors include the serendipity set of edge functions plus some interior functions. Their data shows that, for R/t = 100 and R/t = 1000, reduced integration produces substantially better accuracy than full integration for $p \le 4$ and slightly better accuracy out to p = 7. This result should provide a useful guide for the developers of higher order shell elements.

The design of practical higher order shell elements must cope with the changing realities of finite element analysis. Two points can be emphasized. The first is that practical p elements must allow different values of p on each of their edges in order to accommodate the variation of p over a field of elements, such as might be required to minimize discretization errors efficiently. The second point is that commercial p element codes cannot afford to abandon the lowest two orders, p = 1 and p = 2. In other words, some of the remedies for locking should be put into p elements, even if only for the lowest two orders.

REFERENCES

10.1   W. Flugge, *Stresses in Shells*, 2nd Ed., Springer-Verlag, New York, 1973 (2nd printing 1990).

10.2   J. R. Vinson, *Structural Mechanics: The Behavior of Plates and Shells*, Wiley International, New York, 1974.

10.3   A. L. Goldenveizer, *Theory of Thin Elastic Shells*, Pergamon, Great Britain, 1961.

10.4   S. Utku, "Stiffness Matrices for Thin Triangular Elements of Nonzero Gaussian Curvature," *J. AIAA*, 5, pp 1659-67, 1967.

10.5   S. Ahmad, "Curved Finite Elements in the Analysis of Solid, Shell, and Plate Structures," Ph.D. Thesis, University of Wales, Swansea, 1969.

10.6   B. M. Irons, "Engineering Application of Numerical Integration in Stiffness Methods," *J. AIAA*, 14, pp 2035-7, 1966.

10.7   S. Ahmad, B. M. Irons, and O. C. Zienkiewicz, "Analysis of Thick and Thin Shell Structures by Curved Finite Elements," *Intl. J. Numer. Methods Eng.*, 2, pp 419-51, 1970.

10.8   L. Vu-Quoc and J. A. Mora, "A Class of Simple and Efficient Degenerated Shell Elements—Analysis of Global Spurious Mode Filtering," *Comput. Methods Appl. Mech. Engrg.*, 74, pp 117-75, 1989.

10.9   K. C. Park and G. M. Stanley, "A Curved C^0 Shell Element Based on Assumed Natural-Coordinate Strains," *J. Appl. Mech.*, 53, pp 278-90, 1986.

10.10   R. H. MacNeal, "Specifications for the QUAD8 Quadrilateral Curved Shell Element," MacNeal-Schwendler Corp. Memo RHM-46B, 1980.

10.11   H. V. Lakshminarayana and K. Kailesh, "A Shear Deformable Curved Shell Element of Quadrilateral Shape," *Comput. Struct.*, 33, pp 987-1001, 1989.

10.12   H. Kebari and A. C. Cassell, "Nonconforming Modes Stabilization of a Nine-Node Stress-Resultant Degenerated Shell Element with Drilling Freedom," *Comput. Struct.*, 40, pp 569-80, 1991.

10.13   L. S. D. Morley and A. J. Morris, "Conflict Between Finite Elements and Shell Theory," Royal Aircraft Establishment Report, London, 1978.

10.14   L. S. D. Morley, "The Constant Moment Plate Bending Element," *J. Strain Analysis*, 6, pp 20-4, 1971.

10.15   R. Phaal and C. R. Calladine, "A Simple Class of Finite Elements for Plate and Shell Problems, II: An Element for Thin Shells with Only Translational Degrees of Freedom," *Intl. J. Numer. Methods Eng.*, 35, pp 979-96, 1992.

10.16   R. H. MacNeal and R. L. Harder, "A Proposed Standard Set of Problems to Test Finite Element Accuracy," *Finite Elem. Analysis & Design*, 1, pp 3-20, 1985.

10.17   A. C. Scordelis and K. S. Lo, "Computer Analysis of Cylindrical Shells," *J. Amer. Concr. Inst.*, 61, pp 539-61, 1969.

10.18   W. Flugge, *Stresses in Shells*, 2nd Ed., Springer-Verlag, New York, p. 85, 1973 (2nd printing 1990).

10.19   R. W. Clough and J. L. Tocher, "Finite Element Stiffness Matrices for Analysis of Plates in Bending," *Proc. Conf. Matrix Methods in Struct. Mech.*, Air Force Inst. of Tech., Wright-Patterson AFB, Ohio, 1965.

10.20   R. H. MacNeal and R. L. Harder, "Eight Nodes or Nine?," *Intl. J. Numer. Methods Eng.*, 33, pp 1049-58, 1992.

10.21   R. H. MacNeal, "Derivation of Element Stiffness Matrices by Assumed Strain Distributions," *Nucl. Eng. Design*, 70, pp 3-12, 1982.

10.22   H. Stolarski and T. Belytschko, "Shear and Membrane Locking in Curved C^0 Elements," *Comput. Methods Appl. Mech. Engrg.*, 41, pp 279-96, 1983.

10.23   B. M. Irons, "The Semiloof Shell Element," *Finite Elements Thin Shells & Curved Members* (Eds. D. G. Ashwell and R. H. Gallagher), pp 197-222, John Wiley, London, 1976.

10.24   W. Gilewski and M. Radwańska, "A Survey of Finite Element Models for the Analysis of Moderately Thick Shells," *Finite Elem. Analysis & Design*, 9, pp 1-21, 1991.

10.25   O. C. Zienkiewicz, J. Too, and R. L. Taylor, "Reduced Integration Technique in General Analysis of Plates and Shells," *Intl. J. Numer. Methods Eng.*, 3, pp 275-90, 1971.

10.26   T. J. R. Hughes and M. Cohen, "The 'Heterosis' Finite Element for Plate Bending," *Comput. Struct.*, 9, pp 445-50, 1978.

10.27   H. C. Huang and E. Hinton, "Lagrangian and Serendipity Plate and Shell Elements Through Thick and Thin," *Finite Element Methods for Plate & Shell Struct.*, 1, *Element Technology* (Eds. T. J. R. Hughes and E. Hinton), pp. 62-84, Pineridge Press Int'l., Swansea, U. K., 1986.

10.28   J. J. Rhiu and S. W. Lee, "A New Efficient Mixed Formulation for Thin Shell Finite Element Models," *Intl. J. Numer. Methods Eng.*, 24, pp 581-604, 1987.

10.29 T. Y. Chang, A. F. Saleeb, and W. Graf, "On the Mixed Formulation of a Nine-Node Lagrange Shell Element," *Comput. Methods Appl. Mech. Engrg.* 73, pp 259-81, 1989.

10.30 H. C. Huang and E. Hinton, "A New Nine-Node Degenerated Shell Element with Enhanced Membrane and Shear Interpolation, *Intl. J. Numer. Methods Eng.*, 22, pp 73-92, 1986.

10.31 J. Jang and P. M. Pinsky, "An Assumed Covariant Strain Based Nine-Node Shell Element," *Intl. J. Numer. Methods Eng.*, 24, pp 2389-411, 1987.

10.32 O. C. Zienkiewicz and R. L. Taylor, *The Finite Element Method*, 4th Ed., McGraw Hill, London, pp 243-56, 1989.

10.33 T. Belytschko, W. K. Liu, J. S.-J. Ong, and D. Lam, "Implementation and Application of a Nine-Node Lagrange Shell Element with Spurious Mode Control," *Comput. Struct.*, 20, pp 121-8, 1985.

10.34 D. W. White and J. F. Abel, "Accurate and Efficient Nonlinear Formulation of a Nine-Node Shell Element with Spurious Mode Control," *Comput. Struct.*, 35, pp 621-41, 1990.

10.35 T. Belytschko, B. L. Wong, and H. Stolarski, "Assumed Strain Stabilization Procedure for the Nine-Node Lagrange Shell Element," *Intl. J. Numer. Methods Eng.*, 28, pp 385-414, 1989.

10.36 S. W. Lee, C. C. Dai, and C. H. Yeom, "A Triangular Finite Element for Thin Plates and Shells," *Intl. J. Numer. Methods Eng.*, 21, pp 1813-31, 1985.

10.37 P. Seide and R. A. Chaudhuri, "Triangular Finite Element for Analysis of Thick Laminated Shells," *Intl. J. Numer. Methods Eng.*, 24, pp 1563-79, 1987.

10.38 R. A. Chaudhuri, "A Degenerated Triangular Shell Element with Constant Cross-Section Warping," *Comput. Struct.*, 28, pp 315-26, 1988.

10.39 J. J. Rhiu and S. W. Lee, "A Sixteen-Node Shell Element with a Matrix Stabilization Scheme," *Comput. Mech.*, 3, pp 99-113, 1988.

10.40 J. J. Rhiu, R. M. Russell, and S. W. Lee, "Two Higher-Order Shell Finite Elements with Stabilization Matrix," *J. AIAA*, 28, pp 1517-24, 1990.

10.41 B. A. Szabó and G. J. Sahrmann, "Hierarchic Plate and Shell Models Based on p-Extension," *Intl. J. Numer. Methods Eng.*, 26, pp 1858-81, 1988.

10.42 K. S. Surana and R. M. Sorem, "p-Version Hierarchical Three-Dimensional Curved Shell Elements for Elastostatics," *Intl. J. Numer. Methods Eng.*, 31, pp 649-76, 1991.

10.43 Y. Leino and J. Pitkäranta, "On the Membrane Locking of h-p Finite Elements in a Cylindrical Shell Problem," Helsinki Univ. of Technology, Instit. of Mathematics Research Reports A311, June 1992.

11

Finite Element Design
In Perspective

With the chapters on plates and shells, we have completed the formal treatment of issues in the design of finite elements. Here we will summarize and generalize the material in ways which may lend a fuller perspective to finite element design.

11.1   THE MEDICAL ANALOGY

Throughout our treatment of finite element design we have found many instances where a medical analogy served to sharpen the issues involved. The reader may well wonder why an analogy between such remotely different subjects is so frequently apt. The reason is probably related to the fact that both deal with complex, highly organized objects for which a normal, healthy state of performance is definable. These attributes also fit most complex machines and many institutions which, accordingly, may accommodate a medical analogy. And indeed such usage is common. We sometimes say that our car is in the hospital or that the school system is sick. A common feature of all such "systems" is that they bring order out of the normally chaotic state of their constitutive materials or, to put the matter more succinctly, they locally defy the second law of thermodynamics (or its analog in information theory).

The "illness" of such a system does not usually reveal itself as a small change in the system's efficiency. Rather, it is more likely that some part of the system performs badly or not at all. A machine part breaks, an aircraft wing flutters, a hole develops in a dike. In short, the system or object is sick, with possibly disastrous consequences. What is at work is that the normally chaotic state of nature finds a way to break down the object's layers of orderliness, usually in unexpected ways. After all, if the designer had been able to anticipate the breakdown he would have accounted for it in the design. Once he observes the breakdown the doctor or the engineer or the repairman proposes a cure or a fix, which may work or may not, or which may have undesirable side effects.

All of this applies, as we have seen, to finite elements. The underlying mathematical base is chaotic (could a chimpanzee really type out a valid computer program?); the normal state of health is well defined (as correct answers for patch tests and other simple problems); disorder attacks in dramatic, unexpected ways (particularly in the locking phenomenon); many remedies are proposed (reduced integration, assumed strain, etc.); and they have undesirable side effects (spurious modes, patch test failure). The fact that the medical analogy applies to finite elements is not remarkable. It applies equally well to any complex object or system which shares these qualities.

11.2   REMEDIES FOR FINITE ELEMENT DISORDERS AND THEIR SIDE EFFECTS

The following list identifies six basic disorders which afflict finite elements. Preceding chapters have treated each of them to a greater or lesser degree. Here we summarize their characteristics and look for interrelationships.

>Rigid body failure
>Induced anisotropy
>Interpolation failure
>Integration failure
>Equilibrium failure
>Spurious modes

Rigid body failure is easily the most severe disorder and at the same time the least excusable. Recall Irons' remark[1] quoted in Section 3.5:

"We shall try to kill any element that fails a test for rigid body motions."

Rigid body failure most often occurs because rigid body motion inadvertently shows up in an expression for strain. The cause can be subtle, as is the case for the definition of bending curvature in curved shells (Section 10.5.1), or it can be flagrant, as when superparametric interpolation is used to supply a "correct" representation of geometry (Section 3.5).

The term *induced anisotropy* describes the condition where features of the element design cause its stiffness matrix to depend on the orientation of its coordinate system and hence on the sequence in which its nodes are numbered. As discussed in Section 3.2, the desire to avoid induced anisotropy plays a role in the choice of node patterns and basis functions for elements. Later we saw that selective underintegration and its surrogates (Sections 7.6 and 7.7) will induce anisotropy of the stiffness matrix. A remedy for quadrilateral elements is to arrange the selection of the element coordinate system so that it changes by multiples of 90° when the node sequence is changed (see Figure 7.12). This remedy can be extended with difficulty to hexahedral solid elements, but a better solution is to express the underintegrated terms in a skewed coordinate system tangent to the element's parametric (ξ, η, ζ) axes, either locally or at the center of the element. The only acceptable remedy for induced anisotropy in triangular elements is to make the anisotropic terms invariant to the sequence of the corner nodes; for example, when selecting bubble functions, select one of identical form for each edge.

Interpolation failure is the most difficult disorder to cure. As we saw in Chapter 6, the inability of an element's basis functions to properly interpolate higher order functions from their values at nodes is the immediate cause of locking and shape sensitivity. We have also seen that "standard" isoparametric elements are acutely susceptible to the consequences of interpolation failure and that the remedies proposed for locking frequently cause other types of failure. Figure 11.1 illustrates the linkage between remedies and failure modes starting with a standard isoparametric formulation.

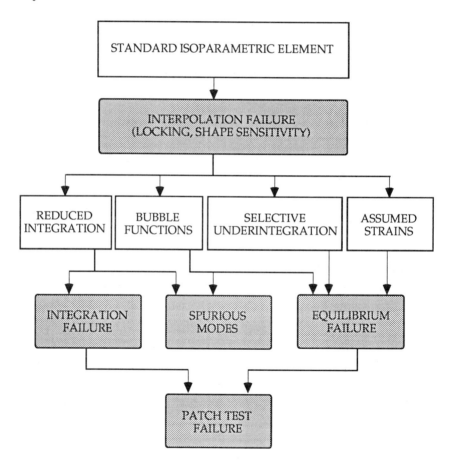

Figure 11.1   The Side Effects of Various Remedies for Locking.

The number of integration points required to produce precise integration depends on the quantity integrated, the shape of the element, and the number of nodes (see

Tables 4.12 and 4.13). We reserve the term *integration failure* for cases which result in the failure of constant strain patch tests. Table 5.1 lists the minimum integration rules which various elements require to pass constant strain patch tests with general shape. We observed that reduced integration causes integration failure for solid brick elements but not for flat two-dimensional elements.

Reduced integration also produces *spurious modes*. As shown in Sections 7.3 and 7.4, reduced integration of Lagrange elements creates global spurious modes which pose an unacceptably large risk for practical application. Underintegration of serendipity elements, on the other hand, yields only relatively benign noncommunicating spurious modes. The net result of these observations is that reduced integration constitutes an attractive remedy for locking in eight-node shell elements and twenty-node brick elements but not in other elements.

Selective underintegration is a simple modification of reduced integration which eliminates spurious modes. It works well for the four-node quadrilateral and, with a little care, for the eight-node brick (see Sections 7.6.1 and 7.6.2). If, however, selective underintegration is tried with the eight- or nine-node quadrilateral or the twenty-node brick, it results in patch test failure through the mechanism of *equilibrium failure*. As explained in Chapter 5, equilibrium failure occurs when the nodal forces computed from constant element strains are not in equilibrium.

Researchers have sought remedies for locking which avoid spurious modes and equilibrium failure since the early 1970s. In this book we have grouped such remedies into assumed strain hybrid formulations (Section 7.7), spurious mode stabilization (Section 7.8), bubble functions (Section 8.1), and *direct* assumed strain formulations (Section 8.4). These classifications are fairly arbitrary and they do not exhaust the possibilities. We have, for example, described an N-bar method which explicitly modifies an element's displacement shape functions (Section 8.3), and methods which employ drilling freedoms (Section 8.2). With the exception of methods which use drilling freedoms, all of these remedies can be classed as B-bar methods;[2] in other words, they accept the element's nodal degrees of freedom as given and replace the standard isoparametric strain-displacement matrix with a new one, $\left[\overline{B}\right]$.

Some of these remedies for locking, such as the assumed strain hybrid formulations and mode stabilization, include patch test satisfaction as an inherent design feature. Others do not and have suffered equilibrium failure. In the case of bubble functions, equilibrium failure is readily avoided by selecting the nonconforming part of the strain-displacement matrix so that it integrates to zero over the element's surface (see Equation 5.18). In the case of direct assumed strain formulations, no general procedure exists (we will show in Section 11.4 that none *can* exist) which will guarantee patch test satisfaction. Direct assumed strain formulations do, however, pass patch tests in some cases, such as when used to compute the transverse shear strains of three- and four-node plate elements (Sections 9.4.3 and 9.4.4).

Because they do not automatically pass patch tests, the direct assumed strain formulations are often regarded as measures of last resort, to be used when other methods fail to remedy locking. Examples of this sort include trapezoidal locking of the four-node quadrilateral (Section 8.4.2) and membrane locking of the six-node triangle (Section 10.5.3). We should also mention the nine-node ANS shell element (Sections 8.4.3 and 10.5.2) as an example of a direct assumed strain formulation which fails patch tests where other elements succeed but which is otherwise exceptionally competitive.

It is possible to take all of these considerations and to construct a designer's guide which indicates the remedies to be preferred in various circumstances. With some trepidation we attempt just that in Table 11.1 using the word *recommended* rather than *preferred* to identify exactly whose preferences are indicated. It will be noted right away that we recommend no locking remedy for membrane action of the three-node triangle even though the element locks quite severely (see Section 6.3). The only known remedy which alleviates locking for this case is to add drilling freedoms. Drilling freedoms are not included in the table because, unlike the other remedies, they add to the nodal degrees of freedom and because their suitability for practical applications remains (December 1992) to be fully validated.

We see that each remedy listed in Table 11.1 has more than one application where it is recommended. In the case of the four-node quadrilateral, selective underintegration has been a preferred remedy for shear locking since 1969.[3] Wilson's incompatible mode method[4] adds the avoidance of dilatation locking,

Table 11.1

Recommended Locking Remedies for Various Elements

ELEMENT	STRAIN COMPONENTS	REDUCED INTEGRATION	SELECTIVE UNDERINTEGRATION	BUBBLE FUNCTIONS	ASSUMED HYBRID STRAIN	MODE STABILIZATION	DIRECT ASSUMED STRAIN
TRIA3	Membrane						
	Transverse Shear						√
QUAD4	Membrane		√	√		√	
	Transverse Shear						√
TRIA6	Membrane						√
	Transverse Shear	— None Required —					
QUAD8	Membrane	√					
	Transverse Shear	√					
QUAD9	Membrane			√	√	√	√
	Transverse Shear			√	√	√	√
HEXA8	All		√	√			
HEXA20	All	√				√	

and mode stabilization[5] improves the computational efficiency. The only effective remedy for transverse shear locking of QUAD4 is the direct assumed strain formulation. This appears also to be the case for membrane locking of the TRIA6.

The eight- and nine node shell elements present an interesting contrast. For the QUAD8, simple reduced integration has worked well since 1971[6] and has only recently been improved by an N-bar correction[7] which allows satisfaction of constant bending patch tests under the same conditions as the nine-node element (Section 8.3). The QUAD9, on the other hand, has received constant attention in recent years from proponents of various advanced formulations. The difference, of course, is that simple reduced integration of QUAD9 causes global spurious modes.

In the case of the eight-node brick element, selective underintegration eliminates shear locking and, if carefully done, avoids equilibrium failure (Section 7.6.2). The addition of bubble functions to the element eliminates dilatation locking (Section 8.1.3).

Finally, we are tempted to recommend reduced integration for the twenty-node brick element, possibly with an N-bar refinement similar to that described in Section 8.3. For those who find the nonsatisfaction of constant strain patch tests objectionable, we recommend full integration of the twenty-node brick with an assumed hybrid strain formulation to remove locking modes (Section 7.7).

TETRA and PENTA solid elements do not appear in Table 11.1. TETRA elements are like triangular membrane elements; they are what they are, perfect in their polynomial completeness, and little if anything can be done to improve them. PENTA solid elements, on the other hand, share some of the characteristics of hexahedral elements and can benefit from the remedies which have been described. For example, Section 7.2 discusses integration patterns for six-node and fifteen-node PENTAs and indicates that a 3×2 pattern for PENTA15 avoids a particular locking state.

11.3   DIAGNOSTIC TESTS AND OTHER TESTS

Anyone who attempts to devise a set of finite element tests soon discovers that the first task is to decide who will use the test results and for what purpose. The author recalls that at the end of the original NASTRAN development (1969) NASA required the developers to submit results for NASA-supplied tests. These demonstration problems dealt mostly with axisymmetric shells and they came with analytically derived solutions. The purposes of the tests were to validate the program and to demonstrate to potential users the kinds of problems that NASTRAN could solve. We, the developers, were not restricted on the type, number, or arrangement of the elements we could use.

Later we modified these tests so that they would validate each of the types of analysis that NASTRAN could solve and they became known as the Holy Thirteen. We ran these problems after each new version of NASTRAN to ensure that no errors had crept in. This particular use of test problems continues to grow. At MSC our test problem library now contains over 2,000 problems which are run by quality assurance engineers to validate that the answers do not change between releases of MSC/NASTRAN and that the results are (virtually) the same for all computers and operating systems. The results are, for the most part, checked by computer. The problems are designed to traverse all the important paths through the system. Inevitably a few important paths are not tested until the program gets into the hands of users. The errors found by users may be important enough to trigger re-releases.

None of this has much to do with finite element design except to note that the quality assurance testers complain when modifications to the elements cause the test results to change, however slightly. It also illustrates that test results have different constituencies, including the Q. A. department, finite element developers, the marketing department and, to be sure, the company's clients. Different kinds of test users require different kinds of tests. The marketing department wants tests that demonstrate the product in a good light, potential new users want benchmarks with which to compare competing software systems, the Q. A. department wants tests which ensure that nothing has changed between systems (or that all changes can be explained), and software developers need tests to diagnose the causes of errors or poor performance.

Finite element designers fit into the last category. They need tests which tell them how well their elements are performing and which reveal the causes, as closely as possible, of substandard performance. We have included many such diagnostic tests in this book, first to identify and to demonstrate the disorders which afflict finite elements, and then to register the improvements provided by remedies. These requirements call for tests which illustrate, and perhaps exaggerate, the disorders in question and which are simple enough to be understood in detail. Such tests can guide the developer to the root causes of the disorders and can also suggest remedies.

Unfortunately, the test problems which developers employ have not always been adequate. The early developers of finite elements were only vaguely aware of such disorders as locking and spurious modes and so they did not test for them. The author recalls that the only diagnostic tests applied to the original plate elements in NASTRAN involved a regular rectangular array of elements with various combinations of lateral loads and boundary conditions (see Figure 6.15 and Table 6.5). These tests reveal nothing about in-plane shear locking, patch test satisfaction, or shape sensitivity. Their only diagnostic value, beyond the ability to detect the outright blunders which any test can provide, is their ability to detect transverse shear locking.

Such was the state of affairs, at least with commercial codes, until the latter half of the 1970s. Until then, Irons' earlier (1966) revelation of the benefits of patch tests [8] went largely unheeded. As the reader will recall, the 1973 version of Wilson's incompatible mode quadrilateral, [4] which failed the constant strain patch test, was corrected and re-released [9] in 1976. Many other elements developed in the 1970s, including some of the author's, might not have appeared or might have appeared minus some of their defects if adequate diagnostic tests had been employed.

The inadequacy of diagnostic element testing was driven home toward the end of the 1970s by John Robinson's publication in his *Finite Element News* [10] and elsewhere [11,12] of benchmark tests comparing commercial elements in simple problems where shear locking, spurious modes, and patch test failures were prominent. Some of the failures revealed by Robinson's tests were spectacular. They made many finite element users aware, for the first time, that the finite element method is not infallible. They caused designers, including the author, to reevaluate their elements, and they led to a demand for action by government and professional organizations to monitor and possibly to regulate finite element codes.

One of the results of this outcry was the formation in the United Kingdom of a National Agency for Finite Element Methods and Standards (NAFEMS). By 1984 NAFEMS was beginning to publish benchmark tests for finite element assemblies.[13] One of their favorite tests involves an elliptical plate with an elliptical hole (Figure 11.2). This example is typical of the benchmark problems which NAFEMS has published.[14] They are good comparative benchmark problems because they appeal to engineers as the sort of problems which finite element programs should be able to solve handily. They do not make good diagnostic tests problems, however, because they do not isolate or emphasize particular finite element disorders.

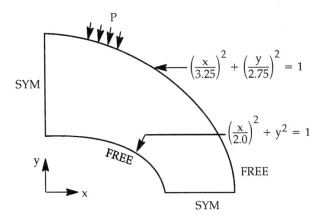

Figure 11.2   NAFEMS Elliptical Plate Problem.   Uniform In-Plane Pressure Applied to Exterior Boundary.   t = 1.0.

Meanwhile, in 1982 or 1983 the American Institute of Aeronautics and Astronautics (AIAA) set up a Standards Committee on Structural Analysis to look into the desirability of issuing standards for finite element analysis. The principal outcome of that effort was the publication in 1985 of a "Proposed Standard Set of Problems to Test Finite Element Accuracy."[15] These so-called MacNeal-Harder test problems have since become a virtual de facto standard for finite element development, in that most papers on new finite elements include one or more of the problems as examples and sometimes the whole lot.[16] Throughout this book we have had occasion to use all of the MacNeal-Harder test problems one or more times. Table 11.2 identifies the problems and the figures where they are first described.

Table 11.2

The MacNeal-Harder Test Problems

PROBLEM NAME	FIGURE WHERE DESCRIBED
Patch Test for Plates	5.1
Patch Test for Solids	5.3
Straight Cantilever Beam	1.4
Curved Beam	8.14
Twisted Beam	10.8
Rectangular Plate	6.15
Scordelis-Lo Roof	1.6
Hemispherical Shell	10.14
Thick-Walled Cylinder	6.5

The MacNeal-Harder test problems were specifically chosen to be a comprehensive set of diagnostic tests, and this they have proven to be. We have, for example, used them to illustrate all of the known types of locking. Whether they constitute a *sufficient* set of problems to validate an element's accuracy remains an open question.

Authors of new finite elements generally include tests problems from other sources, such as the short "obstacle course" for shell elements offered by Belytschko, et al.[17] Two of these three problems are the Scordelis-Lo roof and the Morley hemispherical shell without hole (Figure 10.18). The third is the pinched cylinder shown in Figure 11.3. The salient feature of this problem is a stress concentration under the applied load which decays to insignificance within

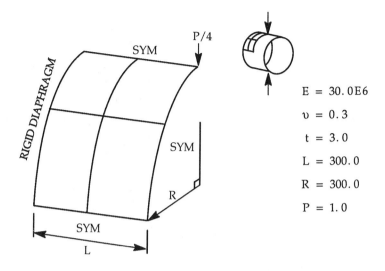

Radial Displacement at Load Point: w = 1.8248E-05.

Figure 11.3   Pinched Cylinder Problem. [17]

about one characteristic length, $\ell = \sqrt{Rt}$. Because of this stress concentration, the pinched cylinder problem tends to favor higher order elements.

Another test problem which has attracted a good deal of attention is the Morley rhombic plate problem [18] illustrated in Figure 11.4. Robinson proposed this problem as a challenge in the June 1983 issue of the *Finite Element News* and later published results for thirty-three elements from nineteen finite element systems.

The challenge has been controversial from the start. A major issue is whether to use a "hard" condition for the simple supports, as prescribed by Robinson, or a "soft" condition. The difference is that the hard condition requires rotation about the in-plane normal, θ_n, to be zero, while the soft condition allows θ_n to be free. The difference should, in any case, be small because

$$\theta_n = w_{,t} - \gamma_{tz} \tag{11:1}$$

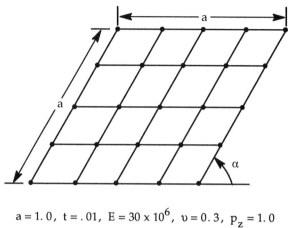

$$a = 1.0, \quad t = .01, \quad E = 30 \times 10^6, \quad \upsilon = 0.3, \quad P_z = 1.0$$

All Edges Simply Supported

Figure 11.4   The Morley Rhombic Plate Problem.

where $w_{,t} = 0$ because $w(t) = 0$ and γ_{tz} is presumably small for a thin plate (in the test problem $t/a = .01$). In actual fact, many of the elements recorded much better results for the soft supports than for the hard supports. This result is related to a phenomenon known as Babuška's paradox,[19] which concerns the simply supported, multi-sided plate shown in Figure 11.5. If the Kirchhoff hypothesis is prescribed, then θ_n will, from Equation 11.1, be zero at all points on the boundary so that, at the intersection of two sides, *both* in-plane components of rotation must be zero. This leads to the paradox that, as the number of sides increases, so does the number of points of fixed support and, in the limit, the simply-supported plate will behave like a built-in plate.

Babuška's paradox can be used to attack the Kirchhoff hypothesis which, among its other eccentricities, predicts a singularity in bending moment at the obtuse corners of the Morley rhombic plate. Babuška and Scapolla have studied the Morley rhombic plate[20] and have observed large differences between the Kirchhoff and Mindlin solutions. They concluded that "the hard simple support seems to be physically incorrect." The hard simple support does, however, have a physical interpretation in Morley's problem. It corresponds to a plane of antisymmetry at the edge of the plate. With this boundary condition, the rhombic

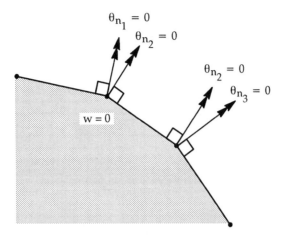

Figure 11.5   Babuška's Paradox. A Multi-Sided Plate with Hard (Kirchhoff) Simple Supports.

plate represents a unit cell in an infinite plane of identical rhombic cells deflected alternately up and down in a checkerboard pattern.

In any event, while the Robinson skew plate challenge has prompted a good deal of valuable research, it would not appear to be an appropriate benchmark to test the accuracy of finite elements, or at least not until the theoretical issues are resolved.

11.4   THE LIMITS OF FINITE ELEMENT PERFECTIBILITY

The task of developing good finite elements never seems to be finished. Designers return, again and again, to the same basic configuration of nodes and find some way to eke out an improvement. Consider the four-node membrane quadrilateral. We showed in Section 6.2 that the standard isoparametric formulation[21] (1961) suffers from shear and dilatation locking, even for rectangular shapes. Selective underintegration,[3] (1969) (Section 7.6.1) eliminates the shear locking but not the dilatation locking. Wilson's incompatible modes[9] (1976) (Section 8.1.1) eliminate both shear locking and dilatation locking for rectangular and parallelogram shapes. The gamma elements of Belytschko, et al.,[22] (1986) (Section 7.8) eliminate locking under the same conditions at lower computation cost. Locking of the four-node membrane element with trapezoidal shape (Section 6.6.2) remained, however, as an outstanding issue. The QUADH element[23] (1982) (Section 8.4.2) eliminates shear locking for trapezoidal shape but fails the constant strain patch test. After a fruitless search for a four-node design which eliminates locking for trapezoidal shape without creating additional external degrees of freedom and which also passes the patch test, the author began to suspect and then was able to prove[24] (1987) that no such element exists.

More recently[25] (1992) the author has generalized this result into a concept of the limited perfectibility of all finite elements. In particular, he has shown that the limited ability of elements to interpolate nodal displacements, i.e., interpolation failure, leads to unsymmetric coupling between low order strain states and high order strain states which cannot be resolved without causing patch test failure in the low order states or locking in the high order states.

The concept of limits on the perfectibility of finite elements would seem to be a fitting, if still speculative, topic on which to end our treatment of finite element design. We will, accordingly, describe the theory in some detail and apply it to a number of examples.

11.4.1 Theory, With Application to the Four-Node Trapezoid

We begin by assuming an elastic continuum, Ω, with linear properties. Within the continuum, deformations are described by an ordered set of displacement states. The first states are rigid body translations and rotations; the next describe constant strains; the ones beyond that represent, in order, linear strain states, quadratic strain states, etc. The independent variables are ordinary Cartesian coordinates, x, y, z. Taken together, the displacement states are complete and can, if enough of them are used, represent any nonsingular deformation state.

We consider next a bounded subdomain, Ω_e, of the elastic continuum which we intend to replace by a finite element (see Figure 11.6). At each point on the boundary, Γ_e, of the subdomain, there are tractions $\{t\}$ and displacements $\{u\}$. The quality of the finite element is measured by its ability to mimic the relationship between $\{t\}$ and $\{u\}$. Thus, given $\{u\}$, the measure of quality is how well the element reproduces $\{t\}$, or vice versa.

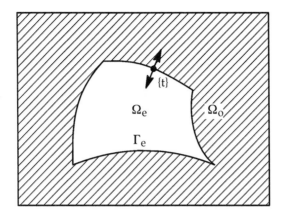

Figure 11.6   Finite Element Embedded in an Elastic Continuum.

Within the portion of the elastic continuum outside the element, Ω_o, the relationship between tractions and displacements is well defined. Thus the strains $\{\varepsilon\} = [L]\{u\}$, where $[L]$ is a linear differential operator, the stresses $\{\sigma\} = [D]\{\varepsilon\}$, where $[D]$ is the modulus matrix, assumed constant and symmetric; and the tractions $\{t\} = \left\{ \sum_j \sigma_{ij} n_j \right\}$ on any surface, Γ, where n_j is the jth component of the outward normal. Thus, given the spatial dependence of $\{u\}$, we can readily compute $\{t\}$ on any surface, including Γ_e.

Since the finite element replacing Ω_e has a finite number of degrees of freedom, it cannot exactly represent the relationship between $\{t\}$ and $\{u\}$ for all possible $\{u\}$. We are led, therefore, to restrict $\{u\}$ to be a linear combination of a finite number of the displacement states, or modes, previously defined. Thus

$$\{u\} = \sum_{m=1}^{M} \{u_m\} \alpha_m \quad , \quad \{t\} = \sum_{n=1}^{M} \{t_n\} \alpha_n \qquad (11:2)$$

where $m = 1, \cdots, M$ includes all displacement states of order S and lower.

Let P_m be the generalized force acting on α_m, i.e., the work done by incremental tractions $\{t\}$ on mode $\{u_m\}$. Thus

$$P_m = \int_{\Gamma_e} \{u_m\}^T \{t\}\, dS = \int_{\Gamma_e} \sum_{n=1}^{M} \{u_m\}^T \{t_n\} \alpha_n\, dS = \sum_{n=1}^{M} P_{mn} \alpha_n \quad (11:3)$$

where

$$P_{mn} = \int_{\Gamma_e} \{u_m\}^T \{t_n\}\, dS \quad (11:4)$$

Considering the generalized forces on all modes to be the vector, $\{P_m\} = [P_{mn}]\{\alpha_n\}$, we note that the coefficient matrix $[P_{mn}]$ is easily computed using the known spatial dependencies of $\{u_m\}$ and $\{u_n\}$. The matrix $[P_{mn}]$ is also symmetric, a fact which can be demonstrated by using the divergence theorem to convert Equation 11:4 to

$$P_{mn} = \int_{\Omega_e} \{\varepsilon_m\}^T [D]\{\varepsilon_n\}\, dV \quad (11:5)$$

We note that for this to hold we must assume that the finite element does not yet occupy Ω_e. Alternatively, we can use the exterior domain Ω_o to demonstrate the symmetry of P_{mn}.

The matrix $[P_{mn}]$ is not necessarily diagonal. It can, however, be made diagonal by Schmidt orthogonalization, i.e., by adding some of mode $\{u_1\}$ to mode $\{u_2\}$, and some of modes $\{u_1\}$ and $\{u_2\}$ to mode $\{u_3\}$, etc. We note also that the elements of $[P_{mn}]$ corresponding to rigid body modes are null.

The finite element design problem can now be stated. We require that the tractions, and hence also the elements of the generalized force vector, $\{P_m\}$, be equal on the two sides of the boundary. On the element's side, the generalized force vector is related to the element's modal coefficients, $\{\beta_n\}$, by the element's modal stiffness matrix, $[K_{mn}]$. Thus

$$\{P_m\} = [P_{mn}]\{\alpha_n\} = [K_{mn}]\{\beta_n\} \tag{11:6}$$

Taking $\{\beta_n\} = \{\alpha_n\}$, we see that the design problem is to make $[K_{mn}] = [P_{mn}]$. If this is done, the finite element will exactly mimic the continuum domain it replaces up to and including the highest mode, α_M in $\{\alpha_n\}$.

If this were all there was to finite element design the subject would have been wrapped up long ago. The complicating factor is that finite elements are based on *nodes*, not *modes*. In other words, the element's degrees of freedom which communicate with the outside world consist of the displacements at boundary nodes and/or the coefficients of functions which describe displacements along the boundary [the so-called *hierarchical* degrees of freedom (see Section 4.5)]. Since hierarchical degrees of freedom are interchangeable with nodal degrees of freedom, we retain only the latter for ease of explanation.

Let the displacement vector at node i be $\{u_i\}$ and let the force vector at node i be $\{P_i\}$. The relationship between nodal displacements and nodal forces can be written as

$$\{P_i\} = \sum_{j=1}^{N} [K_{ij}]\{u_j\} \tag{11:7}$$

where $[K_{ij}]$ is the ij partition of the element's nodal stiffness matrix and N is the number of nodes. We can express nodal displacements in terms of the modal coefficients previously defined and form a weighted sum of the nodal forces to obtain generalized modal forces. Thus

$$\{u_i\} = \sum_{n=1}^{M} \{u_{in}\}\beta_n \quad , \quad \{P_m\} = \sum_{i=1}^{N} [u_{im}]^T \{P_i\} \tag{11:8}$$

where $\left\{u_{in}\right\}$ is the vector of displacement components for mode n at node i and $\left\{P_m\right\}$ is the vector of generalized forces for all modes. We can express the relationship between $\left\{P_m\right\}$ and $\left\{\beta_n\right\}$ *for the element* as

$$\left[K_{mn}\right]\left\{\beta_n\right\} = \left\{P_m\right\} \tag{11:9}$$

where

$$\left[K_{mn}\right] = \sum_{i=1}^{N} \sum_{j=1}^{N} \left[u_{im}\right]^T \left[K_{ij}\right] \left[u_{in}\right] \tag{11:10}$$

The design problem for the element (Equation 11:9) has the same form as before, but the description is incomplete because we have not yet expressed $\left\{P_m\right\}$ in terms of modal coefficients. To do this we first express the nodal forces in terms of modal coefficients, i.e.,

$$\left\{P_i\right\} = \sum_{n=1}^{M} \left\{P_{in}\right\} \alpha_n \tag{11:11}$$

so that

$$\left\{P_m\right\} = \left[P^a_{mn}\right]\left\{\alpha_n\right\} \tag{11:12}$$

where, from Equations 11:8 and 11:11,

$$P^a_{mn} = \sum_{i=1}^{N} \left\{u_{im}\right\}^T \left\{P_{in}\right\} \tag{11:13}$$

The element design problem is now in exactly the same form as before

$$\left[K_{mn}\right]\left\{\beta_n\right\} = \left[P^a_{mn}\right]\left\{\alpha_n\right\} \tag{11:14}$$

where the object is to make $\left[K_{mn}\right] = \left[P^a_{mn}\right]$. As yet we have not specified a method for computing $\left\{P_{in}\right\}$, the nth vector of the modal coefficients for nodal forces. If that selection were arbitrary, we could make the elements of $\left[P^a_{mn}\right]$ anything we pleased. This follows because, if we take $M = NC$ where C is the number of components of motion per node, the number of available coefficients in the $\{P_{in}\}$ vectors will be just equal to the number of elements in $\left[P^a_{mn}\right]$. We could in this way select the $\{P_{in}\}$ vectors so that $\left[P^a_{mn}\right] = \left[P_{mn}\right]$, the generalized modal force matrix computed from properties of the continuum.

The $\{P_{in}\}$ vectors cannot, however, be selected arbitrarily. They must be chosen so that the forces at nodes satisfy equilibrium. This requirement takes priority over equating $\left[P^a_{mn}\right]$ to $\left[P_{mn}\right]$ because satisfaction of the latter condition on an element-by-element basis cannot guarantee correct solutions in a field of elements unless nodal equilibrium is also satisfied.

We begin the construction of $\{P_{in}\}$ vectors which satisfy nodal equilibrium by relating the forces on node i to the tractions on the boundary of the element, i.e.,

$$\{P_i\} = \int_{\Gamma_e} \left[N_i\right]\{t\}dS \qquad (11:15)$$

where $\left[N_i\right]$ is a $C \times C$ matrix of unspecified functions of position. We have not, as yet, given up anything in regard to the arbitrariness of selecting $\{P_i\}$. We know, however, from Section 2.4.3, that $\left[N_i\right]$ must also be the matrix of displacement shape functions for boundary points and, from Section 5.3, that we can assure satisfaction of equilibrium only if the shape functions are conforming. The nonconforming shape functions, or bubble functions, introduced in Section 8.1 do not alter this conclusion because boundary tractions are not applied to them. In addition, finite element designers invariably select $\left[N_i\right]$ to be a diagonal matrix in which the same terms are used for directions with isotropic properties. Thus, for a plate element, $N_v = N_u$ to preserve isotropy but $N_w = N_u$ is not required.

These considerations severely limit the designer's options with respect to $\left[N_i\right]$ and hence with respect to $\left[P^a_{mn}\right]$. The conformability requirement rules out elementary basis functions which are direct functions of metric position coordinates, such as x, y, z, except in triangular and tetrahedral elements with straight edges. The conformability requirement admits functions of the parametric variables ξ, η, ζ as basis functions and, indeed, parametric variables were introduced into finite element analysis specifically to satisfy this requirement. The practical range of choice for basis functions is further restricted by the arrangement of an element's nodes (see Section 3.2). In the case of the four-node quadrilateral, for example, $\lfloor 1, \xi, \eta, \xi\eta \rfloor$ is really the only practical choice.

Returning to consideration of the element design equation, we note that if we assume identical shape functions for all components of motion, Equation 11:13 can be written as

$$P^a_{mn} = \sum_{i=1}^{N} \left\{ u_{im} \right\}^T \int_{\Gamma_e} N_i \left\{ t_n \right\} dS = \int_{\Gamma_e} \left\{ u^a_m \right\}^T \left\{ t_n \right\} dS \qquad (11:16)$$

where

$$\left\{ u^a_m \right\} = \sum_{i=1}^{N} N_i \left\{ u_{im} \right\} \qquad (11:17)$$

is the representation of modal displacement $\left\{ u_m \right\}$, i.e., its *alias*, on the element's boundary. If $\left\{ u^a_m \right\} = \left\{ u_m \right\}$, the design problem returns to its original modal formulation which admits a straightforward solution. In isoparametric elements, we are guaranteed that $\left\{ u^a_m \right\} = \left\{ u_m \right\}$ for all rigid body and constant strain modes. With higher modes there is no guarantee. Thus we can separate the modes into a lower order set, $\left\{ \alpha_\ell \right\}$, where $\left\{ u^a_\ell \right\} = \left\{ u_\ell \right\}$ and a higher-order set, $\left\{ \alpha_h \right\}$, where $\left\{ u^a_h \right\} \neq \left\{ u_h \right\}$ except possibly under special circumstances. Observe what happens to coupling terms $P^a_{\ell h}$ and $P^a_{h\ell}$ defined by Equation 11:16. Since, $\left\{ u^a_\ell \right\} = \left\{ u_\ell \right\}$, $P^a_{\ell h} = P_{\ell h}$ but, since $\left\{ u^a_h \right\} \neq \left\{ u_h \right\}$, $P^a_{h\ell} \neq P_{h\ell}$ except under special circumstances. As a result, $\left[P^a_{mn} \right]$ is *not* guaranteed to be symmetric. The design problem (Equation 11:14) can, in this circumstance, be solved completely only by permitting the element's stiffness matrix to be unsymmetrical. Since this is abhorrent for many reasons, we are left with the conclusion that the design problem is not solvable under all circumstances or, in other words, that there are limits to a finite element's perfectibility.

Before we explore the consequences of this conclusion, the example of the four-node trapezoid shown in Figure 11.7 may reinforce our understanding of it. We assume two modes, a vertical stretching mode and a horizontal bending mode. The properties of these modes are, assuming zero Poisson's ratio:

MODE 1	MODE 2
$u = 0,\ v = y$	$u = xy,\ v = -\frac{1}{2}x^2$
$\varepsilon_x = 0,\ \varepsilon_y = 1,\ \gamma_{xy} = 0$	$\varepsilon_x = y,\ \varepsilon_y = 0,\ \gamma_{xy} = 0$
$\sigma_x = 0,\ \sigma_y = E,\ \tau_{xy} = 0$	$\sigma_x = Ey,\ \sigma_y = 0,\ \tau_{xy} = 0$

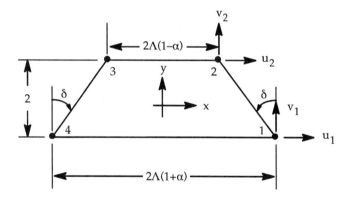

Figure 11.7   Four-Node Trapezoidal Element.

The displacements and forces at nodes 1 and 2 are tabulated below. In transferring tractions to nodes the shape functions are assumed, as is usual for a four-node element, to be linear on each edge. We need make no assumption about shape function form in the element's interior.

DEGREES OF FREEDOM	MODE 1		MODE 2	
	u_{1i}	P_{1i}	u_{2i}	P_{2i}
u_1	0	0	$-\Lambda(1 + \alpha)$	$-E/3$
u_2	0	0	$\Lambda(1 - \alpha)$	$E/3$
v_1	-1	$-\Lambda E$	$-\frac{1}{2}\Lambda^2(1 + \alpha)^2$	0
v_2	1	ΛE	$-\frac{1}{2}\Lambda^2(1 - \alpha)^2$	0

The matrix $\left[P^a_{mn}\right]$ for the two modes, obtained by summing the inner products of modal displacements and modal forces at nodes 1 and 2, is

$$\left[P^a_{mn}\right] = 2E\begin{bmatrix} \Lambda & 0 \\ \hline \alpha\Lambda^3 & \frac{\Lambda}{3} \end{bmatrix} \qquad (11{:}18)$$

Note that, since both modes are symmetric about the y-axis, this result would just be doubled if nodes 3 and 4 were included. The nonzero off-diagonal term P^a_{21} results from the product of the constant vertical traction of mode 1 and the quadratic vertical displacement of mode 2. An accurate integration of the product of traction and modal displacement would, of course, give a zero result because v in mode 2 is the same at corresponding points on the top and bottom surfaces. Thus the inability of the shape functions to represent a general quadratic function can be taken as the cause of the coupling between modes 1 and 2.

Returning to the element design problem, we note that a separation into low-order modes and high-order modes allows it to be stated in the following form, where $\left[P^a_{\ell h}\right] = 0$ and $\left[P^a_{\ell\ell}\right]$ is diagonal by Schmidt orthogonalization.

$$\begin{bmatrix} K_{\ell\ell} & \vdots & K_{\ell h} \\ \cdots & + & \cdots \\ K_{\ell h}^T & \vdots & K_{hh} \end{bmatrix} \begin{Bmatrix} \beta_\ell \\ \cdots \\ \beta_h \end{Bmatrix} = \begin{bmatrix} P_{\ell\ell}^a & \vdots & 0 \\ \cdots & + & \cdots \\ P_{h\ell}^a & \vdots & P_{hh}^a \end{bmatrix} \begin{Bmatrix} \alpha_\ell \\ \cdots \\ \alpha_h \end{Bmatrix} \tag{11:19}$$

Although this design problem cannot be solved perfectly, the designer still has choices. To see what they are more clearly, we put Equation 11:19 in the form

$$\begin{bmatrix} \gamma_{\ell\ell} & \vdots & \gamma_{\ell h} \\ \cdots & + & \cdots \\ \gamma_{h\ell} & \vdots & \gamma_{hh} \end{bmatrix} \begin{Bmatrix} \beta_\ell \\ \cdots \\ \beta_h \end{Bmatrix} = \begin{Bmatrix} \alpha_\ell \\ \cdots \\ \alpha_h \end{Bmatrix} \tag{11:20}$$

where

$$\left[\gamma_{\ell\ell} \right] = \left[P_{\ell\ell}^a \right]^{-1} \left[K_{\ell\ell} \right] \quad ; \quad \left[\gamma_{\ell h} \right] = \left[P_{\ell\ell}^a \right]^{-1} \left[K_{\ell h} \right]$$

$$\left[\gamma_{h\ell} \right] = \left[P_{hh}^a \right]^{-1} \left[K_{\ell h}^T - P_{\ell h}^a \left[P_{\ell\ell}^a \right]^{-1} K_{\ell\ell} \right] \tag{11:21}$$

$$\left[\gamma_{hh} \right] = \left[P_{hh}^a \right]^{-1} \left[K_{hh} - P_{\ell h}^a \left[P_{\ell\ell}^a \right]^{-1} K_{\ell h} \right]$$

and rigid body modes have been removed from α_ℓ and β_ℓ.

The designer has at least two reasonable choices. The first is to choose stiffnesses so that $\left[\gamma_{\ell\ell} \right] = [I]$ and $\left[\gamma_{h\ell} \right] = 0$, i.e.,

$$\begin{bmatrix} I & \vdots & \gamma_{\ell h} \\ \cdots & + & \cdots \\ 0 & \vdots & \gamma_{hh} \end{bmatrix} \begin{Bmatrix} \beta_\ell \\ \cdots \\ \beta_h \end{Bmatrix} = \begin{Bmatrix} \alpha_\ell \\ \cdots \\ \alpha_h \end{Bmatrix} \tag{11:22}$$

This choice satisfies patch tests with the lower modes because, if nodal displacements are selected so that $\{\beta_h\} = 0$, then $\{\alpha_\ell\} = \{\beta_\ell\}$ and $\{\alpha_h\} = 0$; in other words, the modal forces relate correctly to modal displacements for this case. This choice requires $\left[K_{\ell\ell} \right] = \left[P_{\ell\ell}^a \right]$ and $\left[K_{\ell h} \right]^T = \left[P_{h\ell}^a \right]$. It gives $\left[\gamma_{\ell h} \right] = \left[P_{\ell\ell}^a \right]^{-1} \left[P_{h\ell}^a \right]^T$ which is nonzero. As a result, a high-order modal displacement $\{\beta_h\}$ produces a low-order generalized modal force coefficient $\{\alpha_\ell\}$.

The low-order modal force is the cause of the locking of the four-node trapezoidal element described in Section 6.6.2. Inserting values of $\left[P^a_{\ell\ell}\right]$ and $\left[P^a_{h\ell}\right]$ from Equation 11:18 we see that $\{\alpha_\ell\} = \alpha_1 = \beta_1 + \alpha\Lambda^2\beta_2$. The factor $\alpha\Lambda^2 = \Lambda\tan\delta$, where Λ is the element's aspect ratio and δ is the taper angle (see Figure 11.7), was identified in Section 6.6.2 as the magnitude parameter for trapezoidal locking of four-node membrane elements.

The designer's other reasonable choice is to choose stiffnesses so that $\left[\gamma_{\ell\ell}\right]=[I]$, $\left[\gamma_{hh}\right]=[I]$, and $\left[\gamma_{\ell h}\right]=0$, i.e.,

$$\begin{bmatrix} I & 0 \\ \hline \gamma_{h\ell} & I \end{bmatrix}\begin{Bmatrix} \beta_\ell \\ \beta_h \end{Bmatrix} = \begin{Bmatrix} \alpha_\ell \\ \alpha_h \end{Bmatrix} \tag{11:23}$$

This choice reproduces the higher modes exactly; in fact, it satisfies patch tests with them but it fails patch tests for the lower modes. It requires $\left[K_{\ell\ell}\right]=\left[P^a_{\ell\ell}\right]$, $\left[K_{hh}\right]=\left[P^a_{hh}\right]$, and $\left[K_{\ell h}\right]=0$. It gives $\left[\gamma_{h\ell}\right]=-\left[P^a_{hh}\right]^{-1}\left[P^a_{h\ell}\right]$. For the example of the four-node trapezoid we have, from Equation 11:18, $\{\alpha_h\} = \alpha_2 = -3\alpha\Lambda^2\beta_1 + \beta_2$. This result appears to be worse, if anything, than the result of the first choice. The relative "merits" of the two choices are application dependent. The first choice is clearly best if in-plane bending is unimportant. Conversely, the second choice is best for slender beams and, by extension to three dimensions, for thin shells where stretching in the transverse direction is unimportant. The QUADH element, described in Section 8.4.2, uses this choice.

Put in general terms, the choice offered by the offending coupling matrix is to compromise either the accuracy of the lower modes or the accuracy of the higher modes. An apparent escape is provided by selecting the higher modes so that their displacements conform to the element's basis functions, thereby producing a null coupling matrix. (In the case of the four-node quadrilateral, this would imply $u_h = \xi\eta$ or $v_h = \xi\eta$.) Unfortunately this selection merely begs the question. The higher continuum modes are needed to represent important mechanical effects and their suppression through locking is a real concern.

Perhaps the most striking feature of these results is their independence of the element's interior design features. The offending coupling matrix, $\left[\gamma_{\ell h}\right]$ or, $\left[\gamma_{h\ell}\right]$ depends only on the properties of the assumed continuum fields and on the manner of their interpolation on element *boundaries*. Interior design features, such as reduced integration, bubble functions, or the use of mixed energy principles, may be needed to achieve the desired modal stiffnesses, but the values of these stiffnesses are set by exterior considerations.

11.4.2 The Four-Node Parallelogram

Consider the four-node membrane parallelogram sketched in Figure 11.8. Its Cartesian position coordinates are related to its parametric coordinates by

$$x = \Lambda\xi + \eta \tan \delta \qquad y = \eta \qquad (11{:}24)$$

We see that the bending state, $u = xy$, $v = -\frac{1}{2}x^2$, includes terms $\left(\xi^2, \eta^2\right)$ which are not in the element's displacement basis$(1, \xi, \eta, \xi\eta)$. As a result, coupling between the bending state and constant strain states may exist. A symmetry argument can, however, be invoked to show that the coupling is null. As

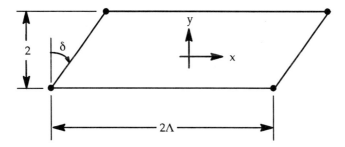

Figure 11.8 **Four-Node Parallelogram Element.**

illustrated in Figure 11.9, a parallelogram has two classes of symmetry and the linear displacement states are in a different class from the quadratic displacement states.

Four-node membrane elements exist which give correct results for skewed shapes. See, for example, Sections 7.8 and 8.1.1.

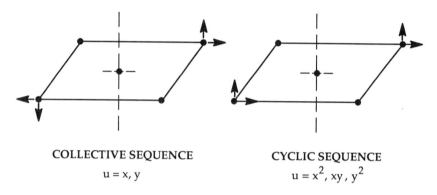

COLLECTIVE SEQUENCE
$u = x, y$

CYCLIC SEQUENCE
$u = x^2, xy, y^2$

Figure 11.9   Symmetry Types of the Parallelogram.

11.4.3   The Four-Node Quadrilateral with Drilling Freedoms

As illustrated in Figure 11.10, drilling freedoms contribute quadratic terms to the normal displacements of an edge. The magnitudes of the added terms are

$$u_n = \left(1 - \xi_e^2\right)\frac{\ell_e}{8}\left(\theta_2 - \theta_1\right) \tag{11:25}$$

where θ_2 and θ_1 are rotations at the adjacent corners, ℓ_e is the length of the edge, and ξ_e is measured form the midpoint of the edge. We note that u_n cannot be independently specified on all edges because equal rotations produce a null result. In the symmetrical example of Figure 11.10, the top and bottom edges use up the available options. In addition, the tangential components of displacement still vary linearly along the edges. Thus the quadratic displacement state $u = xy$, $v = -\frac{1}{2}x^2$ will not be accurately represented along the nonparallel edges. Still, as shown in Figure 8.9, the performance of a four-node trapezoid can be greatly improved by the addition of drilling freedoms.

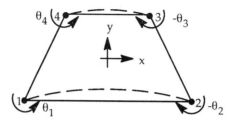

Figure 11.10. Four-Node Quadrilateral with Drilling Freedoms.

11.4.4 Higher-Order Quadrilaterals with Straight Sides and Uniformly Spaced Edge Nodes

As first noted in Section 3.5, some restrictions must be placed on the shapes of elements in order to achieve accurate representations of quadratic and higher-order displacement states. We consider here quadrilateral elements with bilinear shapes such that $x = \left(1, \xi, \eta, \xi\eta\right)$.

With this restriction the quadratic displacement states $u = x^2$, xy, or y^2 include the terms $\left(1, \xi, \eta, \xi\eta\right)^2 = \left(1, \xi, \eta, \xi^2, \xi\eta, \eta^2, \xi^2\eta, \xi\eta^2, \xi^2\eta^2\right)$. The lowest order element which includes all of these terms in its displacement basis is the nine-node Lagrange quadrilateral which can, therefore, under the specified shape restriction, accurately represent any quadratic displacement state.

In the eight-node serendipity element the function $f = \xi^2\eta^2$ is represented by an alias, $f^a = \xi^2 + \eta^2 - 1$. Note that $f^a = f$ at all points on the boundary of the element because $\xi^2\eta^2 = \xi^2 + \eta^2 - 1$ for $\xi = \pm 1$ and for $\eta = \pm 1$. We may conclude from the theory developed in Section 11.4.1 that the eight-node serendipity element should be able to interpolate quadratic displacement states under the same conditions as the nine-node Lagrange element. See Section 8.3 for a particular implementation.

Cubic and higher-order serendipity elements can also be modified to provide the same accuracy as the corresponding Lagrange elements. We know this because conformability requires the shape functions on an edge to depend only on the properties of the edge, i.e., to be independent of the existence of interior nodes.

Thus it can readily be shown that a quadrilateral serendipity (or Lagrange) element with bilinear shape and $p + 1$ equally spaced nodes per edge can be designed to exactly represent any displacement state of degree p in x and y.

The perfectibility troubles of straight-sided quadrilateral elements begin with displacement states of degree $p + 1$. We have already seen that the four-node ($p = 1$) quadrilateral of trapezoidal shape has essential difficulty with uniform in-plane bending. The eight-node, straight-sided trapezoid has similar difficulty with in-plane bending moments that vary linearly. Consider, as an exercise, the element shown in Figure 11.11 and the following two deformation states.

MODE 1	MODE 2
$u = -\frac{1}{2}y^2, \ v = xy$	$u = \frac{1}{2}x^2y, \ v = -\frac{1}{6}x^3$
$\varepsilon_x = 0, \ \varepsilon_y = x, \ \gamma_{xy} = 0$	$\varepsilon_x = xy, \ \varepsilon_y = 0, \ \gamma_{xy} = 0$
$\sigma_x = 0, \ \sigma_y = Ex, \ \tau_{xy} = 0$	$\sigma_x = Exy, \ \sigma_y = 0, \ \tau_{xy} = 0$

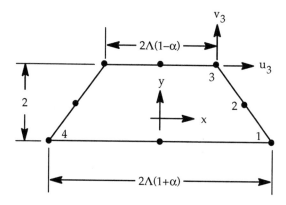

Figure 11.11   Eight-Node Trapezoidal Element, $\upsilon = 0$.

Note that mode 2 requires a distributed load $p_x = -Ey$ to be in equilibrium. Note also the antisymmetry of both modes with respect to a vertical axis.

The modal displacements and forces at nodes 1, 2, and 3 are tabulated below. The element's shape functions were used to convert tractions into forces at nodes.

DEGREES OF FREEDOM	MODE 1		MODE 2	
	u_{1i}	P_{1i}	u_{2i}	P_{2i}
u_1	$-\frac{1}{2}$	0	$-\frac{1}{2}\Lambda^2(1+\alpha)^2$	$\frac{E\Lambda}{3}\left\{\begin{array}{c}-\left(1+\frac{3\alpha}{5}\right)\\ -\frac{4\alpha}{5}\\ \left(1-\frac{3\alpha}{5}\right)\end{array}\right.$
u_2	0	0	0	
u_3	$-\frac{1}{2}$	0	$\frac{1}{2}\Lambda^2(1-\alpha)^2$	
v_1	$-\Lambda(1+\alpha)$	$\frac{\Lambda^2 E}{3}\left\{\begin{array}{c}-(1+\alpha)\\ 4\alpha\\ 1-\alpha\end{array}\right.$	$-\frac{1}{6}\Lambda^3\left\{\begin{array}{c}(1+\alpha)^3\\ 1\\ (1-\alpha)^3\end{array}\right.$	0
v_2	0			0
v_3	$\Lambda(1-\alpha)$			0

The $\left[P^a_{mn}\right]$ matrix, obtained by summing the inner products of modal displacements and modal forces at nodes 1, 2, and 3 is

$$\left[P^a_{mn}\right] = E\left[\begin{array}{c|c} \frac{2}{3}\Lambda^3(1+\alpha^2) & \frac{\Lambda\alpha}{5} \\ \hline \frac{2}{9}\Lambda^5\alpha(1+2\alpha^2) & \frac{\Lambda^3}{3}\left(1+\frac{11}{5}\alpha^2\right) \end{array}\right] \qquad (11{:}26)$$

We see that P^a_{12} is nonzero. As noted earlier we can make it zero by adding some of mode 1 to mode 2. The result is that the value of P^a_{21} is changed to $P^a_{21} - P^a_{12}$. As before, if we wish to have mode 1 satisfy a patch test (in this case a linear-strain patch test) then $K_{11} = P^a_{11}$ and $K_{12} = \overline{P}^a_{21} = P^a_{21} - P^a_{12}$. The resulting value of the generalized force coefficient for mode 1 is

$$\alpha_1 = \beta_1 + \frac{\overline{P}^a_{21}}{P^a_{11}}\beta_2 = \beta_1 + \left(\frac{\Lambda^2\alpha\left(1 + 2\alpha^2\right)}{3\left(1 + \alpha^2\right)} - \frac{3\alpha}{10\Lambda^2\left(1 + \alpha^2\right)} \right)\beta_2 \quad (11:27)$$

It is seen that, for Λ large and α small, the locking parameter is proportional to $\Lambda^2\alpha$, just as in the case of the four-node trapezoid.

It is also easily shown, through a symmetry argument, that coupling of the two modes can be avoided for elements with rectangular or parallelogram shapes. Other locking troubles which are purely internal, such as shear locking, can also be avoided by appropriate design.

11.4.5 Plate and Shell Elements

Plate bending elements require quadratic displacement states to represent even the simplest uniform bending modes. Thus three- and four-node plate bending elements with C_0 continuity must represent displacement states which are not in their basis sets. The difficulty typically manifests itself as locking of the transverse shear strains, particularly for nonrectangular shapes. The theory presented here shows, however, that there should be no essential difficulty in resolving such locking problems because symmetry mandates zero coupling between bending and through-the-thickness extension or between bending and any membrane strain state. Three- and four-node plate bending elements which do not lock for general shapes and which pass patch tests are described in Sections 9.4.3 and 9.4.4.

The symmetry alibi does not apply to curved shell elements and there we indeed find significant coupling between bending states, membrane states, and through-the-thickness extension. For example, we have seen in Table 10.2 that the eight-node brick element with reduced integration performs poorly in the Morley hemispherical shell problem while flat-faceted plate elements perform extremely well. The reason is that curvature causes through-the-thickness taper of the eight-node solid element with effects similar to those of trapezoidal shape in a four-node membrane element. As a consequence, locking is unavoidable for the eight-node brick but is eliminated by the assumption of zero normal stress in the design of four-node flat plate elements.

Curved shell elements include additional opportunities for high-mode to low-mode coupling. Even simple uniform bending requires displacement components that are cubic functions of Cartesian position coordinates. As a result, eight- and nine-node curved shell elements may encounter fundamental limits to their perfectibility.

To summarize, the requirement of nodal equilibrium places restrictions on an element's shape functions which lead, for general element shapes, to essential coupling between low modes whose displacement states can be exactly represented by an element's basis functions and high modes whose displacement states cannot be so represented. The perfection of an element's design consists of finding exact representations of both high and low modes for the special symmetries which cancel the coupling. Progress beyond this limit requires compromise between the integrity of the low modes and the viability of the high modes. Usually, but not always, the element designer's choice comes down on the side of the low modes.

The limit of perfectibility, as defined here, has been achieved for the four-node membrane element. Its achievement for all other constant and linear strain elements is only a matter of time.

REFERENCES

11.1   B. M. Irons and S. Ahmad, *Techniques of Finite Elements*, Ellis Horwood, Chichester, p. 155, 1980.

11.2   T. J. R. Hughes, "Generalization of Selective Integration Procedures to Anisotropic and Nonlinear Media," *Intl. J. Numer. Methods Eng.*, 15, pp 1413-8, 1980.

11.3   W. P. Doherty, E. L. Wilson, and R. L. Taylor, "Stress Analysis of Axisymmetric Solids Using Higher Order Quadrilateral Finite Elements," U. of Calif. Berkeley, Struct. Eng. Lab. Report SESM 69-3, 1969.

11.4   E. L. Wilson, R. L. Taylor, W. P. Doherty, and J. Ghaboussi, "Incompatible Displacement Models," *Numerical & Computer Meth. in Struc. Mech.*, S. T. Fenves, et al. (Eds.), Academic Press, pp 43-57, 1973.

11.5   T. Belytschko, C. S. Tsay, and W. K. Liu, "A Stabilization Matrix for the Bilinear Mindlin Plate Element," *Comput. Methods Appl. Mech. Engrg*, 29, pp 313-27, 1981.

11.6   O. C. Zienkiewicz, J. Too, and R. L. Taylor, "Reduced Integration Technique in General Analysis of Plates and Shells," *Intl. J. Numer. Methods Eng.*, 3, pp 275-90, 1971.

11.7   R. H. MacNeal and R. L. Harder, "Eight Nodes or Nine?," *Intl. J. Numer. Methods Eng.*, 33, pp 1049-58, 1992.

11.8   B. M. Irons, "Numerical Integration Applied to Finite Element Methods," Conf. on Use of Digital Computers in Structural Engineering, Univ. of Newcastle, 1966.

11.9   R. L. Taylor, P. J. Beresford, and E. L. Wilson, "A Nonconforming Element for Stress Analysis," *Intl. J. Numer. Methods Eng.*, 10, pp 1211-9, 1976.

11.10   J. Robinson (Ed.), *The Finite Element News*, Robinson and Associates, Devon, England, 1976 to present.

11.11   J. Robinson and S. Blackham, "An Evaluation of Lower Order Membranes as Contained in the MSC/NASTRAN, ASAS and PAFEC FEM Systems," Robinson and Associates, Dorset, England, 1979.

11.12   J. Robinson and S. Blackham, "An Evaluation of Plate Bending Elements: MSC/NASTRAN, ASAS, PAFEC, ANSYS and SAP4," Robinson and Associates, Dorset, England, 1981.

11.13   A. Kamoulakos, G. A. O. Davies, and D. Hitchings, "Benchmark Tests for Various Finite Element Assemblies," NAFEMS, Nat'l. Eng. Lab., East Kilbride, U. K., 1984.

11.14   NAFEMS Publications, *Benchmark*, April 1988.

11.15   R. H. MacNeal and R. L. Harder, "A Proposed Standard Set of Problems to Test Finite Element Accuracy," *Finite Elem. Analysis & Design*, 1, pp 3-20, 1985.

11.16   D. W. White and J. F. Abel, "Testing of Shell Finite Element Accuracy and Robustness," *Finite Elem. Analysis & Design*, 6, pp. 129-51, 1989.

11.17   T. Belytschko, H. Stolarski, W. K. Liu, N. Carpenter, and J. S.-J. Ong, "Stress Projection for Membrane and Shear Locking in Shell Finite Elements," *Comput. Methods Appl. Mech. Engrg.*, 51, pp 221-58, 1985.

11.18   L. S. D. Morley, *Skew Plates and Structures*, Pergamon Press, London, p. 96, 1963.

11.19   I. Babuška, "The Stability of Domains and the Question of the Formulation of the Plate Problems," *Appl. Math.*, pp 463-67, 1962.

11.20   I. Babuška and T. Scapollo, "Benchmark Computation and Performance Evaluation for a Rhombic Plate Bending Problem," *Intl. J. Numer. Methods Eng.*, 28, pp 155-79, 1989.

11.21 I. C. Taig, "Structural Analysis by the Matrix Displacement Method," Engl. Electric Aviation Report No. 5017, 1961.

11.22 T. Belytschko, W. K. Liu, and B. E. Engelman, "The Gamma-Elements and Related Developments," *Finite Elem. Meth. for Plate & Shell Struct.*, 1, Pineridge Press, Swansea, U. K., pp 316-47, 1986.

11.23 R. H. MacNeal, "Derivation of Element Stiffness Matrices by Assumed Strain Distributions," *Nucl. Eng. Design*, 70, 1, pp 3-12, 1982.

11.24 R. H. MacNeal, "A Theorem Regarding the Locking of Tapered Four-Noded Membrane Elements," *Intl. J. Numer. Methods Eng.*, 24, pp 1793-9, 1987.

11.25 R. H. MacNeal, "On the Limits of Finite Element Perfectibility," *Intl. J. Numer. Methods Eng.*, 35, pp 1589-601, 1992.

Author Index

Numbers in bold type refer to list of references at the end of each chapter.

Subject Index

525